Organic Chemistry

Chemistry 211 / 212 Laboratory Manual

Fifth Edition

Heather J. Corpus • Pamela J. Seaton
University of North Carolina Wilmington
Department of Chemistry and Biochemistry

This lab manual is provided by the UNCW Department of Chemistry and Biochemistry. Royalties from this lab manual are returned to the Chemistry Department to fund the purchase of specialty software, glassware, and supplies used for developmental and instructional purposes for the Organic Chemistry laboratories.

www.kendallhunt.com
Send all inquiries to:
4050 Westmark Drive
Dubuque, IA 52004-1840

ISBN 978-1-5249-3511-5

Published in the United States of America

This lab manual is provided by the UNCW Department of Chemistry and Biochemistry. Royalties from this lab manual are returned to the Chemistry Department to fund the purchase of specialty software, glassware, and supplies used for developmental and instructional purposes for the Organic Chemistry laboratories.

CONTENTS

EXPERIMENTS

APPENDICES

Appendix Title | **Page**

PREFACE

This book was written because, despite the large number of organic lab manuals published, none seemed to fulfill our requirements. We wanted a lab manual that covers the essential techniques of organic chemistry (distillation, recrystallization, etc.), routinely incorporates modern instrumental methods of analysis (gas chromatography, infrared and NMR spectroscopy, etc.), has one or more synthetic sequences (in order to demonstrate the importance of technique, high yield, and product purity), and demonstrates the important physical organic principles such as the tendency of carbocations to rearrange and the effect of the base size on product distribution in dehydrohalogenation reactions. In addition, the experiments should avoid using chemicals that pose unnecessary health hazards, yet still allow training in techniques for the safe use of potentially dangerous substances. We wanted to limit the scale of most reactions for purposes of safety, ease of handling, economy, and to reduce the amount of waste to be disposed. We also sought experiments that were varied and interesting. This book is the result of our quest.

We wish to thank the many students, lab instructors, and graduate teaching assistants who have contributed over the years to the development and improvement of the experiments in this lab manual, and we encourage comments and suggestions for further improvement.

Wilmington, NC
May 2017

Pamela J. Seaton
Christopher J. Halkides
Paulo Almeida
Tom Coombs
Sridhar Varadarajan
Jeremy Morgan
Ralph Mead
Heather Corpus

Laboratory Rules and Safety Guidelines

Laboratory safety is an integral part of this course. It is important both for the immediate safety of you and your fellow students and instructors and as a part of your chemical education. In the future you will need to understand chemical hazards and how to deal with them. Consequently, you are expected to know and understand the safety principles and rules that are presented to you, and your laboratory quizzes may include questions about safety. ***You must sign a safety contract provided by your instructor and return it to your instructor before proceeding with any experiments.***

General Lab Rules

Attitudes and Preparation

- Be sure to be on time to lab. The safety concerns are discussed during the lab briefing, so if you miss the briefing, you cannot perform the experiment, and will receive a zero for the experiment.
- It is ***YOUR*** responsibility to come to lab prepared. This means familiarize yourself with the lab manual and the PowerPoint presentation, as well as with the materials used the hazards associated with them. You are responsible for your own safety.
- Dress properly. Open-toed shoes are not allowed, and shorts are discouraged. Failure to dress properly for lab will result in dismissal from the lab for the day, and therefore result in a grade of zero for the experiment.
- **Safety goggles are a must at all times in the Organic Chemistry lab**. If you forget your safety goggles, they will be supplied to you without penalty once, and then you will be penalized on your technique grade for the day.

Working Environment

- Be careful not to contaminate reagents with spatulas or disposable pipettes. If you take too much reagent/solvent, try to give it to your neighbor. **DO NOT PUT IT BACK INTO THE REAGENT BOTTLE!**
- Do not wander off with the only bottle of a reagent that everyone needs; keep it in the supply hood.
- No book bags on the floor. The aisles must be kept clear at all times.

Glassware

- Never use a thermometer as a stirrer! Always support your thermometer in a beaker or flask with a clamp.
- Round bottom flasks will not stand upright themselves. They must be supported in a beaker or with a clamp.
- All glassware should be cleaned with soap, water, and wash acetone prior to placing it back in the lab drawer.
- Report breakage of glassware to lab TA immediately. **DO NOT ATTEMPT TO CLEAN UP THE BROKEN GLASSWARE YOURSELF!**

Safety Guidelines

Safety Equipment

YOUR INSTRUCTOR WILL SHOW YOU WHERE THE SAFETY EQUIPMENT IS. REMIND YOURSELF FROM TIME TO TIME DURING THE SEMESTER.

a. Fire extinguisher
b. Fire blanket
c. Safety shower
d. Eye wash fountains
e. First aid kit
f. Two exits

Fire Hazards

- Absolutely NO open flames in the laboratory at any time.
- No smoking in lab.
- Know the location and operation of fire extinguishers.
- Know the location and proper use of the safety shower.
- Confine long hair and loose clothing when in the laboratory.
- Most organic compounds are flammable; be cautious.

IF THERE IS A FIRE IN THE LAB IN WHICH YOU ARE WORKING...

A. Shout ***FIRE*** to alert your neighbors and lab TA.
B. If the fire is in a test tube, or other small container, it can usually be contained with a watch glass or book. If it cannot be extinguished by a fire extinguisher, sand, or water, you will need to evacuate the build immediately.
C. If someone's clothes catch on fire, the fire needs to be smothered! Wrap the person in a lab coat, fire blanket, or whatever in order to exclude the oxygen.

Heat Hazards

- Most organic compounds are flammable and may catch fire, even without flames, at high temperatures.
- Make sure you know the location of the nearest fire extinguisher and the nearest exit.
- Reactions that are exothermic or are being heated must be monitored; do not leave them without having someone watch.
- NEVER, EVER heat a closed system. Pressure will build up and break the glass.

Toxic Hazards

- The materials used in lab are the safest we can find to use in order to develop your skills in working with hazardous materials.
- Keep safety goggles on at all times in the lab. If you DO get something in your eye, be sure to flush your eyes out in one of the eye wash fountains for several minutes.
- To prevent inhalation of vapors, do all of your experiments in the fume hoods.
- If you spill a liquid on the bench, immediately soak it up with paper towels, and transfer towels to your hood. Let your instructor deal with the paper towels.

- Know the location and proper use of the shower and eyewash.
- Wash all chemical spills on skin with cold water for several minutes. If the chemical is insoluble in water (as most organic chemicals are), use mild liquid hand soap also.
- Acid spills on the floor or bench should be neutralized cautiously with solid sodium bicarbonate or sodium carbonate, and the residue should then be removed with wet paper towels, a wet mop, or a spill pillow.
- Never touch any chemicals with your fingers.
- All chemical spills, no matter how small, should be cleaned up immediately so that no one will be exposed to them. This applies to spills on or around balances and on or around reagent dispensers.
- Report any spills left by others to your instructor so that they can be cleaned up.

Laboratory Electrical Equipment

- The hot plates that you will use are very powerful, and rarely need to be set higher than 3.
- Electrical heating mantles are to be plugged into a voltage regulator, NEVER directly into the outlet.
- Report frayed cords or nonfunctional equipment to your instructor.
- All equipment is to be unplugged and pushed to the rear of the lab hood before leaving for the day. Failure to do this will result in a deduction from your technique grade.

Waste Disposal

- Directions for waste disposal are located at the end of every laboratory experiment. Never dispose of waste down the laboratory sinks. Be sure to read labels on waste bottles carefully. If there is any question about where to place waste, ask your laboratory instructor.
- Broken glass, melting point capillary tubes, and used TLC spotters should be discarded in the specially marked yellow and white broken glass containers. **NEVER PLACE GLASS OF ANY KIND IN THE TRASHCAN.**

First Aid

- Report every accident or injury to our instructor, no matter how minor.
- Use ice or cold water on heat burns.
- Use bandages on minor cuts after cleaning the wound well.
- Use a gauze compress for larger cuts, and see the University Health Center.
- Wash all chemical spills on skin with cold water for several minutes. If the chemical is water insoluble, use mild liquid hand soap also.

ORGANIC CHEMISTRY LABORATORY (CHML211)

SAFETY CONTRACT

Whenever I am in an area where laboratory reagents are being used, I agree to abide by the following rules:

1. Wear safety goggles.
2. Wear proper clothing, including closed toe shoes and sufficient clothing.
3. Not eat, drink, use tobacco, or apply makeup in the laboratory.
4. Become familiar with actions to be taken in the event of incidents in the laboratory.
5. Be familiar with the location and use of all safety equipment, including fire extinguisher, fire exits, eyewash stations, and safety shower.
6. Become familiar with each laboratory assignment before coming to the laboratory. This includes reading and preparation of PRE LAB assignment, where applicable.
7. Anticipate the common hazards that may be encountered in laboratory.
8. Do only authorized experiments, and work only when the laboratory instructor or another qualified person is present.
9. Dispense reagents carefully, and close reagent bottles immediately after use.
10. Treat laboratory reagents as if they are poisonous and corrosive.
11. Dispose of all laboratory reagents and all laboratory waste as instructed.
12. Clean up work area after use, including lab bench and fume hood.
13. Report all incidents to the laboratory instructor.

Student ID#______________________________

Student signature____ **date**____________

Laboratory instructor____ **date**____________

- ***In the space below, give any health information, such as pregnancy or other circumstance, that might help the laboratory instructor provide a safer environment for you, or that could aid the laboratory instructor in responding to an incident involving you in the laboratory.***

1. **I do / do not** (*circle one*) expect to wear contact lenses during laboratory work. [**NOTE**: Goggles must still be work when contact lenses are worn.]

2. List ***any*** known allergies to medications or other chemicals.

3. **I do / do not** (*circle one*) have a chronic illness and/or a special circumstance that I need to discuss with my lab instructor privately.

ORGANIC CHEMISTRY LABORATORY (CHML212)

SAFETY CONTRACT

Whenever I am in an area where laboratory reagents are being used, I agree to abide by the following rules:

1. Wear safety goggles.

2. Wear proper clothing, including closed toe shoes and sufficient clothing.

3. Not eat, drink, use tobacco, or apply makeup in the laboratory.

4. Become familiar with actions to be taken in the event of incidents in the laboratory.

5. Be familiar with the location and use of all safety equipment, including fire extinguisher, fire exits, eyewash stations, and safety shower.

6. Become familiar with each laboratory assignment before coming to the laboratory. This includes reading and preparation of PRE LAB assignment, where applicable.

7. Anticipate the common hazards that may be encountered in laboratory.

8. Do only authorized experiments, and work only when the laboratory instructor or another qualified person is present.

9. Dispense reagents carefully, and close reagent bottles immediately after use.

10. Treat laboratory reagents as if they are poisonous and corrosive.

11. Dispose of all laboratory reagents and all laboratory waste as instructed.

12. Clean up work area after use, including lab bench and fume hood.

13. Report all incidents to the laboratory instructor.

Student ID#______________________________

Student signature____ **date**____________

Laboratory instructor____ **date**____________

- ***In the space below, give any health information, such as pregnancy or other circumstance, that might help the laboratory instructor provide a safer environment for you, or that could aid the laboratory instructor in responding to an incident involving you in the laboratory.***

1. I **do / do not** (*circle one*) expect to wear contact lenses during laboratory work. [**NOTE**: Goggles must still be work when contact lenses are worn.]

2. List ***any*** known allergies to medications or other chemicals.

3. I **do / do not** (*circle one*) have a chronic illness and/or a special circumstance that I need to discuss with my lab instructor privately.

Organic Chemistry Lab Equipment Checklist (CHML211)

(Leave in Lab Manual for reference until end of semester)

Student Lab Drawers:

Beakers:

_____ (1) 50 mL
_____ (1) 100 mL
_____ (1) 150 mL
_____ (1) 250 mL
_____ (1) 400 mL

Erlenmeyer Flasks:

_____ (1) 25 mL
_____ (3) 50 mL
_____ (3) 125 mL
_____ (1) 250 mL

Filter flasks:

_____ (1) 125 mL

Test tubes:

_____ (6) 18 mm x 150 mm with rack
_____ (20) 13 mm x 75 mm with rack

Graduated cylinders:

_____ (1) 10 mL
_____ (1) 50 mL

Funnels:

_____ (1) plastic stem
_____ (1) plastic powder
_____ (1) Büchner

Filter adapters:

_____ (1) small gray

Other:

_____ (1) glass rod
_____ (1) tweezers
_____ (1) microspatula
_____ (1) scoopula
_____ (1) thermometer
_____ (1) watch glass
_____ (3) TLC chambers
_____ (1) plastic ruler

Student Lab Hoods:

_____(1) hot plate
_____(1) hot plate/stirrer
_____(1) small heating mantle
_____(1) voltage regulator
_____(2) wire gauze pads
_____(2) rubber flask supports
_____(2) ring stands
_____ (2) iron rings
_____ (2) clear tubing
_____ (1) red tubing
_____ (2) clamps

Lab Drawer #___________

Room # ____________

Printed Name______________________________________

Signature____

Instructor Initials_____

Organic Chemistry Lab Equipment Checklist (CHML212)

(Leave in Lab Manual for reference until end of semester)

Student Lab Drawers:

Beakers:

_____ (1) 50 mL
_____ (1) 100 mL
_____ (1) 150 mL
_____ (1) 250 mL
_____ (1) 400 mL

Erlenmeyer Flasks:

_____ (1) 25 mL
_____ (3) 50 mL
_____ (3) 125 mL
_____ (1) 250 mL

Filter flasks:

_____ (1) 125 mL

Test tubes:

_____ (6) 18 mm x 150 mm with rack
_____ (20) 13 mm x 75 mm with rack

Graduated cylinders:

_____ (1) 10 mL
_____ (1) 50 mL

Funnels:

_____ (1) plastic stem
_____ (1) plastic powder
_____ (1) Büchner

Filter adapters:

_____ (1) small gray

Other:

_____ (1) glass rod
_____ (1) tweezers
_____ (1) microspatula
_____ (1) scoopula
_____ (1) thermometer
_____ (1) watch glass
_____ (3) TLC chambers
_____ (1) plastic ruler

Student Lab Hoods:

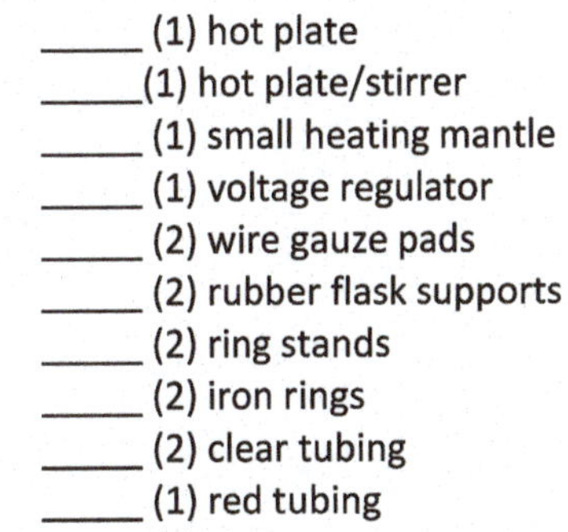

_____ (1) hot plate
_____(1) hot plate/stirrer
_____ (1) small heating mantle
_____ (1) voltage regulator
_____ (2) wire gauze pads
_____ (2) rubber flask supports
_____ (2) ring stands
_____ (2) iron rings
_____ (2) clear tubing
_____ (1) red tubing
_____ (2) clamps

Lab Drawer #___________

Room # ____________

Printed Name_______________________________________

Signature____

Instructor Initials_____

Common Laboratory Glassware and Equipment

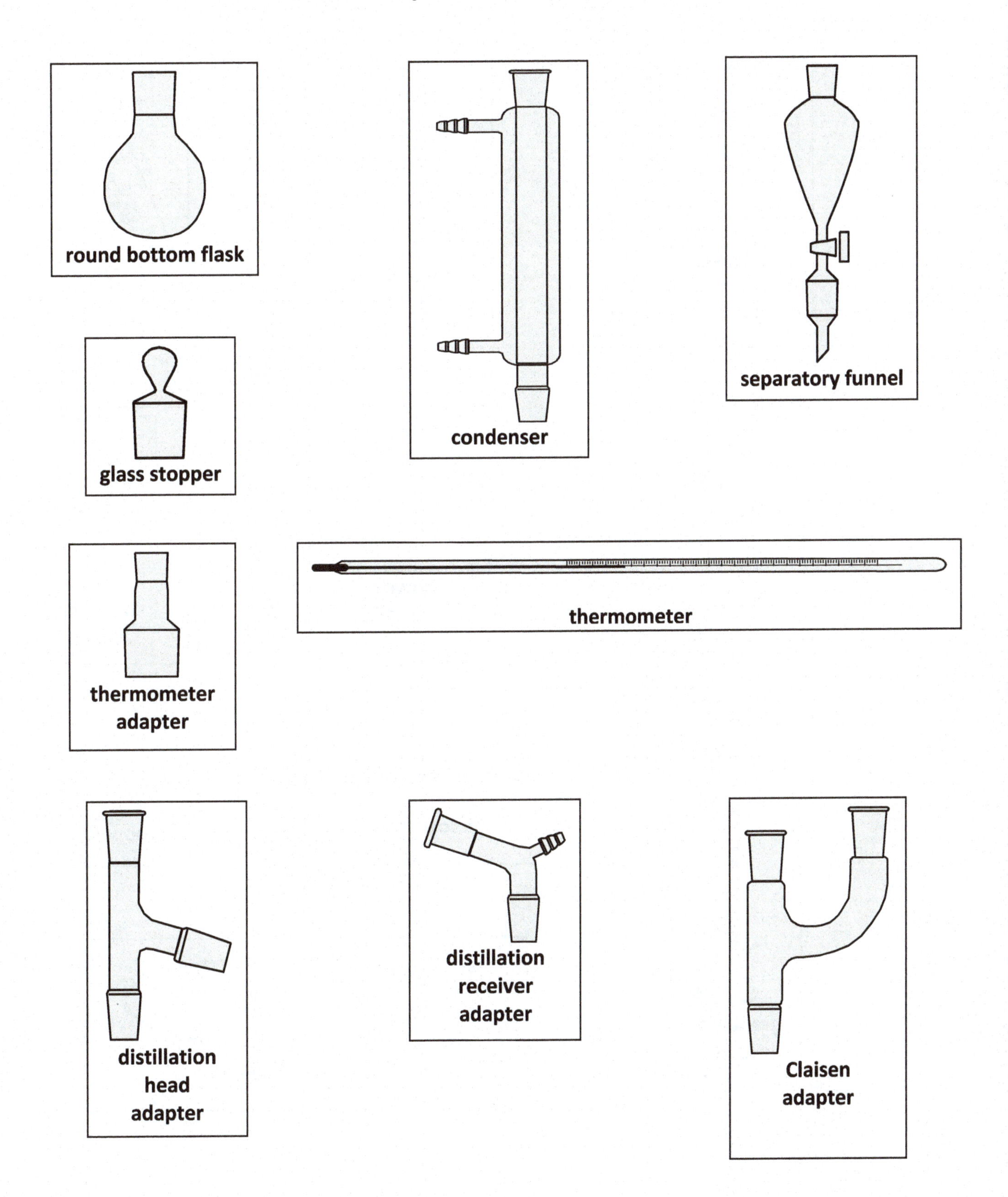

Common Laboratory Glassware and Equipment

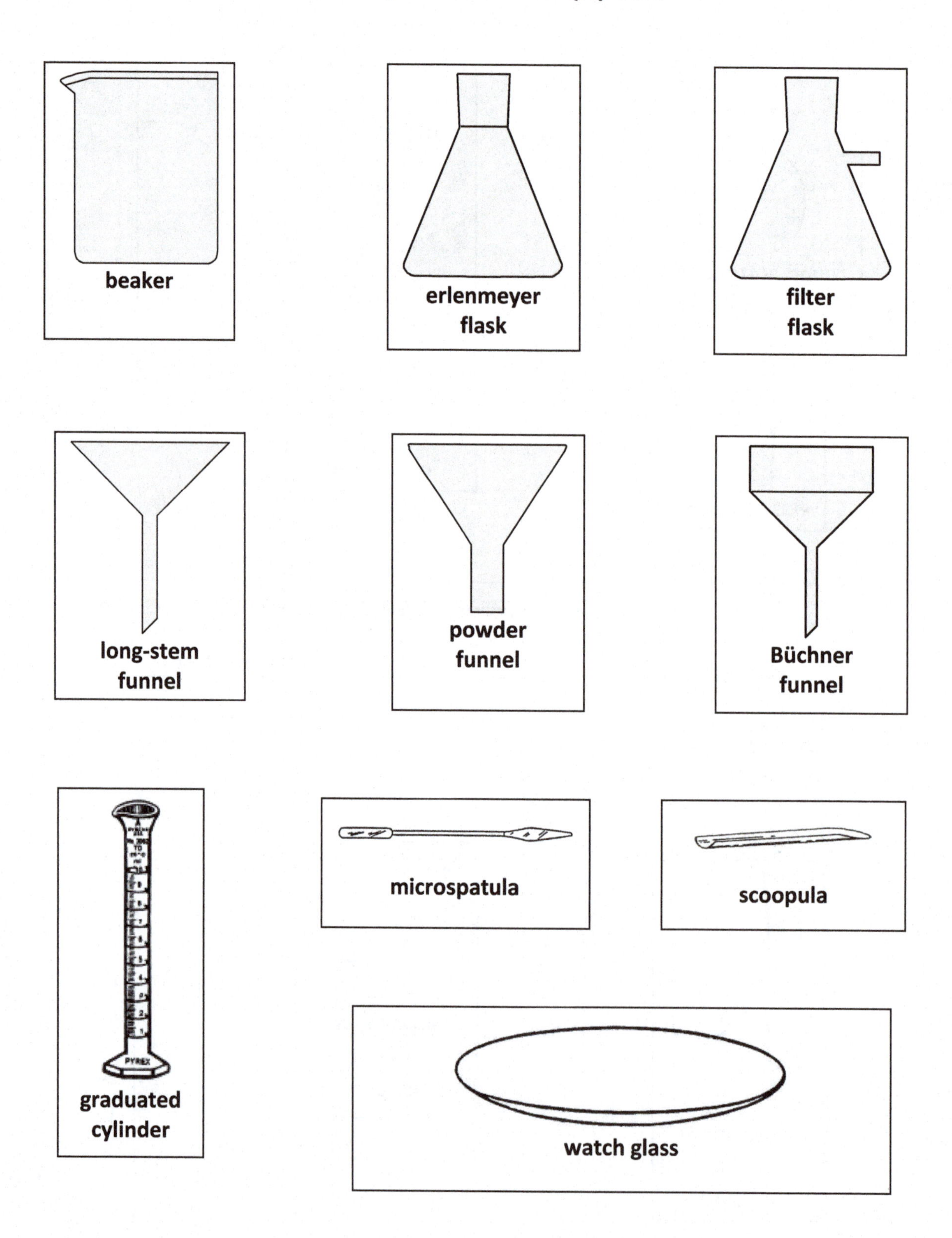

Experiment 1

Structure, Intermolecular Forces, and Solubility

Introduction

In this experiment you will explore the relationship between molecular structure, intermolecular forces, and solubility of various important organic solvents. You will investigate the interaction of organic solvent molecules with other solvent molecules, including water. You will also look at the solubility of neutral and ionic organic solids in an organic solvent, water, aqueous acid, and aqueous base.

Importance of Molecular Structure

The electronic structure of a molecule determines the three-dimensional shape (molecular structure) of the molecule. The molecular shape and the polarity of individual bonds determine whether the molecule will be polar or nonpolar, and whether it will be capable of hydrogen bonding. Thus, the molecular structure determines what kind of intermolecular forces are present between molecules. Intermolecular forces are manifested in the physical state of a compound (solid vs. liquid vs. gas), its physical properties (boiling or melting point and solubility), as well as chromatographic properties. Understanding molecular structure is also very important in understanding chemical characteristics of compounds, including biochemical compounds. For example, the structure and function of proteins is due to intermolecular forces between amino acids. DNA and DNA-RNA base pairing is due in large part to intermolecular forces (in particular, hydrogen bonding) between base pairs.

Intermolecular Forces

Several types of intermolecular forces can be present between interacting molecules. In ionic solids, very strong electrostatic attractive forces exist between the positively and negatively charged ions. These very strong attractive forces result in very high melting points for ionic solids. For neutral molecules, there are three major types of intermolecular attractive forces, collectively known as van der Waals forces: **London dispersion** forces, **dipole–dipole** forces, and **hydrogen bonding.**

London Dispersion Forces

London dispersion forces are weak attractive forces caused by instantaneous (or induced) dipoles in molecules. **All molecules are capable of London dispersion forces**. London dispersion forces are the only forces of attraction between nonpolar molecules, such as alkanes. In general, larger molecules have greater London attractive forces between each other. Methane (CH_4), ethane (C_2H_6), propane (C_3H_8), and butane (C_4H_{10}) are gases; pentane (C_5H_{12}) is a very low boiling liquid (bp = 35–36°C), while the larger alkanes have increasingly greater boiling points. Hexane (C_6H_{14}), a commonly used organic solvent, has a boiling point of 69°C.

Dipole–Dipole Forces

Dipole–dipole forces are due to electrostatic attraction between polar molecules. A bond or molecule is said to be polar if the electron density within the molecule is asymmetrically distributed. The electron density in a bond or molecule is unevenly distributed due to the attraction that highly electronegative atoms have for electrons. For example, chlorine is much more electronegative than hydrogen. The unsymmetrical distribution of electrons in H-Cl leads to the development of partially positive (δ^+) and partially negative (δ^-) ends of the bond. A quantitative measure of polarity is the dipole moment (μ), measured in Debye (D). The dipole moment is represented as an arrow going from the positive to the negative, as shown for H-Cl (Figure 1.1). Oxygen is more electronegative than carbon, and acetone therefore has a dipole. Chlorine is also more electronegative than carbon so that dichloromethane also has a dipole.

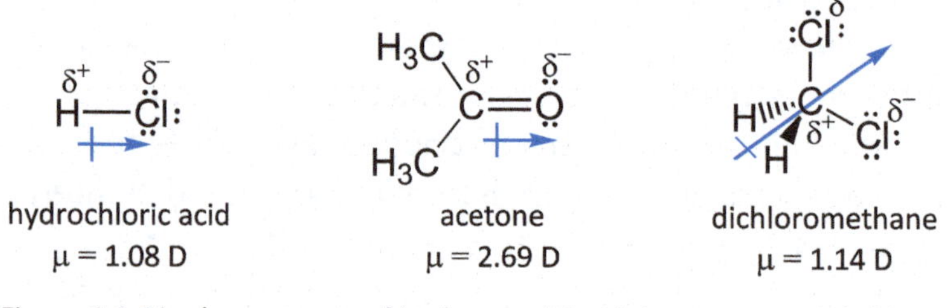

Figure 1.1 Dipole moments of hydrogen chloride, acetone and dichloromethane.

Hydrogen Bonding

Hydrogen bonding is the electrostatic attraction between the partially positive H of -O-H and -N-H bonds with the negative end of the dipoles of adjacent nitrogen- and oxygen- containing dipole molecules. Of the elements most commonly found in organic compounds (C, H, N, O, Cl, Br, S, P), oxygen is the most electronegative and hydrogen is the least electronegative. The large difference in electronegativity between hydrogen and oxygen makes the O-H bond very polar. The positive hydrogen end of the bond is attracted to the partially negative nitrogen or oxygen of another molecule. This strong dipole–dipole attraction is called a hydrogen bond (*H-bond*). The molecule that provides the hydrogen is called the *H-bond donor* and the one that accepts the hydrogen is called the *H-bond acceptor* (Figure 1.2).

Not all H-bond acceptors are H-bond donors. For example, the partially negatively charged oxygen atom of an aldehyde, ketone, ether, ester, and nitro-compound can act as an acceptor of H-bonds, but not as a donor. The ability to H-bond has far-reaching consequences in organic chemistry. Compounds that can donate and accept H-bonds (those with N-H or O-H bonds) have higher boiling and melting points than compounds with similar molecular mass, but without N-H or O-H bonds. Also, N-H or O-H containing compounds are generally more soluble in water and less soluble in nonpolar solvents like hexane than compounds without these bonds. H-bonding is also critical in helping to determine the structure and function of numerous biomolecules, such as proteins, DNA, and RNA, to mention a few.

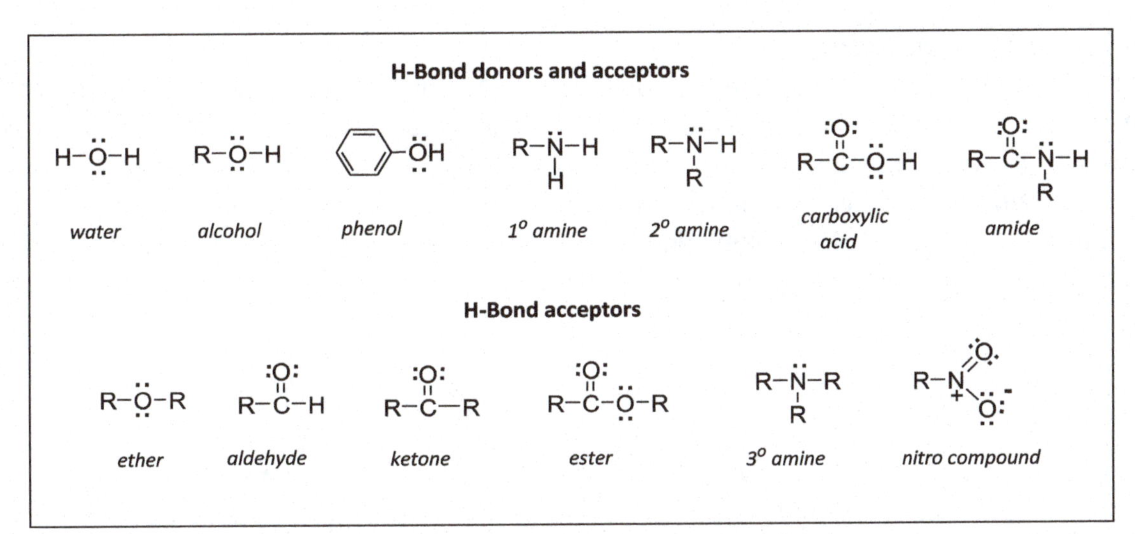

Figure 1.2 Compounds with hydrogen bonding capabilities.

Ion–Dipole Forces

One intermolecular force that is important in the solubility of many solids and liquids are ion–dipole forces. Ion–dipole forces are the attraction between an ion (an electrically charged species) and a polar solvent. The more polar a solvent is, the stronger its attraction to ions. Ion–dipole attraction explains why sodium chloride (NaCl) is soluble in water, but is insoluble in hexane. When an ionic solid is placed into water, it will dissolve if the ion–dipole forces between the ions and the partial charges (δ^+ and δ^-) of water molecules are greater than the electrostatic attractive forces between the ions (Figure 1.3).

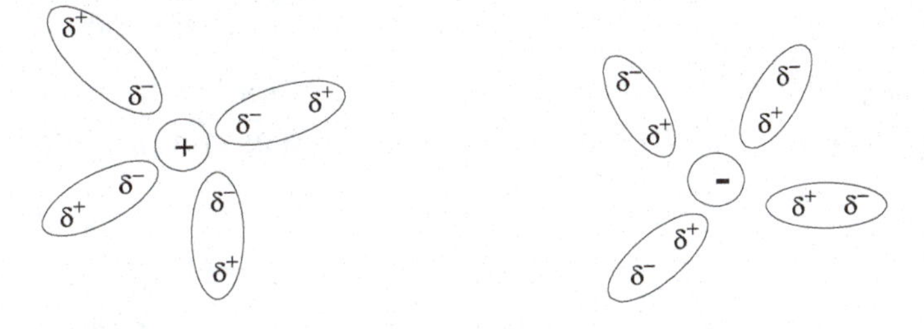

Figure 1.3 Ion–dipole forces between ions and a polar solvent.

The oxygen atoms of water bear a partial negative charge that is attracted to the positively charged sodium atom (Na^+) in solution, and the hydrogen atoms of water are similarly attracted to the chlorine atom (Cl^-). These attractions compensate for the loss of attractive ionic forces when sodium and chloride ions in a crystal are separated by forming a solution. **Ionic solids will not dissolve in solvents that are not capable of strong ion–dipole interactions.** Hexane does not have a partial positive or a partial negative charge, and therefore, sodium chloride remains a solid when mixed with hexane. Thus, sodium chloride does not dissolve in most organic solvents, which are relatively nonpolar compared to water.

Organic Acids and Bases

Many organic solids are insoluble in water, but soluble in organic liquids (solvents). However, sometimes the conjugate acid or the conjugate base forms of the organic solids are soluble in water, but insoluble in organic liquids. The explanation lies within the ion–dipole forces. When an acid is neutral, the conjugate base must be negatively charged. For example, benzoic acid (Figure 1.4) is soluble in organic solvents but is only slightly soluble in water. The benzoate ion, on the other hand, is soluble in water, because of strong ion–dipole attractions.

Figure 1.4 Benzoic acid and its conjugate base.

Likewise, when a base is electrically neutral, its conjugate acid is positively charged (Figure 1.5). Aniline is only slightly soluble in water. However, the anilium ion, its conjugate acid, is much more water soluble because of ion–dipole attractions.

Figure 1.5 Aniline and its conjugate acid.

Acids can be defined (Brønsted-Lowry definition) as molecules or ions that donate an H^+ and bases (which are usually amines) can be defined as compounds that accept an H^+.

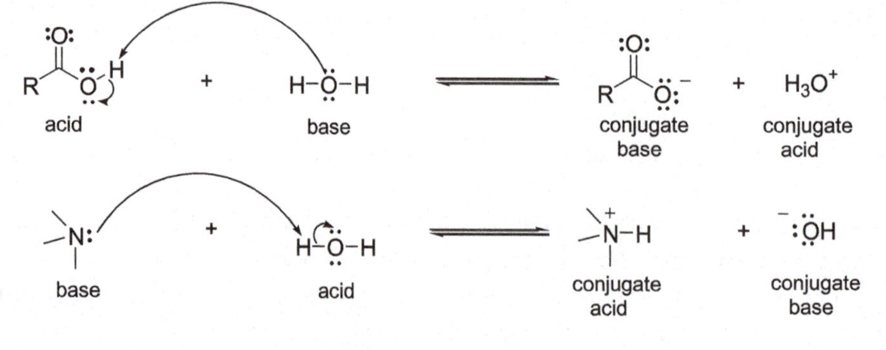

Figure 1.6 Acid-base chemistry.

The O-H bond of an organic acid more readily dissociates than the O-H bond of water or alcohol. Carboxylic acids, such as acetic acid and benzoic acid, are relatively weak acids (pK_a = 4.76 and 4.19) but can be completely deprotonated with $NaHCO_3$ (Figure 1.6). Although acetic acid is completely soluble in water, benzoic acid is only very slightly soluble in water because of the lipophilic (nonpolar) benzene ring. However, upon reaction with $NaHCO_3$, the conjugate base formed—sodium benzoate—is a salt and is soluble in water due to strong ion–dipole interactions

(Figure 1.7). The ionic character of sodium benzoate now prevents the salt from dissolving in relatively nonpolar solvents like hexane, diethyl ether, or ethyl acetate since these solvents are capable of only very weak ion–dipole interactions.

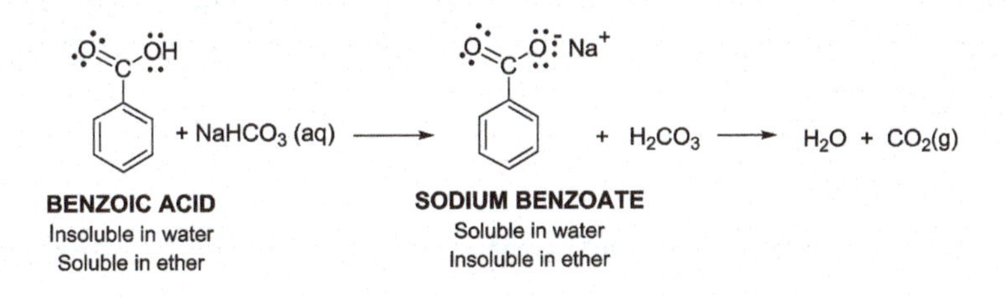

Figure 1.7 Acid-base reaction with an organic acid.

Organic bases are generally amines and accept a proton from a strong acid to form a conjugate acid salt, as shown below for aniline (Figure 1.8). Although aniline is only slightly soluble in water, the hydrochloride salt formed by reaction with HCl (phenylammonium chloride) is very soluble in water due to ion dipole interactions. Its ionic character now prevents the salt from dissolving in relatively nonpolar solvents like hexane, diethyl ether, or ethyl acetate.

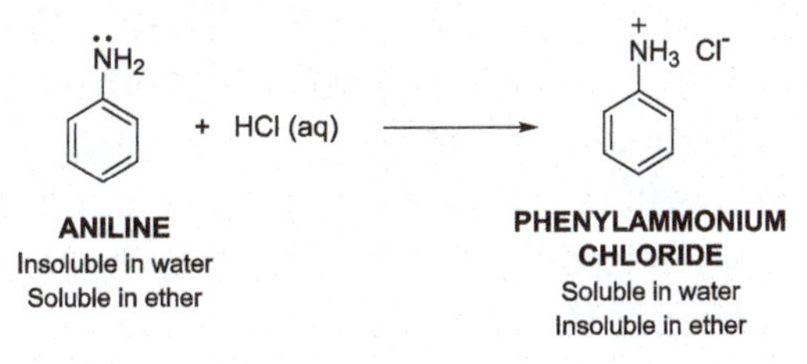

Figure 1.8 Acid-base reaction with an organic base.

Polarity

Several experimental measures of solvent polarity exist. One such measure is the dielectric constant, or the ability of a solvent to screen charges on the solutes from interacting with each other. ***In practice the polarity of a liquid is dependent of at least four factors:***

1. The polarity of individual chemical bonds, arising from the differences in electronegativity
2. The geometry of the molecule
3. The presence of hydrogen bond donating or accepting atoms
4. The relative proportion of polar versus nonpolar bonds in the substance

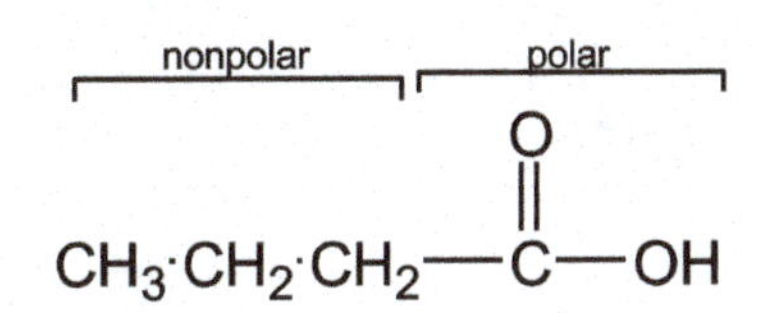

Figure 1.9 Proportion of polar to nonpolar bonds in butanoic acid.

To illustrate this point, let us examine butanoic acid, a carboxylic acid (Figure 1.9).The carboxylic acid functional group on the right-hand side of the molecule is polar. There is a large electronegativity difference between carbon and oxygen, making the carbon-oxygen bonds polar. In addition, a carboxylic acid functional group is both a hydrogen bond donor and a hydrogen bond acceptor. However, the propyl group on the left-hand side of the molecule is made of C-C and C-H bonds, which are not polar. If we added more $–CH_2–$ groups to butanoic acid, it would become **less** polar, but if we subtracted $–CH_2–$ groups from butanoic acid, the molecule would become **more** polar.

Miscibility and Solubility

If two liquids are mixed and two layers form, the liquids are **immiscible** in each other. If two liquids are mixed and only a single layer forms, the two liquids are **miscible** in each other. If a turbid solution forms, the two liquids are not miscible, but they are not as immiscible as when two layers form. The maximum amount of a solute that dissolves in a solvent is the **solubility** of a solute. This can be thought of as how much solute can fit into the solvent until it becomes saturated. The solute may be a liquid or a solid, whereas the solvent is usually a liquid. Even when two liquids **A** and **B** are immiscible, there is always a very small amount of liquid B in the liquid A layer, and vice versa. In other words, two immiscible liquids A and B are **slightly soluble** in each other.

What determines how soluble a solute is in a solvent? There are intermolecular forces (IMFs) between solute molecules, IMFs between solvent molecules, and IMFs between solute and solvent molecules. The IMFs between the solute and solvent must be as favorable as the other IMFs between molecules of solute and between molecules of solvent in order for the solute to be highly soluble. Likewise, when two liquids are miscible, the forces between the two liquids must be as favorable as each solvent's IMFs are with itself. For example, hexane and pentane are miscible because any given hexane molecule experiences London dispersion forces with a pentane molecule almost as easily as with another hexane molecule. Water and methanol are miscible because any given water molecule can form hydrogen bonds with methanol very effectively. But water and pentane are not miscible because water interacts more strongly with itself, via hydrogen bonds, than it interacts with pentane.

Objectives

In this experiment solubility tests will be used to investigate the effect of intermolecular forces (IMF) on the miscibility of various commonly used solvents in water. The relative polarity of various carbon chain length alcohols will be determined based on solubility testing with a polar solvent vs. a nonpolar solvent. Finally, the solubility of neutral and ionic organic solids will be investigated in hexane, water, aqueous acid, and aqueous base in an effort to understand acid-base chemistry.

Experimental Procedure

Miscibility of organic liquids in water

- Test the miscibility of organic liquids in water. Label 5 small test tubes **A-E** and place in test tube rack.
- Add 1 mL of the following solvents to the appropriate test tube:
    - **A**: methanol
    - **B**: ethyl acetate
    - **C**: dichloromethane
    - **D**: acetone
    - **E**: hexane
- Add 1 mL **BLUE WATER** (blue-tinted deionized water) to each test tube. Shake each tube gently. Record which solvents are miscible with water (*they form a single layer*) and which solvents are immiscible with water (*they form two layers*) in Table 1.1 on the POST LAB Assignment (provided online).
- Compare the solvent densities and predict the identity of each layer (higher density solvents will sink, forming the bottom layer).
- Use the solvent structure to identify which IMF(s) the compounds are capable of. Be sure to distinguish between those which can donate hydrogen bonds and those which can accept hydrogen bonds only.

Miscibility of alcohols in hexane vs. water

Miscibility of alcohols in hexane

- Label two small test tubes **F** (methanol), and two small test tubes **G** (1-butanol).
- Add 1 mL of hexane to one **F** test tube, and 1 mL of hexane to one **G** test tube.
- Add 1 mL of methanol to the **F** test tube. Agitate the tube and record whether the liquids are miscible or immiscible with each other in Table 1.2 on the POST LAB Assignment (provided online).
- Add 1 mL of 1-butanol to the **G** test tube. Agitate the tube and record whether the liquids are miscible or immiscible with each other in Table 1.2.

Miscibility of alcohols in water

- Add 1 mL of BLUE WATER to an **F** test tube, and 1 mL of BLUE WATER to a **G** test tube.
- Add 1 mL of methanol to an **F** test tube. Agitate the tube, and record whether the liquids are miscible or immiscible with each other in Table 1.2.
- Add 1 mL of 1-butanol to a **G** test tube. Agitate the tube and record whether the liquids are miscible or immiscible with each other in Table 1.2.

Solubility of organic solids

- Test the miscibility of a neutral organic acid and a charged organic base in hexane, water, aqueous acid and aqueous base.
- Label 4 test tubes **H** (benzoic acid) and 4 test tubes **I** (sodium benzoate).
- ***For the following solubility tests, no specific mass of the solid is necessary, however only an amount equivalent the inside of the circle shown should be used (—O).***

Solubility of benzoic acid

- Add a small amount of benzoic acid to all of the test tubes labeled **H**.
- Add 3 mL of hexane to the first tube, 3 mL of **COLORLESS DEIONIZED WATER** to the second tube, 3 mL of 1M HCl to the third tube, and 3 mL of 10% $NaHCO_3$ to the fourth tube.
- Once the solvents are added, agitate the contents of the tube by holding a micro spatula between two fingers and spinning rapidly inside the tube. Observe the remaining solid, if any. Record whether all of the crystals have dissolved (**SOL**), if only a small amount of the solid remains (**SL SOL**), or if none of the solid dissolved (**INSOL**) into Table 1.3 on the POST LAB Assignment (provided online).

Solubility of sodium benzoate

- Add a small amount of sodium benzoate to all of the test tubes labeled **I**.
- Add 3 mL of hexane to the first tube, 3 mL of COLORLESS DEIONIZED WATER to the second tube, 3 mL of 1M HCl to the third tube, and 3 mL of 10% $NaHCO_3$ to the fourth tube.
- Once the solvents are added, agitate the contents of the tube by holding a micro spatula between two fingers and spinning rapidly inside the tube. Observe the remaining solid, if any. Record whether all of the crystals have dissolved (**SOL**), if only a small amount of the solid remains (**SL SOL**), or if none of the solid dissolved (**INSOL**) into Table 1.3.

SAFETY

All experiments should be performed in a fume hood with appropriate safety glasses. Gloves will be provided by request.

All organic solvents used in this experiment are flammable, irritants, and can be toxic if ingested or absorbed through skin. Dichloromethane is a suspected carcinogen. Hexane is a possible teratogen. Benzoic acid, sodium benzoate, and sodium bicarbonate are all toxic if ingested. Hydrochloric acid is extremely corrosive.

WASTE MANAGEMENT

After recording your results, pour all liquid waste into the container labeled "LIQUID ORGANIC WASTE—IMF" in the waste hood. Do not pour ANY solvents down the drain!

References

Klein, David. (2015). *Organic Chemistry*, 2nd ed. Hoboken: John Wiley and Sons.
Palleros, Daniel R. (2000). *Experimental Organic Chemistry*. New York: John Wiley and Sons.

Name:			
	Max	Score	Total Grade
Tech	10		
Pre	20		
Post	70		

Exp. 1 Structure, Intermolecular Forces, and Solubility

PRE-LAB ASSIGNMENT: *(**EACH STUDENT** will complete and submit an original copy at the beginning of the lab period. **Without a complete pre-lab assignment, you will not be allowed to perform the experiment, and will receive a zero for the lab.**) .....max score = 20 pts.*

1. **Objective** *(Write a brief purpose of the experiment in **complete** sentences, addressing all of the following points)*
   - What effect will be investigated in this experiment using solubility testing?
   - How will the relative polarity of alcohols be determined in this experiment?
   - How will acid-base chemistry be introduced in this experiment?

2. **Physical Data** *(Complete the following table before coming to lab.)*

Compound	**bp** (C°)	**d** (g/mL)
methanol		
1-butanol		
ethyl acetate		
dichloromethane		
acetone		
hexane		

3. **Experimental Outline** *(Give a brief description of the procedure that will be followed in this experiment in 5 lines or less.)*

1	
2	
3	
4	
5	

4. Pre-Lab Questions *(Answer the following questions prior to lab.)*

A. From the following list, circle the ***STRONGEST*** intermolecular force. Mark an "X" through the ***WEAKEST*** intermolecular force.

DIPOLE-DIPOLE FORCE **HYDROGEN BONDING** **LONDON DISPERSION FORCE**

B. Circle the ***most polar*** molecule from the list below:

HEXANE **WATER** **ETHANOL**

I have read and understood the experimental procedure for this experiment. I am familiar with the hazards and the required disposal procedures for this experiment.

Sign here: ____

Experiment 2

Stereochemistry of Cyclohexanes

Introduction

Substituted cyclohexanes are very common in nature. Many natural products contain cyclohexane-like structures as their main component, such as steroids and pharmaceutical agents, making the study of cyclohexane extremely important. The objective of this experiment is to practice drawing cyclohexane rings in chair conformations, and learn how to recognize axial and equatorial positions. This experiment will provide practice drawing monosubstituted and disubstituted cyclohexane rings in both chair conformations that exist. Finally, the stability of chair conformations will be determined in an effort to determine which conformer is favored at equilibrium.

Cyclohexane and the Chair Conformation

There are twelve possible positions on a chair structure. Six of the positions, known as ***axial*** positions, are in vertical positions. The other six positions, known as ***equatorial*** positions, point outward from the center of the ring. Three of the axial positions point upward, the other three point downward. For the equatorial positions, it is the same. Three of the equatorial positions point upward, the other three point downward, though not quite as obvious (Figure 2.1).

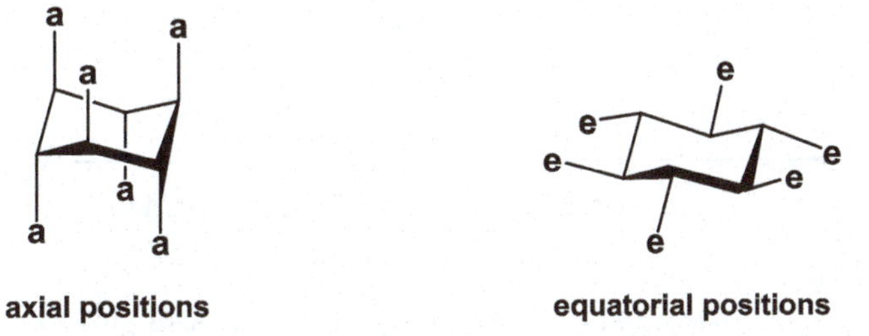

Figure 2.1 Axial and equatorial positions on a cyclohexane ring.

Chair conformations are conformationally mobile. Though not quite as flexible as straight chain alkanes, because of their flexibility chair conformations can "flip" from one possible chair conformation to another. It is important to note that when this chair flip (also called ring flip) occurs, the substituents originally in the axial position become equatorial, and the substituents originally in the equatorial position become axial. However, equally important is the fact that substituents that are pointing upward, remain pointing up after the ring flip occurs, and vice versa. In summary, up stays up and down stays down, but equatorial groups become axial groups and axial groups become equatorial groups (Figure 2.2).

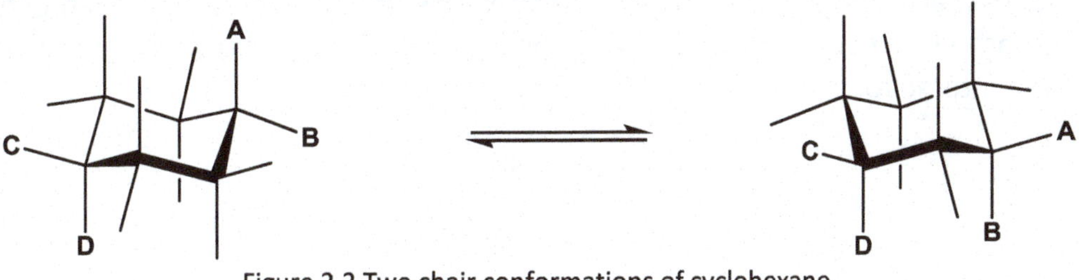

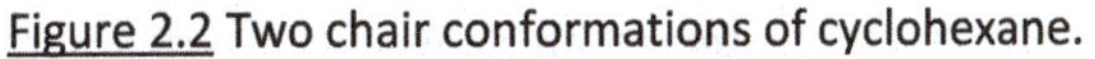

Figure 2.2 Two chair conformations of cyclohexane.

Drawing Chair Conformations

You will frequently be required to draw three-dimensional representations of chair conformations of cyclohexane, and use these representations to show relationships between atoms or groups of atoms that may be bonded to the ring. Molecular models are the best tools to visualize cyclohexane rings, but it is important to practice drawing cyclohexane rings, in the event that models are not available. If drawn correctly, the relationships between groups on a cyclohexane ring are easily identifiable. Stepwise instructions are shown below for drawing a basic cyclohexane chair conformation, along with adding axial and equatorial groups properly (Figure 2.3 and Figure 2.4).

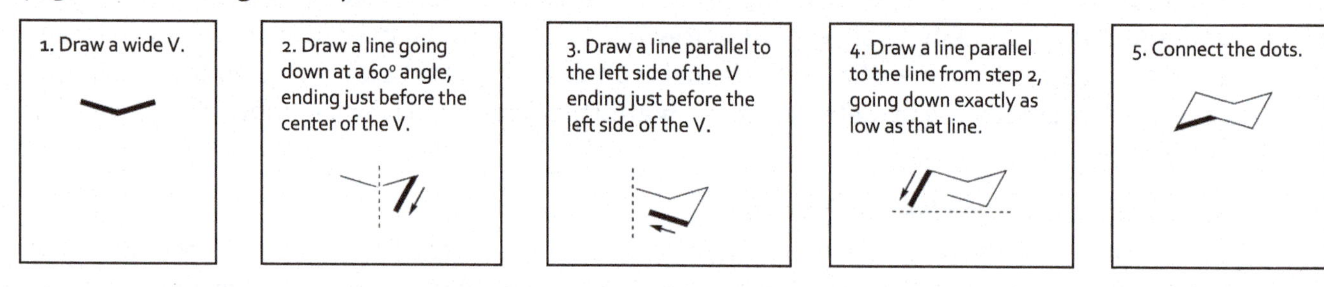

Figure 2.3 Drawing a chair conformation of cyclohexane.

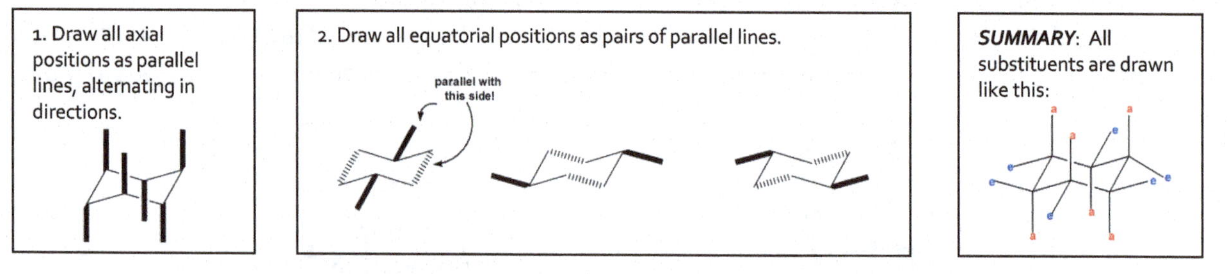

Figure 2.4 Drawing axial and equatorial positions.

Substituted Cyclohexane

The cyclohexane structure will continuously flip from one chair conformation to another at room temperature. Cyclohexane without any substituents has two equivalent chair conformations, both of equal energy, which exist in a 50:50 mixture. Once substituents are added to the ring, however, the two forms may not be equal in energy. Axial substituents have steric interactions with other groups in axial positions. These steric interactions, called ***1,3-diaxial interactions***, occur due to the steric strain which results from a substituent on one carbon being too close to axial hydrogen atoms three carbons away (Figure 2.5). For this reason, a substituent is almost always more stable in the equatorial position than in the axial position. The exact amount of steric strain in the substituted cyclohexane varies depending on the nature and size of the substituent. Generally a larger substituent gives rise to a larger difference in energy between the axial and equatorial conformations. Stepwise instructions are given in Figure 2.6 for drawing both chair conformations of a monosubstituted cyclohexane ring, and identifying axial hydrogen interactions.

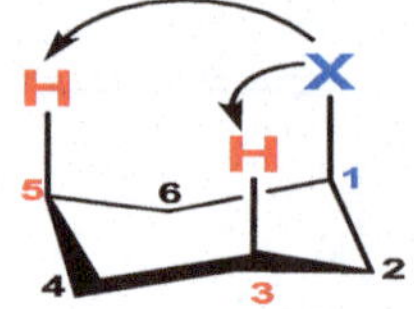

Figure 2.5 Example of 1,3-diaxial interactions between substituent and axial hydrogens.

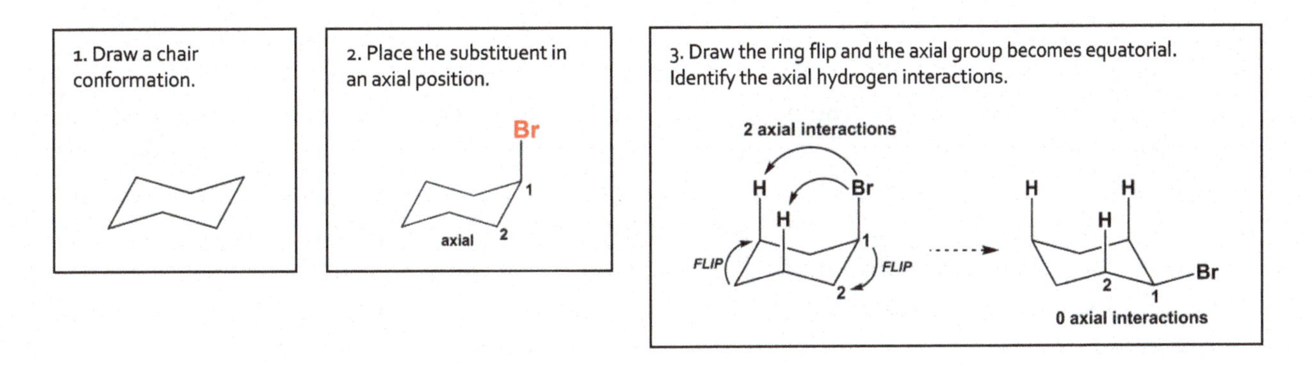

Figure 2.6 Drawing both chair conformations of a monosubstituted cyclohexane ring.

Conformation Analysis of Chair Conformations

A monosubstituted cyclohexane ring is always more stable when the substituent is in the equatorial position, since there are no axial hydrogen interactions. When the ring has more than one substituent, the steric interactions of all substituents must be accounted for. Once the steric interactions for both chair conformations have been analyzed, the favored chair conformation can be determined. Stepwise instructions are given in Figure 2.7 for drawing both chair conformations of a disubstituted cyclohexane ring, identifying axial hydrogen interactions, and calculating total 1,3-diaxial strain for each.

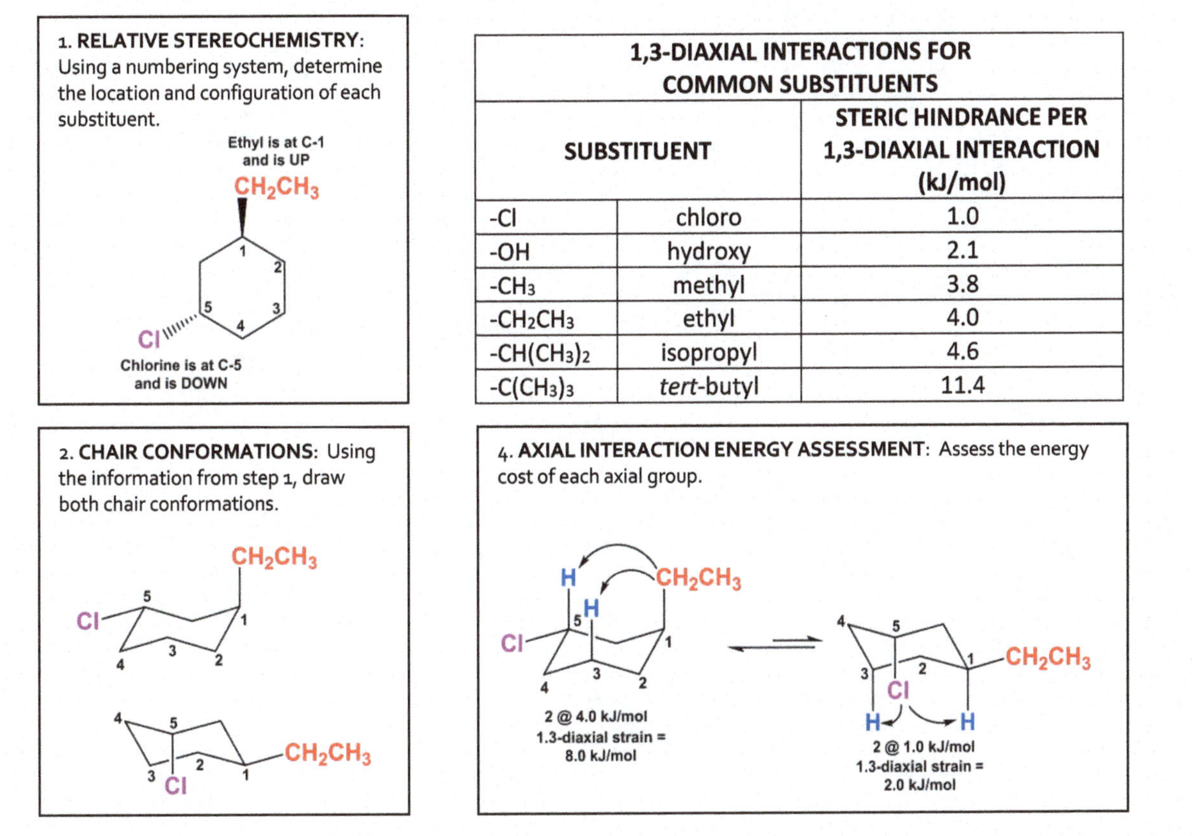

1,3-DIAXIAL INTERACTIONS FOR COMMON SUBSTITUENTS		
SUBSTITUENT		STERIC HINDRANCE PER 1,3-DIAXIAL INTERACTION (kJ/mol)
-Cl	chloro	1.0
-OH	hydroxy	2.1
$-CH_3$	methyl	3.8
$-CH_2CH_3$	ethyl	4.0
$-CH(CH_3)_2$	isopropyl	4.6
$-C(CH_3)_3$	*tert*-butyl	11.4

Figure 2.7 Conformational analysis of a disubstituted cyclohexane ring.

Objectives

On the provided lab worksheet, you will practice sketching various monosubstituted and disubstituted cyclohexane rings. For the disubstituted conformers, the two-dimensional representation will be sketched initially in order to identify the stereochemical relationship between the substituents (*cis* vs. *trans* relationship). This two-dimensional representation will be used to draw both chair conformations three-dimensionally. Conformational analysis of the chair conformations will be used to compare the stability of chair conformations, and to identify the most stable conformer.

References

Klein, David. (2015). *Organic Chemistry*, 2nd ed. Hoboken: John Wiley and Sons.
Brown, Willam, Poon, Thomas (2014). *Introduction to Organic Chemistry,* 5th ed. New Jersey: John Wiley and Sons.

Exp. 2: *Stereochemistry of Cyclohexanes*

Name:			
	Max	Score	Total Grade
Tech	5		
Pre	15		
Post	80		

PRE-LAB ASSIGNMENT: *(**EACH STUDENT** will complete and submit an original copy at the beginning of the lab period. **Without a complete pre-lab assignment, you will not be allowed to perform the experiment, and will receive a zero for the lab.**) .....max score = 15 pts.*

Part 1: Review of Concepts

For the following statements, ***FILL IN THE BLANKS*** using a term from the list below.

chair conformation	axial	angle	1,3-diaxial interactions		
steric	equatorial	torsional	tetrahedral	ring flip	equilibrium

1. ________________________ strain is due to expansion or compression of bond angles, resulting in a deviation of the preferred 109.5° angle preferred in a ________________________ carbon.
2. ________________________ strain is due to eclipsing of bonds on neighboring atoms.
3. ________________________ strain is due to the repulsive interactions which occur when atoms approach each other too closely.
4. Cyclohexane is found in the ________________________________ most of the time, as this form has no torsional strain and very little angle strain.
5. Each carbon atom in a cyclohexane ring can bear two substituents. One substituent is said to occupy an ________________________ position, and the other substituent is said to occupy an ________________________position.
6. When a ring bears one or more substituents, the substituents can occupy either axial or equatorial positions, and these two conformations are at ________________________ with each other.
7. The term ________________________is used to describe the conversion of one chair conformation into the other.
8. Axial substituents generate ________________________________, a form of steric hindrance; therefore bulkier substituents generally prefer the equatorial position.

EXPERIMENTAL RESULTS

(*Complete in pencil or pen!*)

Part 2: Drawing a Chair Conformation of Cyclohexane

9. In the provided space, ***draw a chair conformation*** showing all six axial positions and all six equatorial positions **LABELED**.

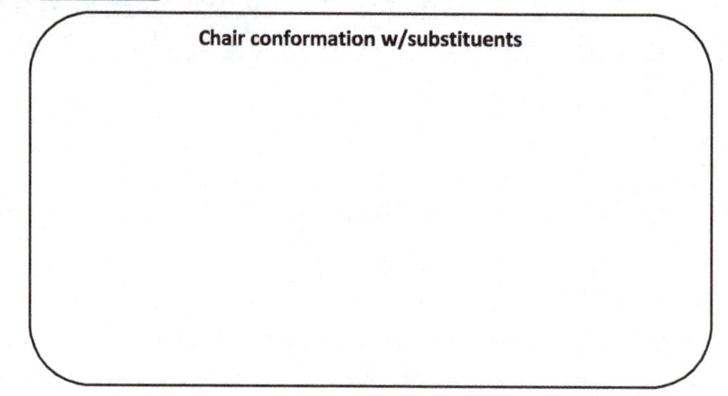

Part 3: Drawing Both Chair Conformations of a Monosubstituted Cyclohexane

10. In the provided space, ***draw both chair conformations*** of *tert*-butylcyclohexane.

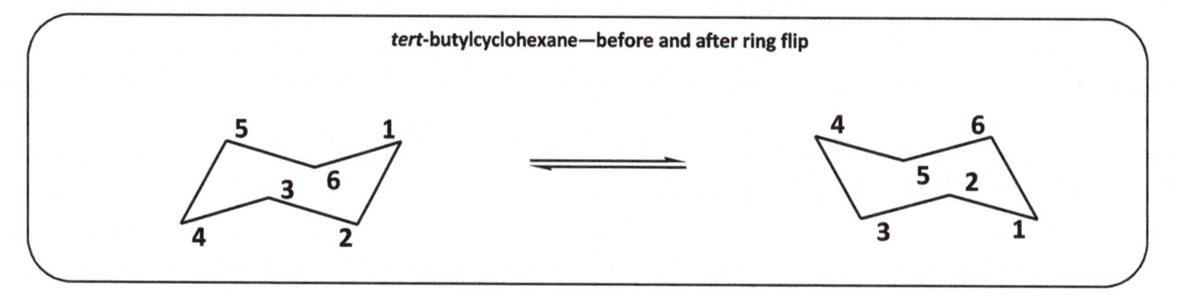

Part 4: Conformational Analysis of Disubstituted Cyclohexanes

11. In the provided space, ***draw both chair conformations*** of *cis*-1-hydroxy-2-methylcyclohexane using the provided numbering system. Show the ***full calculation*** for the 1,3-diaxial strain of each conformation in the provided box, and then ***circle*** the most stable conformer.

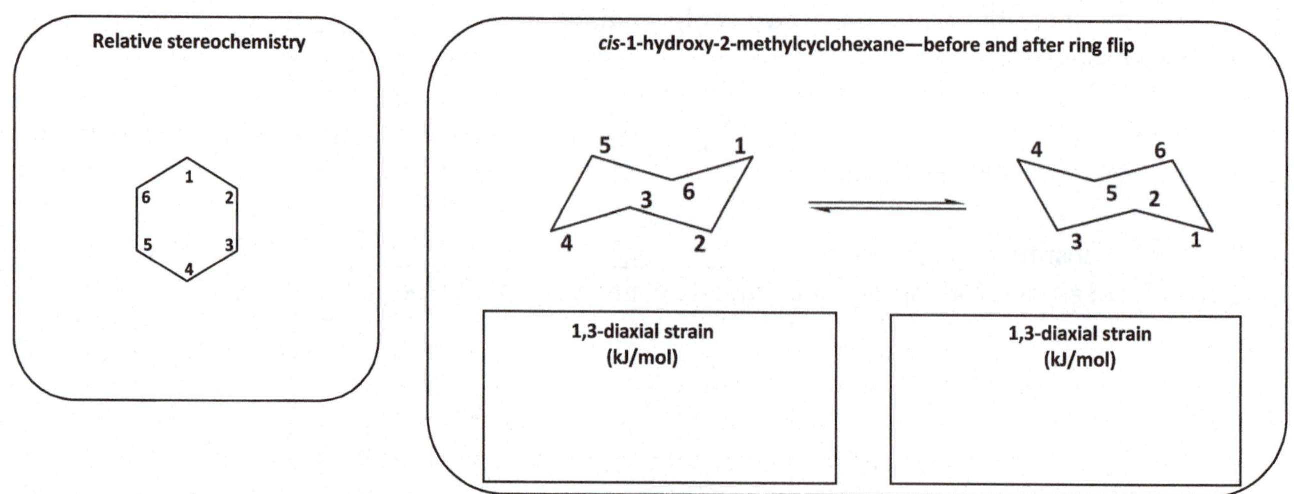

12. In the provided space, **draw both chair conformations** of *trans*-1-isopropyl-3-methylcyclohexane using the provided numbering system. Show the ***full calculation*** for the 1,3-diaxial strain of each conformation in the provided box, and then ***circle*** the most stable conformer.

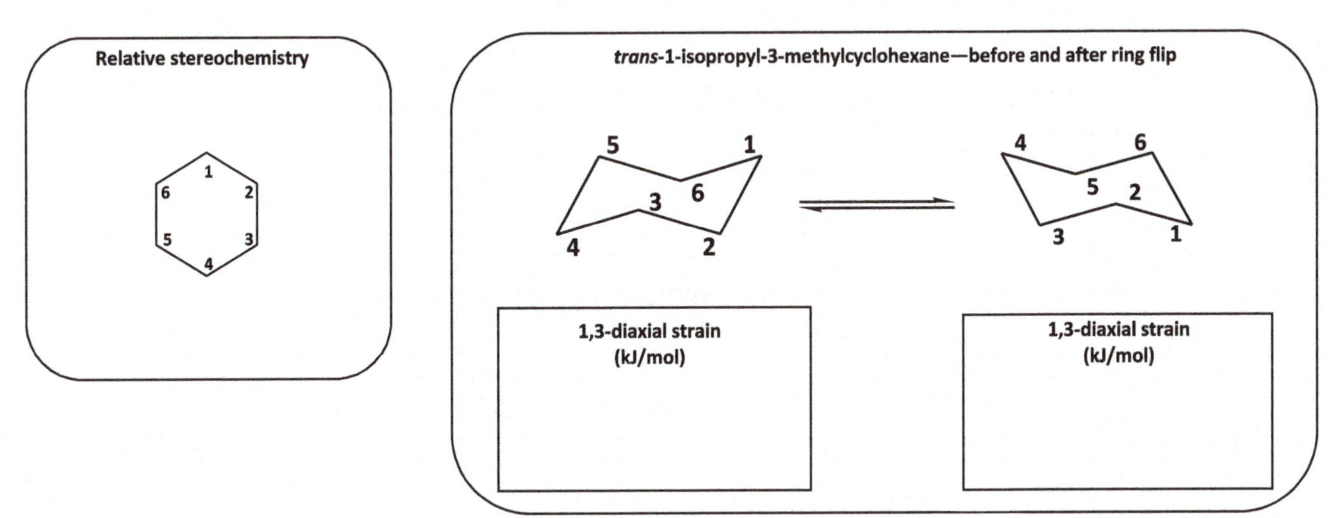

13. In the provided space, ***draw both chair conformations*** of *cis*-1-*tert*-butyl-4-ethylcyclohexane using the provided numbering system. Show the ***full calculation*** for the 1,3-diaxial strain of each conformation in the provided box, and then ***circle*** the most stable conformer.

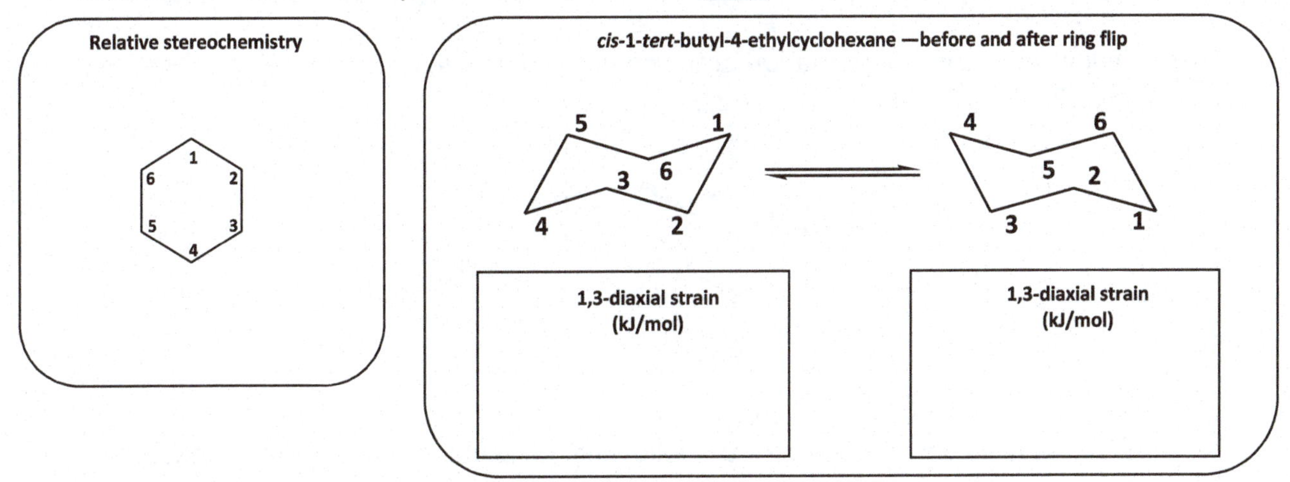

Part 5: General Questions

14. Which of the statements below ***correctly*** describes the chair conformations of *trans*-1,4-dimethylcyclohexane?

a. The two chair conformations are of equal energy
b. The higher energy chair conformation contains one axial methyl group and one equatorial methyl group
c. The higher energy chair conformation contains two axial methyl groups
d. The lower energy chair conformation contains one axial methyl group and one equatorial methyl group

15. Draw the ***most stable*** conformer of *cis*-1-ethyl-3-methylcyclohexane:

16. Draw the ***most stable*** conformer of *cis*-1-ethyl-4-isopropylcyclohexane:

17. In the lowest energy conformation of the compound below, how many alkyl substituents are ***axial***?

a. 0
b. 1
c. 2
d. 3

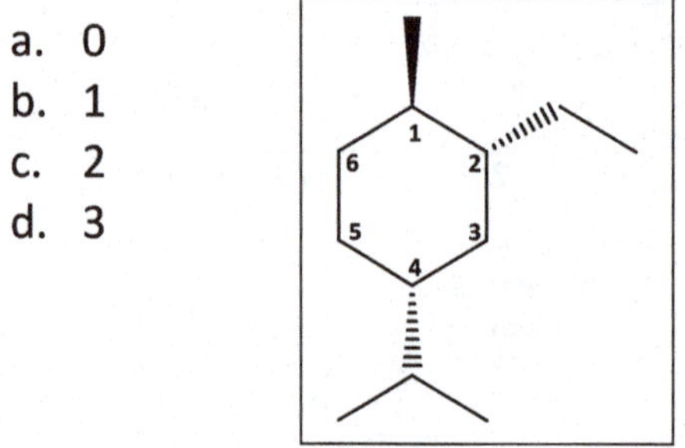

18. Draw both chair conformations for the following molecule, and ***circle*** the most stable conformer.

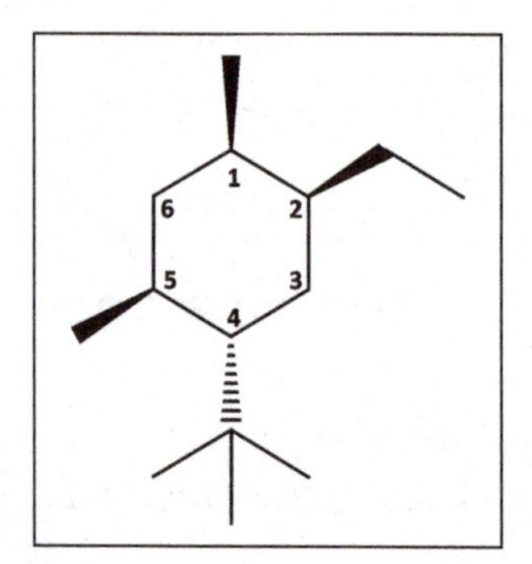

Experiment 3

Stereochemistry: Enantiomers and Diastereomers

Introduction

The importance of stereoisomers in chemistry is illustrated by the different properties compounds have which differ only in their stereochemistry. Understanding how to recognize stereocenters and assign the absolute configuration of an enantiomer or diastereomer is critical in organic chemistry (see Appendix G). In this experiment you will construct molecular models in order to help understand the concepts of chirality, stereogenic carbon atoms, enantiomers, diastereomers, and meso forms. You will learn how to use R-S conventions for designating the configurations of chiral molecules. Finally, you will learn how to convert between line diagrams and Fischer projections. ***It may be helpful to bring your text with you to this lab.***

Importance of Stereochemistry

In biological systems, such as the human body, subtle differences in 3 dimensional molecular structures can have serious implications. One example of the significance of stereochemistry is demonstrated in a drug named Thalidomide (Figure 3.1), which was used in several European countries in the 1950s to suppress the morning sickness experienced by expectant mothers. The drug was dispensed as a 50:50 mixture of mirror image molecules, which is referred to as a racemic mixture. While one stereoisomer of the drug functioned as expected to control morning sickness, the other stereoisomer caused severe birth defects. Over 10,000 children were born with severe deformities such as very short, if not missing, limbs. Due to insufficient pharmaceutical testing, the FDA never approved the sale of Thalidomide in the United States. Through extensive testing, it was discovered that Thalidomide actually racemizes in the body, so administering only one enantiomer of the drug would not prevent the teratogenic effect on humans. In reaction to this tragedy, the drug was pulled from the European market and congress enacted laws requiring more extensive testing of drugs before reaching the market. Today, synthetic organic chemists attempt to develop stereoselective syntheses in which compounds are formed enantiomerically pure.

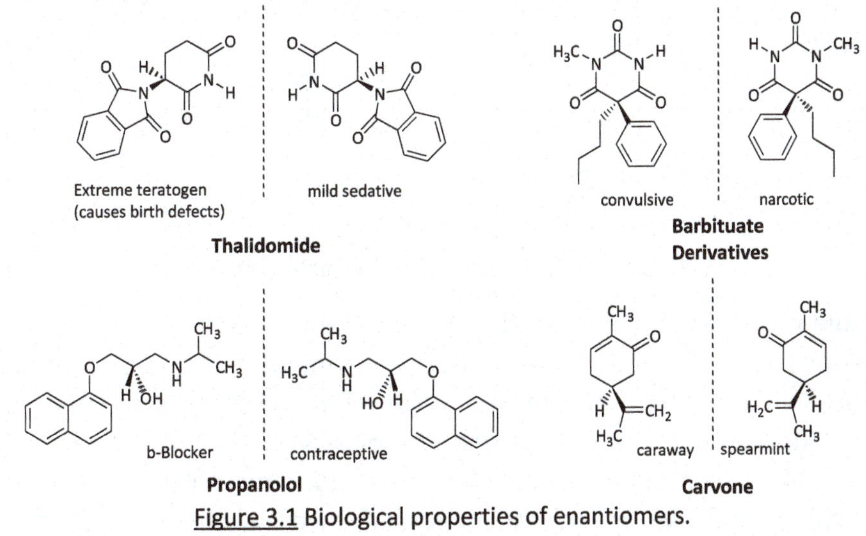

Figure 3.1 Biological properties of enantiomers.

Of course, the importance extends to more than just drugs. All of the proteins that make up our brain, organs, tissues, skin and hair are composed of a single stereoisomer of amino acids. Our bodies can digest starch, but not cellulose, both of which are polymers of glucose which differ only in stereochemistry at the sugar linkage. These are just of a few examples of the importance of stereochemistry.

Chirality and Enantiomers

A **stereogenic center** imparts the property of handedness or **chirality** to a molecule. A carbon that has four different atoms or groups of atoms attached to it is a stereogenic center, a stereogenic carbon atom, or a chiral center. These terms are used interchangeably. The carbon marked with an asterisk in 3-methylhexane (Figure 3.2) is a stereogenic center. Stepwise instructions are shown below for locating chirality centers within a compound (Figure 3.3).

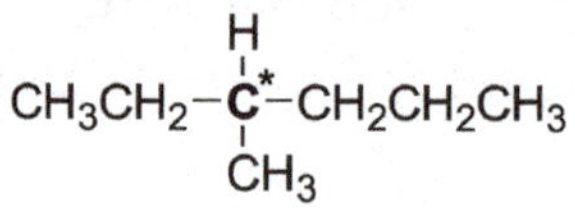

H
$CH_3CH_2-C^*-CH_2CH_2CH_3$
CH_3

Figure 3.2 Structure of 3-methylhexane.

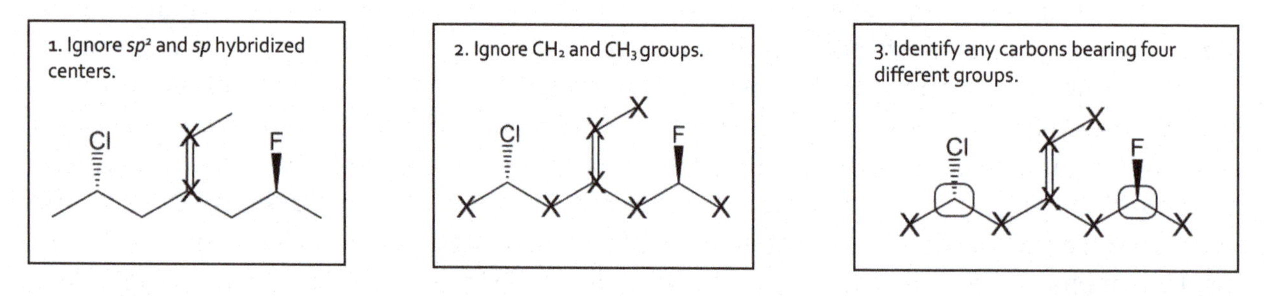

Figure 3.3 Locating chirality centers.

A molecule that is **achiral** has a mirror image that is identical to it. Two molecules are said to be superimposable if they can be placed on top of each other and the three-dimensional position of each atom of one molecule coincides with the equivalent atom of the other atom. We shall see that any molecule with a plane or center of symmetry is achiral (Figure 3.4). Any molecule that has an internal plane of symmetry cannot be chiral, even though it may contain asymmetric carbons.

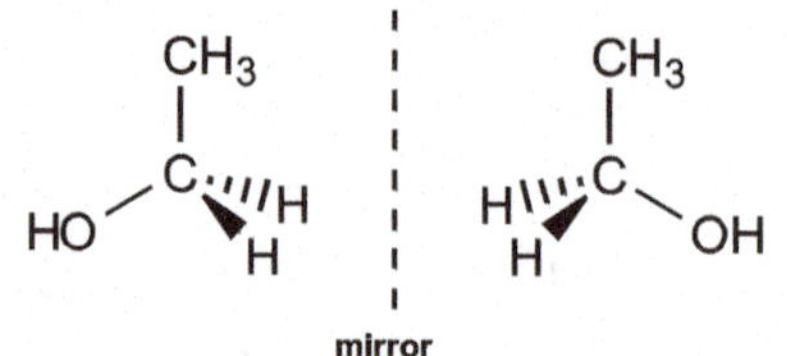

Figure 3.4 Superimposable mirror images of ethanol.

A molecule is said to be **chiral** (that is, having the property of handedness) if its mirror image is ***not*** identical to it. The mirror image of a left hand, for example is a right hand. The left and right hand are non-superimposable mirror images. Two substances with the same molecular structure, which differ only in their 3d arrangement of atoms, are not superimposable on their mirror images, and are called **enantiomers**. The 2-bromobutane molecule shown in Figure 3.5 is an example of a chiral molecule. The mirror images differ from each other *only* in properties that have a direction or "handedness," as, for example, the *direction* (clockwise or counterclockwise) in which they rotate a beam of plane-polarized light. Because of this latter property, such substances are sometimes called **optical isomers**. They are **optically active**.

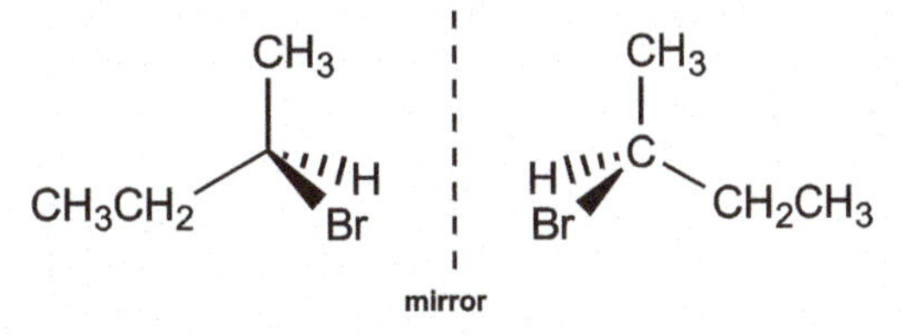

Figure 3.5 Non-superimposable mirror images of 2-bromobutane.

The R-S Convention and Absolute Configuration

The thalidomide structure shown in Figure 3.6 has an asymmetric carbon, and exists as two enantiomeric forms. These mirror images are different, and as previously mentioned, these differences can have serious implications. Both structures are thalidomide; however there is a simple way to distinguish between the enantiomers and give them a unique name.

The difference between the two enantiomers lies in the three dimensional arrangement of the four groups around the asymmetric carbon. The asymmetric carbon has two possible spatial arrangements, called configurations. By naming the two configurations of the asymmetric carbon, specifying and naming the enantiomers can be accomplished.

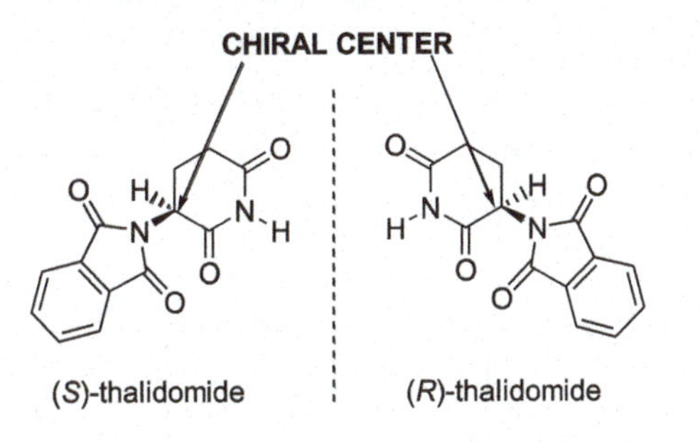

Figure 3.6 Enantiomeric forms of thalidomide.

The Cahn-Ingold-Prelog convention is the most widely used method for naming the configurations of chiral centers. Either the letter ***R*** (*rectus*, or right) or ***S*** (*sinister,* or left) is used to designate the configuration at a stereogenic center. The four atoms or groups attached to the stereogenic center are arranged in a *priority order according to atomic number: the higher the atomic number, the higher the priority.* If two atoms have the same atomic number, we move to the next atoms out from the stereogenic center, or even farther, until we observe a difference in atomic number. We then view the molecule from the side *opposite* the group with the lowest priority. If the remaining three groups in order from highest to lowest priority form a clockwise array, the configuration is ***R***; if counterclockwise, the configuration is ***S***. Stepwise instructions are shown below (Figure 3.7) for assigning configuration to a chiral center. Note that to change **configuration**, we must disconnect and remake bonds. This cannot be accomplished by simply rotating the molecule. However, we can change **conformation** by rotation of groups around single bonds.

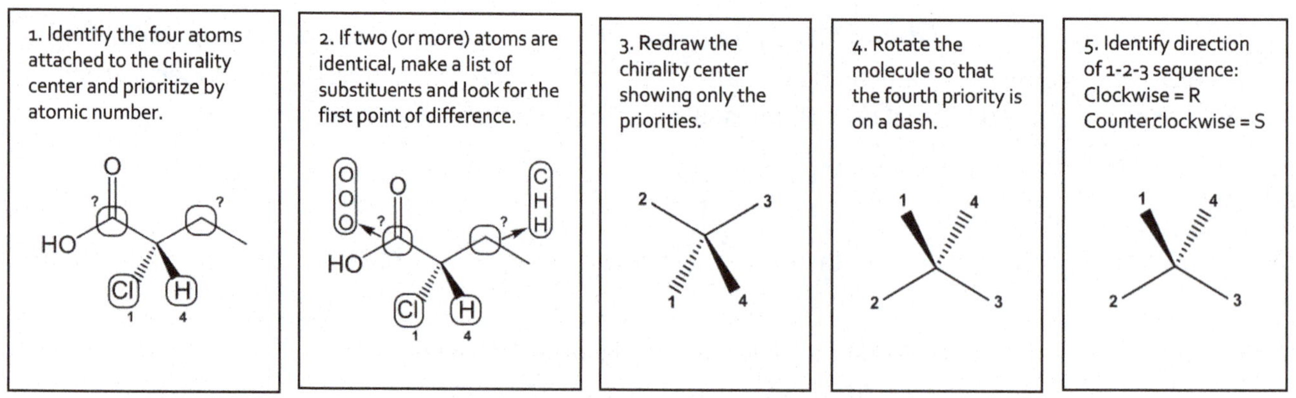

Figure 3.7 Assigning configuration to stereocenters.

Diastereomers and Meso Forms

For any molecule that has *two or more* stereogenic centers, it is possible to have stereoisomers that are not mirror images. A compound with *n* asymmetric carbons can have as many as 2^n stereoisomers, where *n* is the number of chiral centers. The compound 1-hydroxy-2-methylcyclohexane is an example of a compound that has two chiral (or asymmetric) carbons (Figure 3.8). This compound would have 4 stereoisomers. These structures are stereoisomers, since they differ in the 3d arrangement of their atoms, but they are not mirror images of one another, therefore are not enantiomers. Stereoisomers that are not related as enantiomers are called **diastereomers.** Notice that the enantiomers have opposite absolute configuration at both stereocenters, whereas the diastereomers have opposite absolute configuration at one stereocenter, and the same at the other.

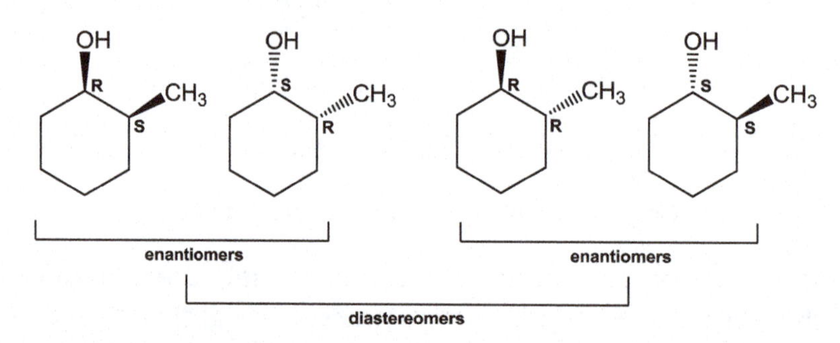

Figure 3.8 Stereoisomers of 1-hydroxy-2-methylcyclohexane.

As previously mentioned, a molecule which contains a plane of symmetry cannot be chiral, even though it may contain chiral centers. These forms are called **meso forms.** This situation arises when a molecule has two *identically substituted* stereogenic centers. Because the molecule has a readily accessible conformation with a plane of symmetry, it is achiral and optically inactive. The structure of 1,2-dichlorocyclohexane is an example of such a compound (Figure 3.9).

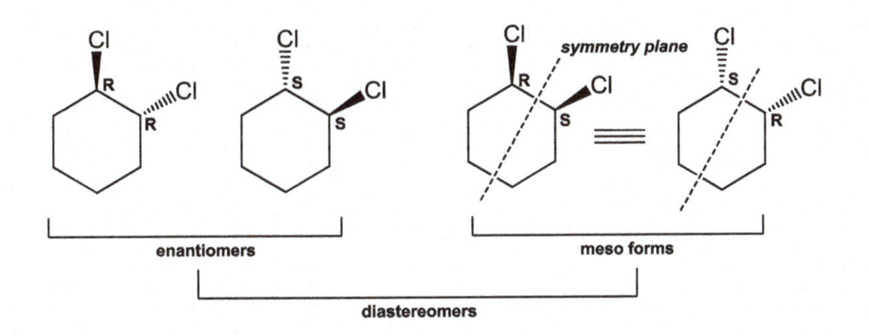

Figure 3.9 Stereoisomers of 1,2-dichlorocyclohexane.

Fischer Projections

Fischer projections are another drawing style that is useful to depict 3-dimensional structures in a 2-dimensional plane. In a Fischer projection, each cross represents a chiral center. The horizontal bonds are considered to be coming out of the page (wedges), while the vertical bonds are considered to be going back into the page (dashes), as shown in Figure 3.10. This method of representing compounds with chiral carbons is particularly useful when investigating compounds with multiple chiral centers, such as sugars, and are helpful for the determination of stereochemical relationships between a pair of stereoisomers.

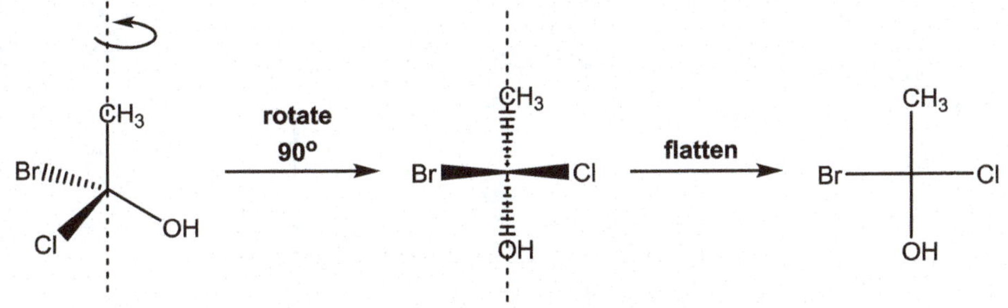

Figure 3.10 Line diagram compared to Fischer projection.

In some cases, it may be necessary to convert a line diagram to a Fischer projection, or convert a Fischer projection to a line diagram. The steps to doing so are outlined in Figure 3.11 and Figure 3.12.

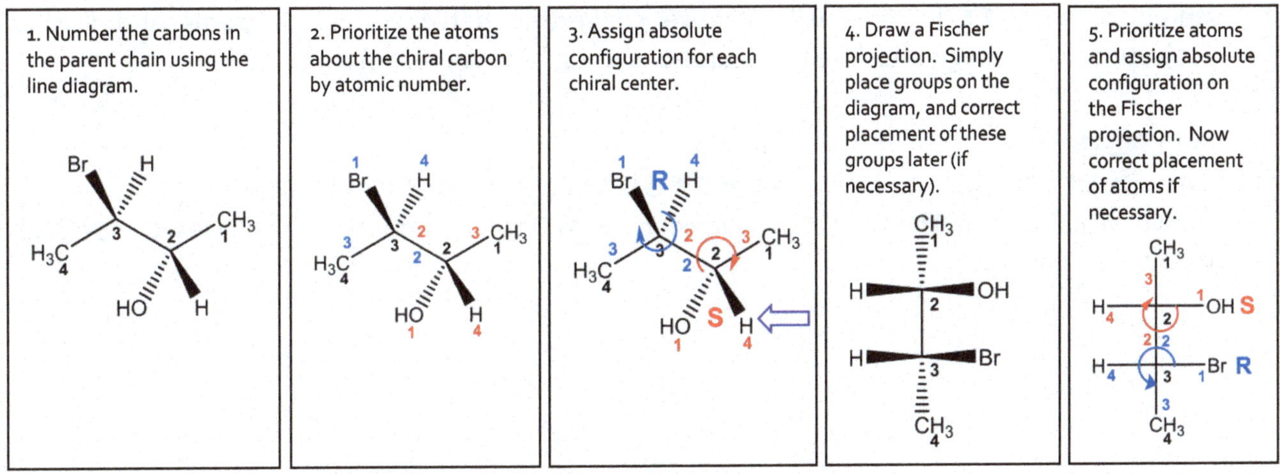

Figure 3.11 Converting a line diagram to a Fischer projection.

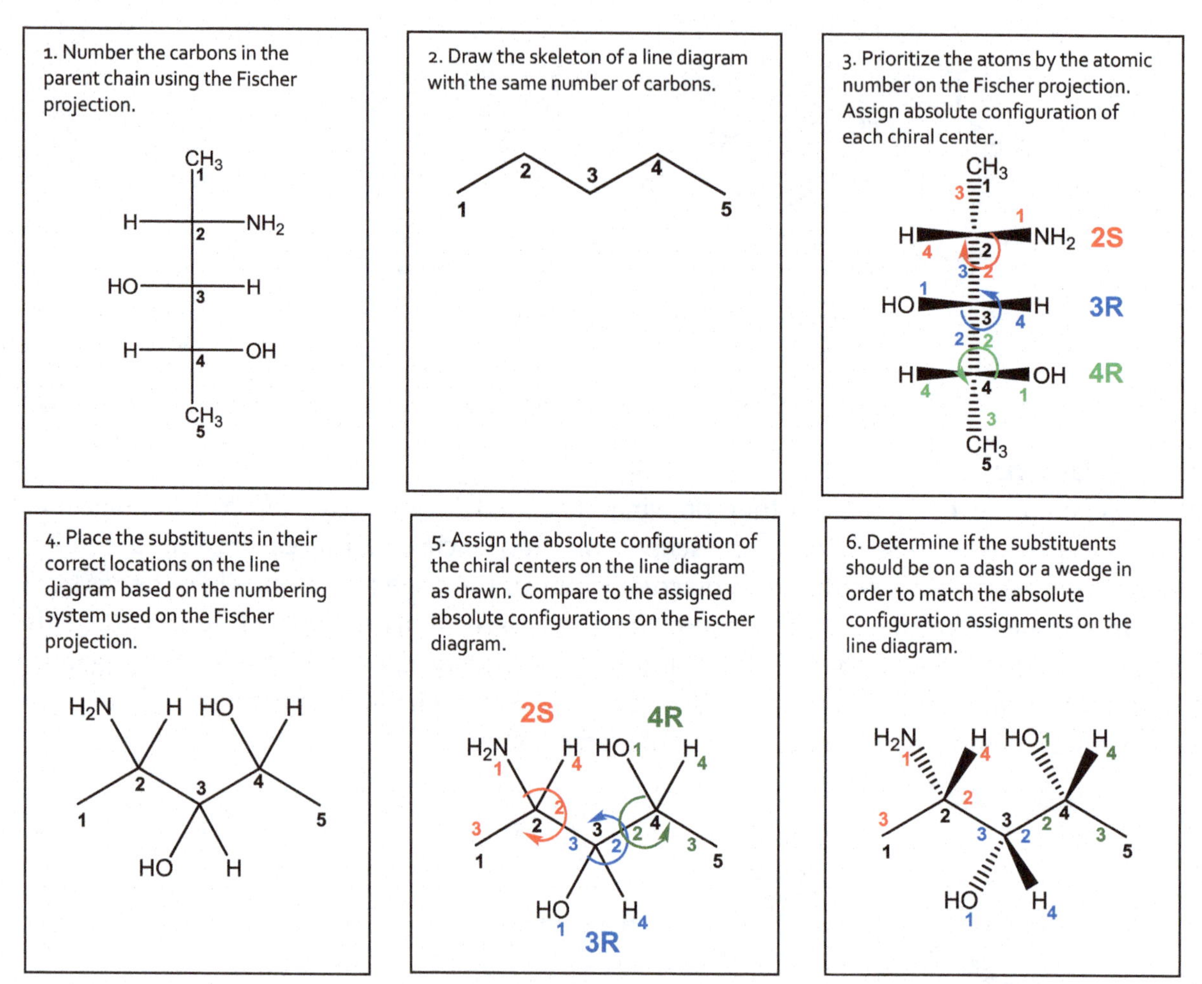

Figure 3.12 Converting a Fischer projection to a line diagram.

Objectives

In this experiment, you will learn how to identify stereocenters using molecular models. You will practice how to assign an order of priority to atoms, and assign absolute configurations to chiral centers using line diagrams, and determine stereochemical relationships between stereoisomers. Finally, you will practice converting between line diagrams and Fischer projections, and assign absolute configurations using both representations.

References

Klein, David. (2015). *Organic Chemistry*, 2nd ed. Hoboken: John Wiley and Sons.
Brown, Willam, Poon, Thomas (2014). *Introduction to Organic Chemistry,* 5th ed. New Jersey: John Wiley and Sons.

Name:			
	Max	Score	Total Grade
Tech	5		
Pre	15		
Post	90		

Exp. 3: Stereochemistry: Enantiomers and Diastereomers

PRE-LAB ASSIGNMENT: *(**EACH STUDENT** will complete and submit an original copy at the beginning of the lab period. **Without a complete pre-lab assignment, you will not be allowed to perform the experiment, and will receive a zero for the lab.**) .....max score = 15 pts.*

Part 1: Review of Concepts

For the following statements, **FILL IN THE BLANKS** using a term from the list below. Refer to *Appendix G* for definitions.

Stereoisomers	meso	polarized light	enantiomer	optically
mirror images	stereogenic carbon atom	diastereomers	absolute configuration	

1. ________________________ have the same connectivity of atoms, but have different 3d arrangement of atoms.

2. **Chiral** objects are not **superimposable** on their _______________ ____________. The most common source of molecular chirality is the presence of a ________________ __________ _____, a carbon bearing four different groups.

3. A compound with one chirality center will have one non-superimposable mirror image, called its ________________________.

4. The Cahn-Ingold-Prelog system is used to assign the _________________ _______________ of a chirality center.

5. A **polarimeter** is a device used to measure the ability of chiral organic compounds to rotate the plane of __________ __________________ __________. Such compounds are said to be _________________________ **active**.

6. For a compound with multiple chirality centers, a family of stereoisomers exists. Each stereoisomer will have at most one enantiomer, with the remaining members of the family being ________________________.

7. A _________________________ **compound** contains multiple stereogenic centers but is nevertheless achiral because it possesses an internal plane of symmetry.

EXPERIMENTAL RESULTS

(*Complete in pencil or pen!*)

Part 2: Locating Chirality Centers & Assigning Priorities to Substituents

8. Identify the chirality centers in each of the following compounds. Mark each chirality center with an asterisk (*):

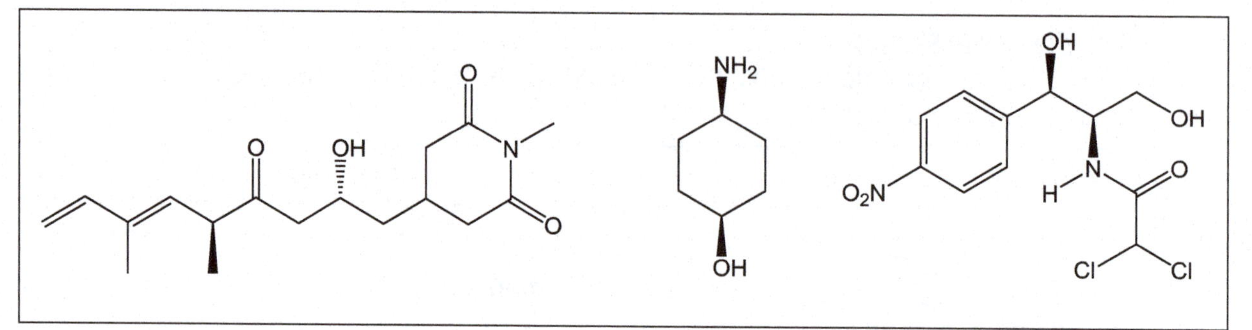

9. Rank the following sets of substituents (1 = highest, 4 = lowest).

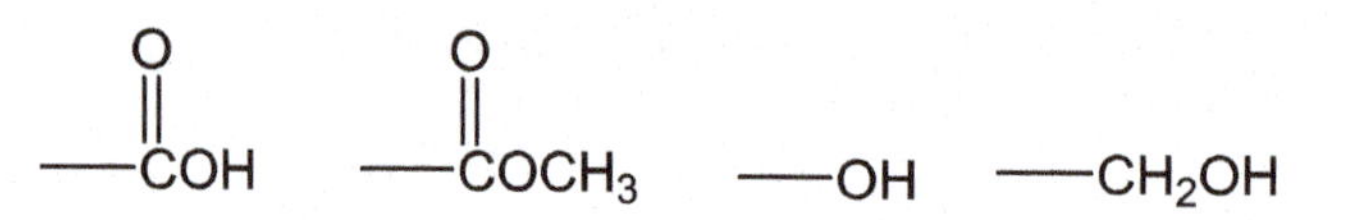

Part 3: Assigning R/S Configuration

10. For the molecules below, number the groups in order of priority. Assign the stereocenter(s) as ***R*** or ***S***.

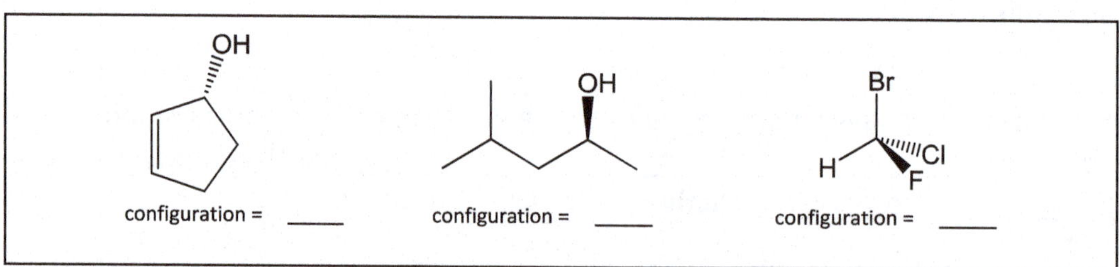

11. For the molecules below, number the groups in order of priority. Assign the stereocenter(s) as ***R*** or ***S***.

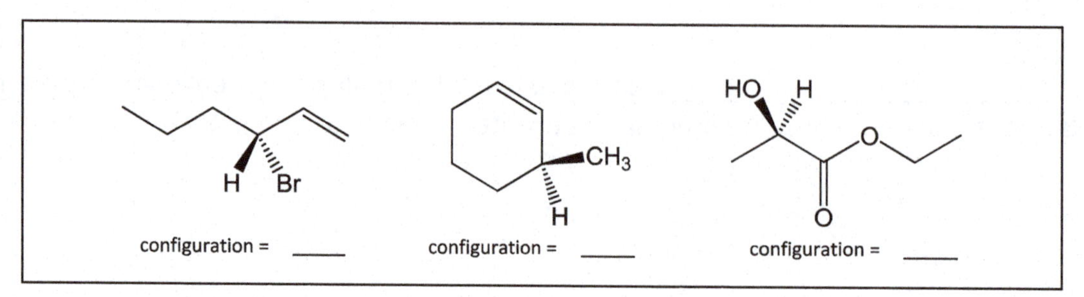

Part 4: Determining the Stereoisomeric Relationship between Two Compounds

12. Assign the absolute configuration for each stereocenter. Which of the following terms best describes the pair of compounds shown: ***enantiomers (E)***, ***diastereomers (D)***, or ***identical (I)***?

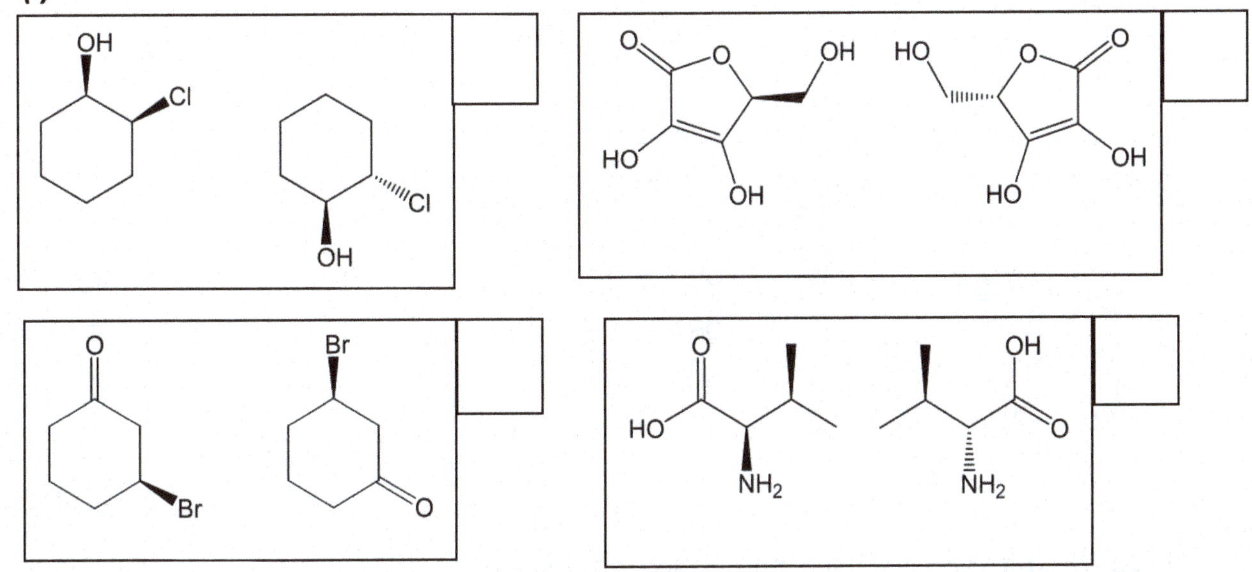

Part 5: Identifying Meso Compounds

13. Each of the following compounds possesses a plane of symmetry. Identify the plane of symmetry in each compound. (***HINT***: In some cases, you may need to rotate a single bond to make it easier to identify the plane of symmetry.)

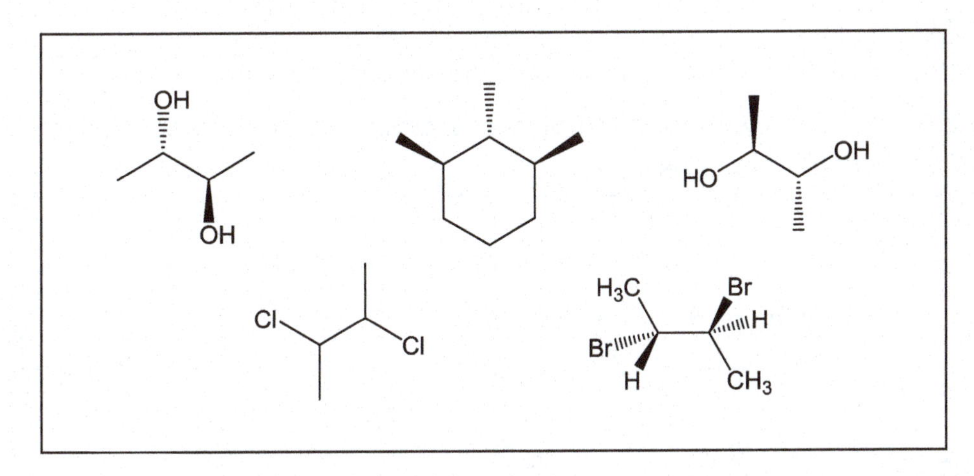

Part 6: Fischer Projections

14. Draw the Fischer projection of the following compound:

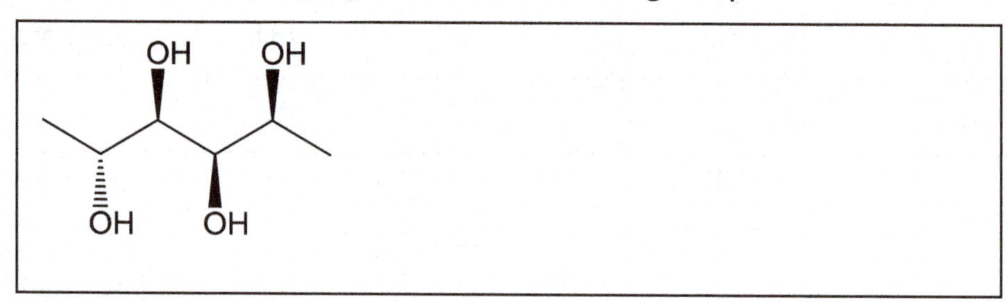

15. Which of the following is a correct Fischer projection of the following compound?

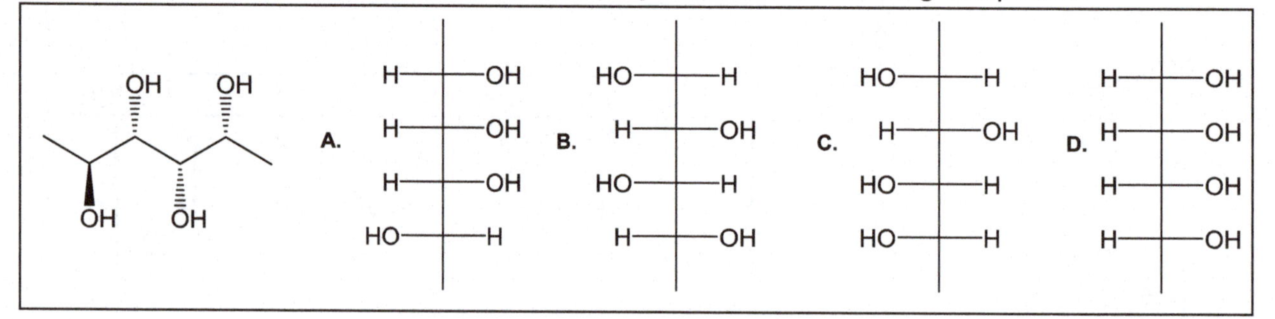

16. Convert the following Fischer projection into a line diagram

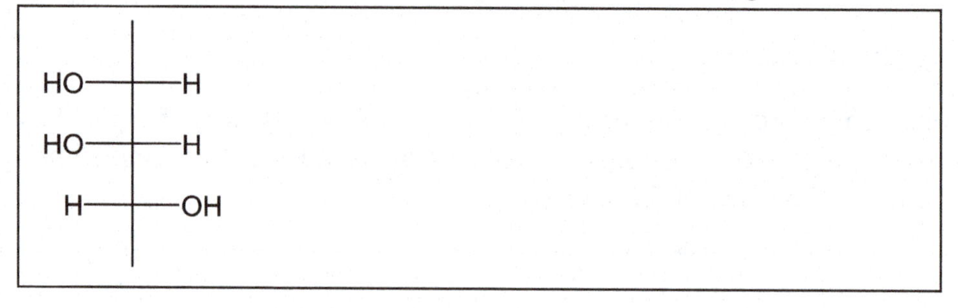

17. For each of the following Fischer projections, assign the configuration at each stereocenter. Then, fill in the blanks below:

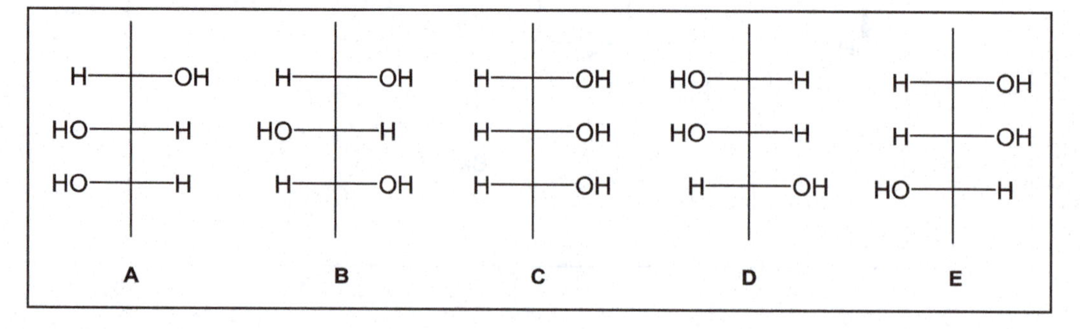

- ***Identical*** to isomer A? _______________
- ***Enantiomeric*** to isomer A? _______________
- ***Diastereomeric*** to isomer A? _______________

Experiment 4

Determination of the Mutarotation Constant of Glucose

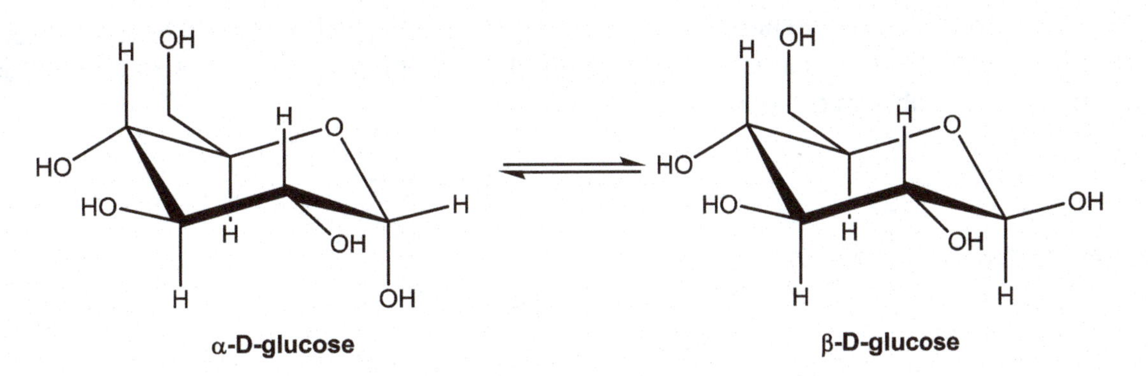

Introduction:

Chiral molecules rotate the plane of plane-polarized light. The extent of this rotation is the specific rotation, a physical characteristic of the compound, which is defined as the number of degrees plane polarized light is rotated upon passing through a 10 cm path of a solution containing 1.0 g of the compound per mL. Since it is not always possible to have these conditions, the specific rotation can be determined by the use of the following formula:

$$[\alpha]_D = \frac{\alpha}{(C)(1)}$$

where:

$[\alpha]_D$ = the specific rotation at the wavelength of the D line of sodium (589.2 nm)
α = the measured rotation, in degrees ((+) is used to denote clockwise rotation, (-) is for counterclockwise rotation.)
C = the concentration of the solute in g/mL
1 = the cell length in dm (10 cm = 1 dm)

The specific rotation can also be determined by measuring the rotation (α) at several different concentrations (C), then by plotting α vs C, establishing the best straight line through the points, and determining the value of α where C = 1.0 g/mL. This method is useful when solubility limitations prevent having a 1.0 g/mL solution, as in the case of glucose.

In this experiment, you will measure the optical rotation of glucose solutions using a polarimeter. You will use this data to determine the equilibrium constant of a 1.0 g/mL glucose solution. Finally, you will use the data collected to determine the concentration of an unknown glucose solution.

Mutarotation of Glucose

Glucose is a chiral molecule and will therefore rotate the plane of polarized light. It can exist in several different forms, which, because of their different structures will rotate plane polarized differently. The two hemiacetal forms interconvert readily in aqueous solution via the open-chain aldehyde. This process is called mutarotation. The purpose of this experiment is to determine the equilibrium constant for the mutarotation:

$$\alpha\text{-D-glucose} \rightleftharpoons \beta\text{-D-glucose}$$

The equilibrium constant has the form:

$$K_{eq} = \frac{C_B}{C_A}$$

where:

K_{eq} = the equilibrium constant,

C_B = the concentration of β-D-glucose in g/mL

C_A = the concentration of α-D-glucose in g/mL

Because the α-form and β-form each have different specific rotations, the concentration of each in a mixture can be determined by measuring the rotation of plane polarized light by their solution.

Objectives

The objective of this experiment is to determine the equilibrium constant for the mutarotation of a 1.0 g/mL solution of glucose. This will be accomplished by measuring the optical rotation of various glucose solutions of known concentrations using a polarimeter. The data collected will then be used to generate a calibration curve, which will be used to extrapolate the angle of rotation of a 1.0 g/mL glucose solution. Once this value is determined, the equilibrium constant will be calculated. Finally, the generated calibration curve will be used to determine the concentration of an unknown glucose solution based on its measured angle of rotation.

* This experiment was developed by Dr. Murray A. Gibas, Chemistry Department, Indiana University-Purdue University at Fort Wayne, and is used with his permission.

Experimental Procedure

- Place a clean empty polarimeter tube in the polarimeter instrument. Look into the eyepiece and turn the eyepiece until the image becomes darkest. Read the number of degrees of rotation from the scale around the eyepiece. The reading should be about 0°. Record it as a blank for that polarimeter; this value must be accounted for in the optical rotation values of each solution.
- Replace the empty tube with one containing a glucose solution, filled to a 10 cm (1dm) path length, of 0.1 g/mL, 0.2 g/mL, 0.4 g/mL, 0.6 g/mL, or a solution of unknown glucose concentration. Each polarimeter has a different glucose solution; you must go to each station, determine its blank reading, and then measure the rotation of the glucose solution at that polarimeter.
- Record the *observed angle of rotation* and the *blank correction* for EACH polarimeter in Table 4.1 on the POST LAB assignment (available online). Apply the *blank correction* from the *observed angle of rotation* to determine the *corrected angle of rotation.*
- Using a computer and *Microsoft Excel*, enter the rotation data in a spreadsheet and prepare a graph of the *corrected angle of rotation* vs. the *known concentrations* of glucose.
- Determine the best-fit straight line, and from the equation of the line, obtain α at 1.0 g/mL. This equation can also be used to calculate values of the concentration that correspond to a given α value, so that the concentration of the unknown glucose solution can be determined.

Calculations:

Because the length of the tube used in this experiment is 1 dm, we can simplify the calculations by omitting the number 1 from the following equation:

$$[\alpha]_D = \frac{\alpha}{C}$$

which may be arranged to give:

$$C = \frac{\alpha}{[\alpha]_D}$$

and:

$$\alpha = C\,[\alpha]_D$$

The total rotation of a binary mixture in which the two components do not interact is the sum of the rotations of the individual components:

$$\alpha = C_A\,[\alpha_A]_D + C_B\,[\alpha_B]_D$$

where the subscripts A and B refer to the α- and β- forms of D-glucose respectively.

It is known that for the pure anomeric forms of D-glucose:

$[\alpha_A]D$ for α–D-glucose = **+ 112.2°**

$[\alpha_B]D$ for β–D-glucose = **+ 18.7°**

Using *Microsoft Excel*, perform a least-squares analysis on the data. To determine α, the rotation at 1.0 g/mL, use the equation of the best fit line that the computer determines. At this point the total glucose concentration is 1.0 g/mL, or:

$$C_A + C_B = 1$$

which may be written:

$$C_B = 1 - C_A$$

This expression for C_B may be put into equation below along with the known [α] values:

$$\alpha = 112.2\, C_A + 18.7\,(1 - C_A)$$

Now that α has been determined, equation has a single unknown: C_A. Solve this equation for C_A, and then use it in equation to determine C_B. Finally, equation, on the previous page, may be used to determine K_{eq}:

$$K_{eq} = \frac{C_B}{C_A}$$

WASTE MANAGEMENT

Be sure to cover polarimeter tubes with Parafilm immediately after performing measurements to prevent evaporation.

Reference

Klein, David. (2015). *Organic Chemistry*, 2nd ed. Hoboken: John Wiley and Sons.

Name:			
	Max	Score	Total Grade
Tech	10		
Pre	20		
In	20		
Post	50		

Exp. 4 Mutarotation of Glucose

PRE-LAB ASSIGNMENT: *(**EACH STUDENT** will complete and submit an original copy at the beginning of the lab period. **Without a complete pre-lab assignment, you will not be allowed to perform the experiment, and will receive a zero for the lab.**) .....max score = 20 pts.*

1. **Objective** *(Write a brief purpose of the experiment in **complete** sentences, addressing all of the following points)*
   - What instrument will be used during this experiment, and what will it be used for?
   - How will the concentration of the unknown solution be determined from this?
   - Once determined, what value will be calculated using this information?

2. **Chemical Structures** *(Draw the chemical structures of the compounds below)*

α-D-glucose	β-D-glucose

3. **Physical Data**

Compound	$[\alpha]_D$
α-D-glucose	
β-D-glucose	

4. Experimental Outline *(Give a brief description of the procedure that will be followed in this experiment in 5 lines or less.)*

1	
2	
3	
4	
5	

5. Pre-Lab Question *(Answer the following question prior to lab.)*

A. Define ***mutarotation***:

B. Why can't the angle of rotation of a 1.0 g/mL solution of glucose be measured directly?

I have read and understood the experimental procedure for this experiment. I am familiar with the hazards and the required disposal procedures for this experiment.

Sign here: ______________________________

Experiment 5

TLC and HPLC Analysis of Analgesics

Introduction

In this experiment you will identify an over-the-counter analgesic using thin layer chromatography (TLC) and high performance liquid chromatography (HPLC). These chromatographic techniques will allow you to separate the components present in the analgesics so that they can be compared to standards and identified.

Analgesics

Aspirin (Figure 5.1) is one of the most popular over the counter drugs available. Aspirin acts as an analgesic (pain reliever), an antipyretic (fever reducer) and an anti-inflammatory (reduces swelling) drug. The discovery of aspirin originates from the centuries old use of willow bark extracts as a pain reliever. In the late 1800s the active ingredient from willow bark was isolated and identified as salicylic acid. The usefulness of salicylic acid was limited by its acidic properties, which resulted in severe irritation of the mouth and stomach. In 1893 Felix Hoffman, a chemist for Bayer Chemical company in Germany, synthesized a derivative of salicylic acid, acetylsalicylic acid, which had few of the undesirable side effects. This drug was first sold commercially as Bayer aspirin in 1899.

Although aspirin has low toxicity, certain individuals have either allergic reactions to aspirin or experience other undesirable side effects. Other analgesics developed over the past 50 years have also gained wide acceptance as over-the-counter drugs. Acetaminophen and phenacetin were developed after it was accidentally discovered (in 1886) that acetanilide had antipyretic properties. Although acetanilide proved to be toxic, its discovery led to development of acetaminophen (the active ingredient in Tylenol) and phenacetin, which is rarely used now, due to its toxicity when taken in high doses. Acetaminophen is an effective analgesic and potent antipyretic, but is not an effective anti-inflammatory drug. Caffeine is also commonly added to analgesics to improve efficacy of the drug.

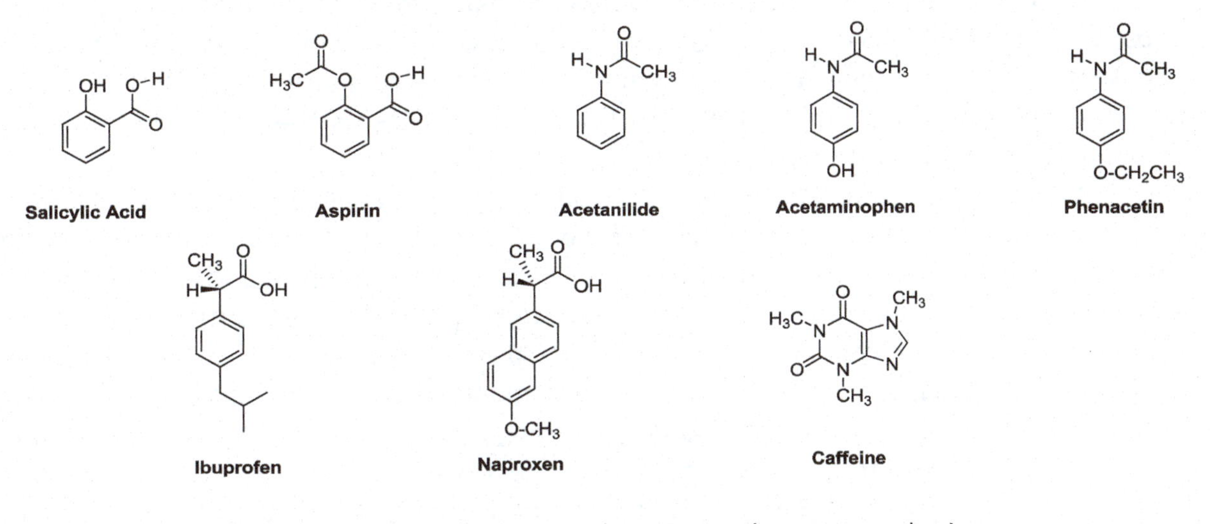

Figure 5.1 Structures of active ingredients in over the counter analgesics.

More recently developed over-the-counter analgesics are ibuprofen (the active ingredient of Advil©) and naproxen (the active ingredient of Aleve©). Both have analgesic, antipyretic and anti-inflammatory properties.

Certain over-the-counter analgesics contain several ingredients. Table 5.1 lists the active ingredients present in some popular over the counter analgesics. Given the different molecular structure of these compounds, it is possible to separate and identify the components using modern chromatographic techniques. Identification of drugs, either in tablets or powders, or in the body, is a very important undertaking, whether for quality control in industry, or to try to determine what is present in an overdose victim. Several chromatographic methods have proven to be powerful techniques for the identification and quantification of drugs. We will use TLC and HPLC to identify both the components and the brand of an unknown over-the-counter analgesic.

Active Ingredient/tablet				
Product	**Aspirin**	**Acetaminophen**	**Ibuprofen**	**Caffeine**
Anacin©	400 mg	---	---	30 mg
Bayer© Aspirin	325 mg	---	---	---
Excedrin© (Extra Strength)	250 mg	250 mg	---	65 mg
Tylenol©	---	500 mg	---	---
Goody's© (Body Pain)	500 mg	325 mg	---	---
Advil	---	---	200 mg	---

Table 5.1 Composition of analgesic tablets.

Thin Layer Chromatography (see also Appendices E and F)

Thin layer chromatography (TLC) is an extremely important and powerful technique in organic chemistry. It can be used to follow the course of a reaction, analyze fractions collected from a chromatographic purification or analyze purity of a compound. TLC can help to identify components present in a mixture, if standards are available.

All forms of chromatography depend on distributing, or partitioning, a dissolved material between a ***mobile phase*** and a ***stationary phase***. In thin layer chromatography the *stationary phase* consists of a thin layer of silica gel (partially hydrated SiO_2) that is supported by a plastic, aluminum, or glass backing. There are many other stationary phases used for TLC, such as alumina, cellulose, reverse phase silica gel, etc., but silica gel is the most commonly used stationary phase in organic chemistry.

As pictured in Figure 5.2, compounds to be analyzed are applied in a row, ~1 cm from the bottom of the TLC plate (the ***origin line***), and the TLC plate is placed in a chamber containing a developing solvent or solvent mixture (the *mobile phase*). The developing chamber also contains a filter paper, which sits in the solvent and serves to keep the chamber saturated with solvent vapors. As the solvent comes in contact with the TLC plate, the solvent rises up the silica layer by capillary action. As the solvent passes through the compounds applied to the plate, the individual components dissolve in the solvent and begin moving up the TLC plate. When the solvent has migrated within 1–2 cm from the top of the TLC plate, the p is removed from the chamber, and the solvent front is marked. Compounds can be visualized by use of an ultraviolet lamp, by staining with iodine vapors, or by spraying with reagents that react with the compounds to produce colored spots.

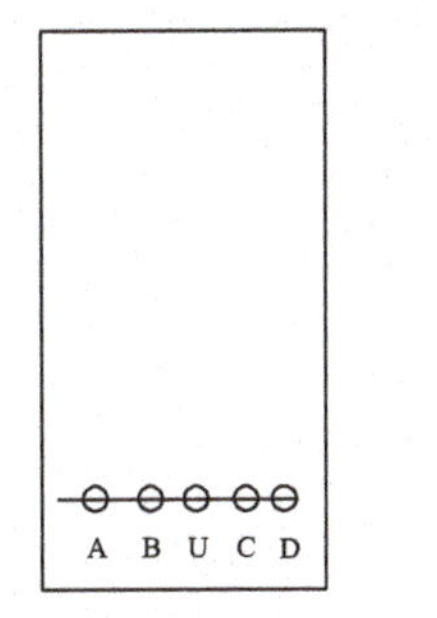

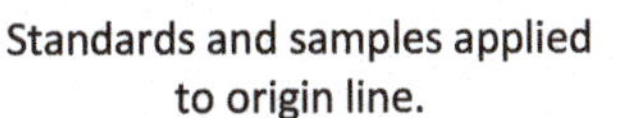
Standards and samples applied to origin line.

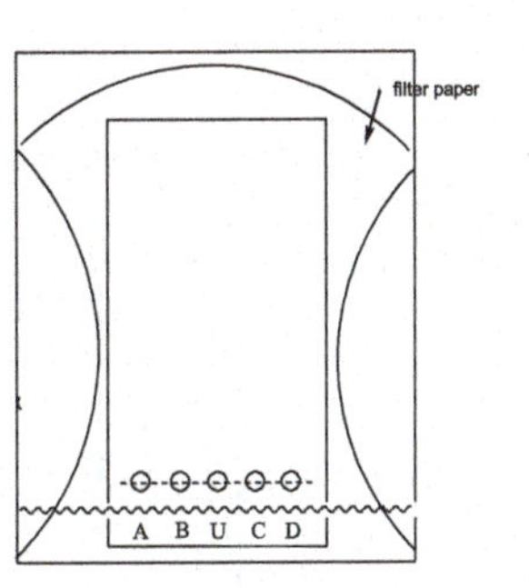

TLC plate placed in developing chamber containing mobile phase.

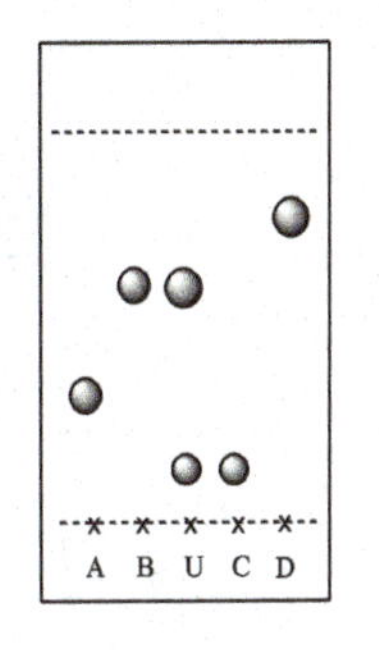

Resulting TLC plate after visualization.

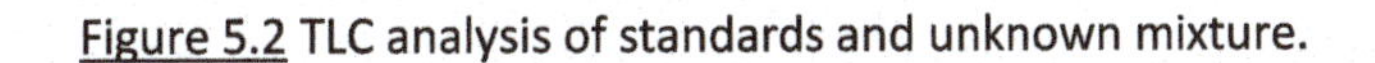
Figure 5.2 TLC analysis of standards and unknown mixture.

Identification of Compounds Based on Retention Factor (R_f)

The TLC R_f (ratio to the front) value of a compound is a ratio of the distance a compound (or analyte) travels up the TLC plate to the distance the solvent traveled (both measured from the origin line). Since the R_f value is a ratio, it is unit less, and usually has two significant figures (Figure 5.3).

A TLC R_f value can be used to aid in the identification of a substance by comparison to standards. The R_f value is ***not*** a physical constant, and comparison should be made only between spots on the same sheet, run at the same time. Two substances that have the same TLC R_f value may be identical; those with different R_f values are not identical. Thus, TLC can serve as a rapid and simple tool for identification of various compounds if appropriate standards are available.

Analyte Polarity vs. R_f Value

Different compounds have differing affinities for the polar silica gel and the nonpolar (relative to the silica gel) developing solvent. The molecular structure and intermolecular forces between the compound and the developing solvent and silica gel on the TLC plate determine how far a compound will travel up the TLC plate. More polar compounds tend to associate more strongly with the very polar silica gel, whereas the less polar components interact weakly with the silica and stay dissolved in the solvent. Thus, the less polar components rise more quickly up the TLC plate, whereas the more polar compounds spend more time "stuck" to the silica gel and travel very slowly up the chromatogram.

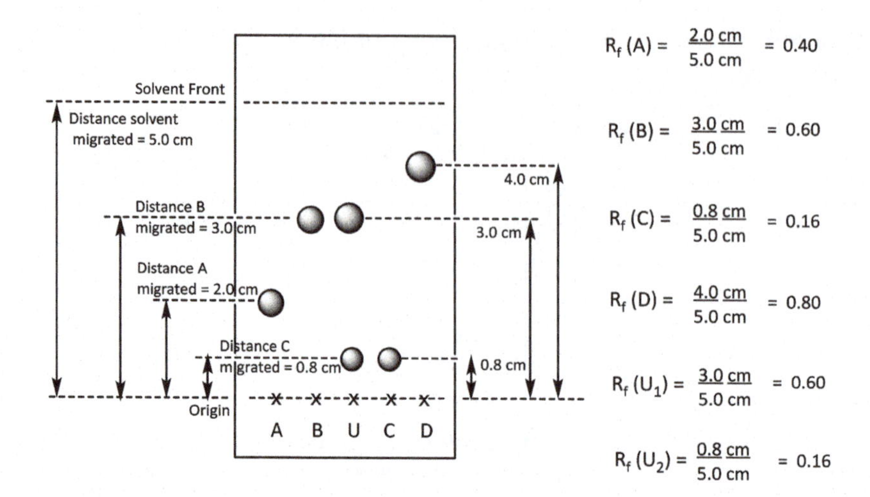

Figure 5.3 Calculation of TLC R_f values. Based on R_f values, unknown U contains compounds B and C.

If a compound travels with the solvent front, the TLC R_f is close to or equal to 1.00. If the compound does not travel at all and stays strongly adsorbed to the TLC plate, the R_f is 0.00. More strongly adsorbed compounds (***more polar*** compounds) don't travel very far up the TLC plate and thus have ***lower R_f*** values. ***Less polar*** compounds have ***larger R_f*** values. The adsorption strength of compounds increases with increasing polarity of functional groups, as shown below:

$-CH=CH_2$, $-X$, $-OR$, $-CHO$, $-CO_2R$, $-NR_2$, $-NH_2$, $-OH$, $-CO_2H$, $-CONR_2$.

(weakly adsorbed, nonpolar) (strongly adsorbed, polar)

Developing Solvent Polarity vs. R_f Value

By changing the developing solvent system, separation and TLC R_f values of components can be manipulated. In *any* solvent system, more polar compounds will be more attracted to the polar silica (the stationary phase) than less polar compounds. The **more polar compounds** will migrate less far up the TLC sheet and thus **have lower R_f values** than less polar compounds.

Changing the developing solvent to a **less polar developing solvent** (a less strongly eluting solvent mixture) will cause virtually **all** of the components to migrate less far up the TLC plate, so that **all of the components will have decreased R_f values.** In less polar solvents, most of the components will be less soluble in the solvent and remain "stuck" to the silica gel on the plate, and will therefore have lower R_f values. However, **more polar developing solvents** will compete with the polar silica gel on the plate, and the components stay dissolved longer in a more polar solvent. Virtually **all** of the components will therefore travel further up the TLC plate so that **all of the components will have higher R_f values.** There are some exceptions to these general trends, but these exceptions are usually based on specific intermolecular forces (especially hydrogen bonding) that can occur between the solvents and the analyte.

There are several measures of the relative elution strengths of different solvents. A commonly used parameter is the **elution strength, ε**, which is based on experimental results with silica gel as the stationary phase (see Table 5.2). As you can see from the values for different solvents, H-bond ***donating*** solvents (such as methanol and ethanol) have the highest elution strengths and nonpolar solvents (like hexane) have low elution strengths. In this experiment you will perform TLC analyses on standards and an unknown mixture in **three** different solvent systems with differing polarities to explore the effect of solvent polarity on TLC R_f values.

High Performance Liquid Chromatography (see also Appendices E and F)

One of the most widely used analytical methods available to the organic chemist today is high performance liquid chromatography (HPLC). HPLC is used routinely to analyze pharmaceuticals or other organic compounds for purity, and to identify substances in complex mixtures. Because analysis is usually performed at room (ambient) temperature or below, even sensitive samples can be analyzed successfully by HPLC.

In principle, HPLC is similar to all forms of chromatography; a mobile phase passes through a stationary phase and components present in a mixture separate due to their different intermolecular forces with the mobile phase vs. the stationary phase. In HPLC the mobile phase (the ***eluting solvent***) is forced through the column by high pressure, typically between 100–1000 psi. Such high pressure requires some specialized instrumentation and typical HPLC tubing and columns are made of small-bore stainless steel. One major difference between TLC (as well as column chromatography) and HPLC is that in HPLC the stationary phase particle size is much smaller. This allows for more surface area and leads to better separation, but the pressure drop from the start of the column to the end is also much greater. Hence, high-pressure pumps are needed to force the eluting solvent through the column.

As in TLC, a variety of stationary phases are available for HPLC. When a polar material, like silica gel, is used as the stationary phase, it is called *normal phase* HPLC. One of the most valuable and commonly used stationary phases is a C-18 modified silica gel, in which organic groups have been covalently bonded to the surface of the silica. A common organic group is the C-18 group; this means that an 18-carbon hydrocarbon chain (nonpolar) is attached covalently to the surface of the silica. This effectively converts the silica from a very polar surface to a nonpolar one, and it causes nonpolar analytes to be more attracted to the C-18 modified silica than polar analytes. Thus, we would expect polar compounds to elute first, followed by increasingly less polar compounds, with nonpolar substances eluting last (Figure 5.5). This is the opposite elution order from that observed on silica gel TLC or ordinary column chromatography and is called *reverse phase* HPLC. ***We will use a <u>normal phase</u> HPLC column for this experiment.***

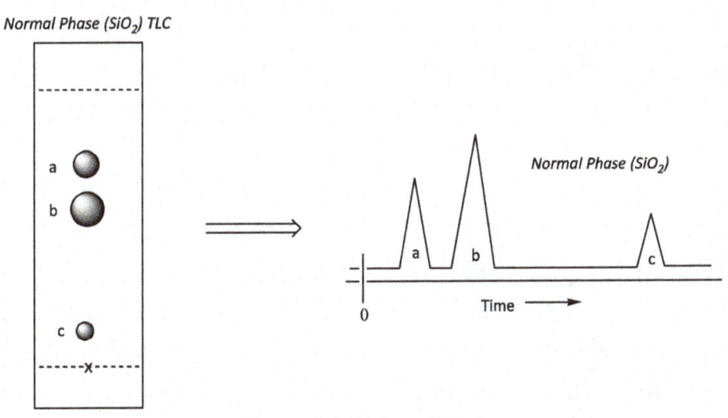

Figure 5.5 TLC vs. HPLC.

Compound Detection

The method of detection used for TLC analysis in this experiment is a hand held UV lamp. The HPLC is equipped with a very sensitive UV-Vis absorption detector. Any sample that absorbs in the 220–800 nm wavelength range can be detected by UV; the absorption is converted to a peak on a chromatogram. This detector relies upon the absorbance of ultraviolet (UV) or visible (Vis) light by the analyte. If an analyte is colored, it must absorb visible light (400–700 nm). If it is colorless, it may still absorb in the UV range (200–400 nm). Most compounds that absorb UV/Vis light have aromatic rings or other multiple bonds. Most solvents used in lab have little ability to absorb UV/Vis light and are therefore not detected, such as ethyl acetate, ethanol, water, and hexane. Thus, ***it is unusual to see a solvent peak in the HPLC chromatogram or on a TLC plate***.

Objectives

In this experiment you will learn the separation techniques of Thin Layer Chromatography and High Pressure Liquid Chromatography. You will use these techniques to separate and identify active ingredients present in an unknown analgesic sample. You will then identify the unknown analgesic based on the active ingredients present in the sample.

Experimental Procedure

Product Analysis:

TLC Analysis

- To prepare a TLC developing chamber, insert a truncated circle of filter paper around the inside of the jar.
- Add ~5 mL of the first developing solvent (15:85 hexane/ethyl acetate) in the jar. The flat side of the filter paper should be in the solvent. Close the jar. The solvent will then wick up the filter paper, saturating the chamber with solvent vapors.
- With a soft pencil, lightly draw a line ~1 cm from the bottom of the TLC sheet. This will be the origin line on which the solutions will be applied. Make 5 small marks, equidistant apart, on the origin line with a pencil. Label these marks A, B, U, C, and D under the origin line.
- You will be provided with standard solutions of aspirin (A), acetaminophen (B), caffeine (C), and ibuprofen (D). The lane marked (U) will represent your unknown analgesic solution.
- Apply each of the provided standard solutions to the appropriate lane, along with the sample solution, to the origin line, using a clean capillary tube for each standard solution.
- Prior to developing the TLC sheet in the chamber, view the plate under a UV lamp to ensure the spots are concentrated enough. If a dark spot is visible on the origin line before development of the TLC sheet, the sample will be visible after the development.
- Using tweezers, place the TLC sheet in the TLC chamber carefully, so that the solvent reaches the origin line slowly and evenly. The TLC plate should lean against the side of the jar. Once the TLC plate is in the solvent, DO NOT MOVE THE TLC CHAMBER.

- Allow the solvent to migrate up the plate until it is ~1 cm from the top of the plate. Remove the plate and quickly mark the solvent front with a faint pencil mark before the solvent evaporates.
- Visualize the TLC plate using a UV lamp (short wave). Circle the spots; measure the migration of the compounds from the origin line to the center of the spot.
- Draw a diagram of the TLC plate and enter all cm measurements in the laboratory notebook. Complete Table 5.1 using this data. Calculate TLC R_f values for each standard and any component present in the sample mixture.
- Perform two more TLC analyses using the procedure described above with *100% ethyl acetate* and *15:85 methanol/ethyl acetate* as the developing solvents. Be sure to sketch all TLC plates in the laboratory notebook, and in Table 5.1 (provided online).

HPLC Analysis

- Standard HPLC chromatograms, along with unknown sample chromatograms, will be provided online. Use these chromatograms to complete Table 5.2.

SAFETY

****All experiments should be performed in a fume hood with appropriate safety glasses. Gloves will be provided by request.****

All organic solvents used in this experiment are flammable, irritants, and can be toxic if ingested or absorbed through skin.

WASTE MANAGEMENT

Place all liquid waste from TLC sample preparation and TLC developing chambers into container labeled "LIQUID WASTE—TLC." Place all used TLC capillary tubes in the broken glass container, and TLC plates in the yellow solid waste trashcan under the supply hood. Leave TLC chambers in the drawer with the cap off. Do ***not*** clean with soap and water.

References

Klein, David. (2015). *Organic Chemistry*, 2nd ed. Hoboken: John Wiley and Sons.
Palleros, Daniel R. (2000). *Experimental Organic Chemistry*. New York: John Wiley and Sons.
Pavia, D. L., Lampman, G. M., & Kriz, G. S. (1988). *Introduction to Organic Laboratory Techniques, a Contemporary Approach*. New York: Saunders College Publishing.

Solvent	MF MW (g/mol)	bp (°C) Density (g/mL)	Hazards	Dipole Moment (μ)	Elution Strength (ε)
Hexane $CH_3(CH_2)_2CH_3$	C_6H_{12} 86.17	68.7 0.659	Flammable Toxic	0.08	0.01
Toluene $C_6H_5CH_3$	C_7H_8 92.13	110.6 0.867	Flammable Toxic	0.31	0.22
Diethyl Ether $CH_3CH_2OCH_2CH_3$	$C_2H_{10}O$ 72.12	32.6 0.713	Flammable Toxic CNS depressant	1.15	0.29
Dichloromethane CH_2Cl_2	CH_2Cl_2 82.92	39.8 1.326	Toxic Irritant Cancer suspect	1.12	0.32
Ethyl Acetate $CH_3CO_2CH_2CH_3$	$C_2H_8O_2$ 88.10	77.1 0.901	Flammable Irritant	1.88	0.25
Acetone CH_3COCH_3	C_3H_6O 58.08	56.3 0.790	Flammable Irritant	2.69	0.23
1-butanol $CH_3CH_2CH_2CH_2OH$	$C_2H_{10}O$ 72.12	117.7 0.810	Flammable Irritant	1.75	0.27
Propanol $CH_3CH_2CH_2OH$	C_3H_8O 60.09	97.0 0.802	Flammable Irritant	1.72	0.63
Ethanol CH_3CH_2OH	C_2H_6O 26.07	78.5 0.789	Flammable Irritant	1.70	0.68
Methanol CH_3OH	CH_2O 32.02	62.7 0.791	Flammable Toxic	1.70	0.73
Water HOH	H_2O 18.02	100.0 0.998		1.87	>1

Table 5.2 Properties of commonly used solvents for normal phase (on SiO_2) chromatography.

Name:			
	Max	Score	Total Grade
Tech	10		
Pre	20		
In	20		
Post	50		

Exp. 5 TLC and HPLC Analysis of Analgesics

PRE-LAB ASSIGNMENT: *(**EACH STUDENT** will complete and submit an original copy at the beginning of the lab period.* ***Without a complete pre-lab assignment, you will not be allowed to perform the experiment, and will receive a zero for the lab.****)* *.....max score = 20 pts.*

1. **Objective** *(Write a brief purpose of the experiment in **complete** sentences, addressing all of the following points)*
   - What techniques will be introduced in this experiment?
   - What goal will be accomplished using these techniques?
   - What effect will be changed that could possibly affect the outcome of the experiment?

2. **Compound Structures** *(Complete the following table before coming to lab.)*

aspirin	acetaminophen	caffeine	ibuprofen
ethyl acetate	**hexane**	**methanol**	

3. Experimental Outline *(Give a brief description of the procedure that will be followed in this experiment in 5 lines or less.)*

1	
2	
3	
4	
5	

4. Pre-Lab Questions *(Answer the following questions prior to lab.)*

A. TLC plates separate molecules on the basis of:

a. Polarity
b. Color
c. Size
d. Bond orientation

B. Define ***stationary phase***:

C. Define ***mobile phase***:

I have read and understood the experimental procedure for this experiment. I am familiar with the hazards and the required disposal procedures for this experiment.

Sign here: ______________________________

Experiment 6

Extraction of Analgesics

Introduction

In this experiment you will separate aspirin from acetaminophen, the two analgesics present in "Goody's Pain Formula" powders© (Figure 6.1). Goody's powders are over-the-counter analgesics that contain aspirin (500 mg/powder) and acetaminophen (325 mg/powder). You will rely on the fact that aspirin is a stronger acid than acetaminophen, allowing you to separate the two compounds using the technique of *acid-base extraction*. You will then evaluate the effectiveness of your separation by HPLC. In Experiment 7, you will purify your aspirin and acetaminophen by recrystallization and evaluate your crystals by melting point and HPLC analysis.

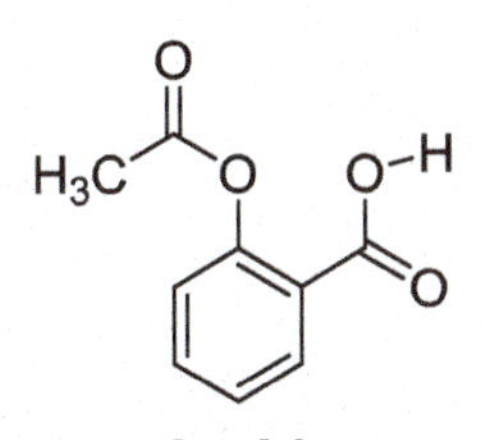

Aspirin
Acetylsalicylic acid
Solubility in H_2O:
1 g/100 mL at 37 °C
1 g/300 mL 21 °C
1 g/400 mL 15 °C
~1 g/550 mL at 0 °C

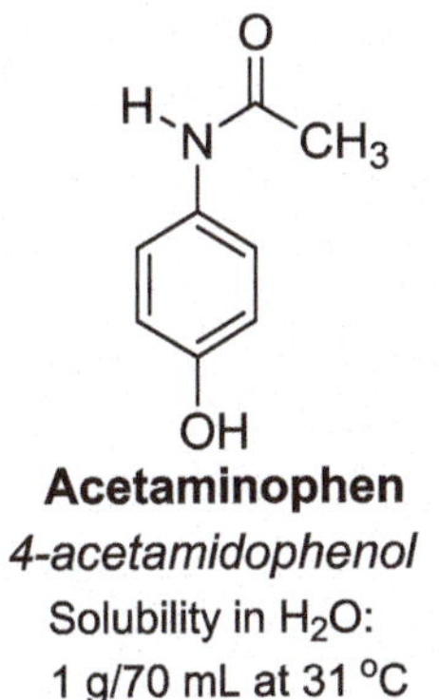

Acetaminophen
4-acetamidophenol
Solubility in H_2O:
1 g/70 mL at 31 °C
1 g/150 mL at 21 °C

Figure 6.1 Solubility of aspirin and acetaminophen at various temperatures.

Extraction

Extraction is the separation of a substance from one phase by another phase. When brewing coffee, caffeine and other compounds are extracted from the solid coffee grounds by hot water. There are numerous other types of extractions. In organic chemistry, liquid–liquid extraction is an extremely important technique. The phases are two immiscible solvents, such as water and ethyl acetate. When a compound is shaken in a separatory funnel with water and an immiscible organic solvent, some of the compound dissolves in the organic layer and some of it dissolves in the aqueous layer. It *partitions* between the two layers. Relatively nonpolar organic compounds are more soluble in the organic layer, whereas very polar compounds and salts are more soluble in the aqueous layer.

Separatory Funnel Technique

In the laboratory, extraction is carried out using a separatory funnel. The separatory funnel is supported between agitations in an iron ring on a ring stand. The iron ring should be cushioned by a rubber flask support to avoid breaking the glass funnel. Remember to tighten the

Teflon stopcock before adding any liquid, since it should have been stored with the stopcock loose. It is best to fill the separatory funnel no more than 3/4 full to allow room for adequate agitation. After the immiscible solvents or mixtures have been transferred to the funnel, the upper opening of the separatory funnel is sealed tightly using a ground glass stopper. The separatory funnel should be held in one hand so that your fingers hold the stopper in place. Gently invert the funnel and open the stopcock to release the pressure that builds up. Be sure to point the stem of the funnel into the hood, away from yourself and others nearby, since the pressure that builds up can cause some liquid to spurt out. Gradually the funnel can be agitated more and more vigorously, venting less and less frequently, as the pressure no longer builds. Vigorous agitation should be carried out continuously for 30-60 seconds. Then the funnel is returned to the ring stand and the layers are allowed to separate. After removing the glass stopper, the lower layer is drawn off the bottom through the stopcock into an Erlenmeyer flask, and then the top layer is ***poured*** out through the upper neck of the funnel (to avoid contamination).

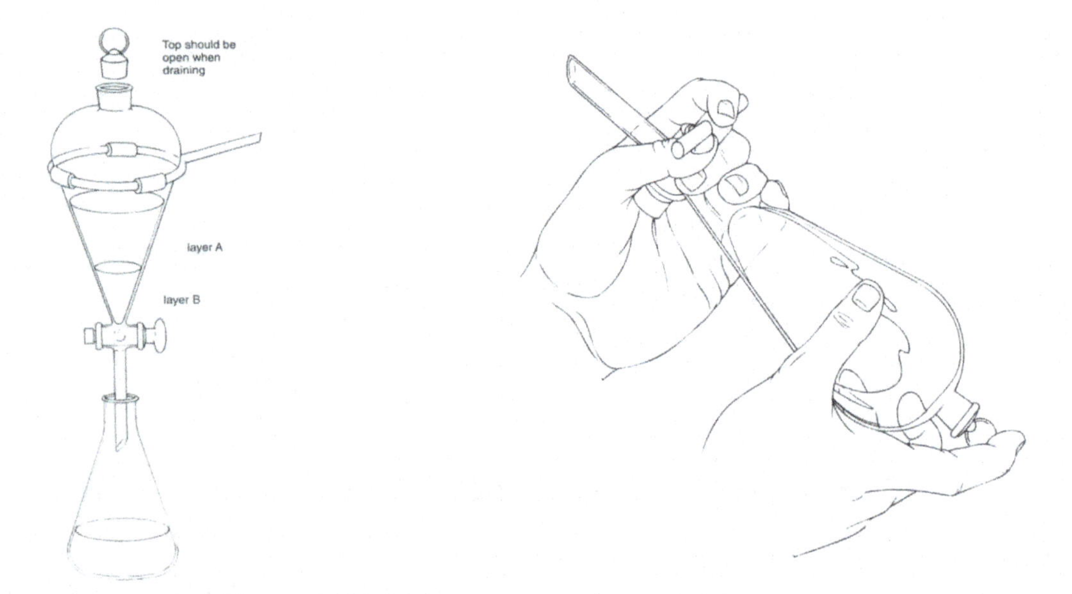

Figure 6.2 How to suspend and hold a separatory funnel.

Properties of Extraction Solvents

The solvent used for extraction should have many of the following properties to be satisfactory. The solvent should readily dissolve the substance to be extracted; it should not react with the solute or another solvent; it should have a low boiling point so that it can be readily removed by evaporation; and it should be relatively nontoxic (refer to Appendix J). Finally, it should not be miscible with water (typically the second phase). No solvent meets all of the criteria, but several come close.

Emulsions

Sometimes a clean separation between layers does not occur; instead, a cloudy emulsion is formed. Several techniques are available to break an emulsion. The simplest, although sometimes time-consuming, is to wait and allow the phases to separate. Alternatively, the emulsion can often be broken by adding a salt such as NaCl or NaBr, either as a solid or a saturated aqueous solution. Lastly, the emulsion can be filtered with suction through a Büchner funnel to obtain a two-phase filtrate, free of emulsion. If only a small amount of emulsion appears at the interface between phases, don't be concerned about getting rid of it. If part of an aqueous phase is mistakenly transferred with the organic phase, it will be removed during subsequent washing, or during drying and filtration.

Washing

Often, after several extractions with an organic solvent, the combined organic phase is returned to the separatory funnel and is "washed" with water or a saturated salt (NaCl) solution. This procedure simply involves extracting the recombined organic extracts with an aqueous solution, and is done to remove water-soluble impurities prior to drying. ***The difference between "washing" and "extracting" depends on the fate of the solvent added.*** The extracting solvent is usually retained, whereas the wash solvent is usually discarded.

Identifying the Phases

In order to determine which phase is organic and which is aqueous, one can be guided initially by the relative densities of the solvents used. For example, water (1.00 g/mL) is more dense than diethyl ether (0.71 g/mL), but less dense than methylene chloride (1.34 g/mL). However, a mixture has a different density than the pure solvent, so it is best to check the identity of the phases by adding a milliliter of water to a milliliter of the phase in question in a test tube. If the water forms a separate phase (either floating or on the bottom), the phase in question is organic. If only one phase is formed, the phase in question is aqueous. It is good procedure to save all phases (clearly labeled) until the final product has been isolated, to avoid the mistake of discarding the wrong phase.

Drying Organic Solvents

Whenever organic solvents contact water, some water dissolves in them, even though they are considered immiscible with water. The amount of water that will dissolve in an organic solvent is dependent on the intermolecular forces between the solvent molecules and water molecules, as you discovered in Experiment 1. For example, the solubility of water in hexane is 0.01 g per 100 g of hexane, whereas the solubility of water in diethyl ether is 1.26 g per 100 g of ether and the solubility of water in ethyl acetate is 3.30 g per 100 g. Therefore, it is necessary to remove the water from an organic phase after an extraction. This is called *drying* the solvent and can be done by adding a ***drying agent***, which is usually an anhydrous salt that readily forms a hydrate with water molecules. Commonly used drying agents are listed in Table 6.1.

The amount of drying agent required depends on the solvent to be dried, the volume of solvent, and the drying agent selected. Ordinarily the amount is not measured accurately, but rather the drying agent is added in small increments, over a 5 to 10 minute time interval, to the Erlenmeyer flask containing the solution to be dried. The flask should be swirled after addition of the drying agent. If all of the drying agent particles clump or stick to the bottom of the flask, additional drying agent should be added. The solvent is dry when drying agent particles swirl freely (like a snow storm). The drying agent is removed by decanting or by filtration. Subsequent evaporation of the volatile solvent from a tared (preweighed) flask allows recovery of the compound isolated by extraction as well as determination of its weight.

Drying agent	Hydrate	Efficiency	Speed	Capacity
$MgSO_4$	$MgSO_4 \cdot 7\ H_2O$	Medium	Fast	High
$CaCl_2$	$CaCl_2 \cdot 2\ H_2O$	High	Fast	Low
$CaSO_4$	$CaSO_4 \cdot 1/2\ H_2O$	High	Fast	Low
Na_2SO_4	$Na_2SO_4 \cdot 7\ H_2O$	Low	Slow	High

Table 6.1 *Commonly used drying agents.*

Acid–Base Extraction

Acid–base extraction relies on changing the solubility of a substance by chemically altering the compound; changing an organic acid to its conjugate base or converting an organic base to its conjugate acid (Figure 6.3). Aspirin has two functional groups, a neutral ester and a carboxylic acid, which is a weak acid ($pK_a = 3.5$). Carboxylic acids are converted to their *sodium carboxylate salts* when treated with a base, such as sodium hydroxide or sodium bicarbonate, as shown below. The resulting *carboxylate salt* is now ionic and soluble in water, even though the original neutral compound was not soluble in water (as you observed for benzoic acid in Experiment 1). Thus, if a carboxylic acid was initially dissolved in an organic solvent, such as ethyl acetate, *extraction* of that solution with aqueous sodium bicarbonate would convert the acid to its conjugate base and *extract* it from the organic layer and into the aqueous layer. Treating the aqueous layer with HCl will result in re-protonation of the carboxylate group, thereby converting it back to the carboxylic acid, which is insoluble in water and will precipitate from the solution.

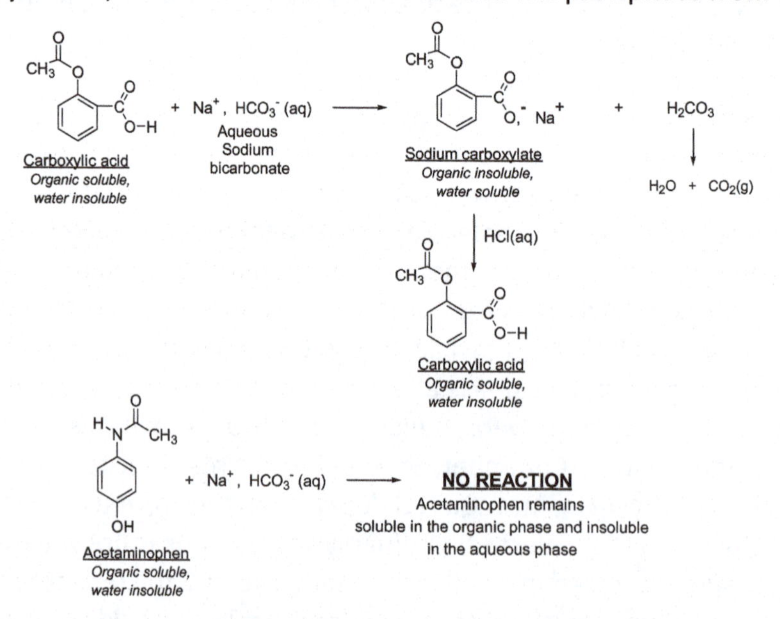

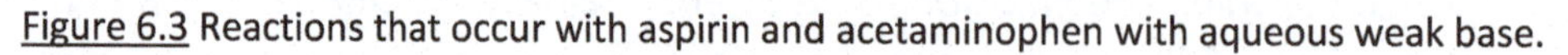

Figure 6.3 Reactions that occur with aspirin and acetaminophen with aqueous weak base.

Acetaminophen contains a phenol and a neutral amide functional group. Phenols are very weak acids (pK_a ~10) and will only be deprotonated by a strong base, such as aqueous NaOH (pH > 11). Acetaminophen will ***not*** be converted to its conjugate base with the weaker base, such as sodium bicarbonate, but ***will*** be deprotonated if a stronger base is used. Thus, since aspirin is more acidic than acetaminophen, these two chemicals can be separated by *acid–base extraction* using sodium bicarbonate, a weak base, as shown in the following flowchart (Figure 6.4). The aspirin can be converted to its carboxylate salt with a weak base, and isolated in the aqueous layer. The acetaminophen phenol will not be converted to a phenoxide ion with a weak base, thus will remain dissolved in the organic layer.

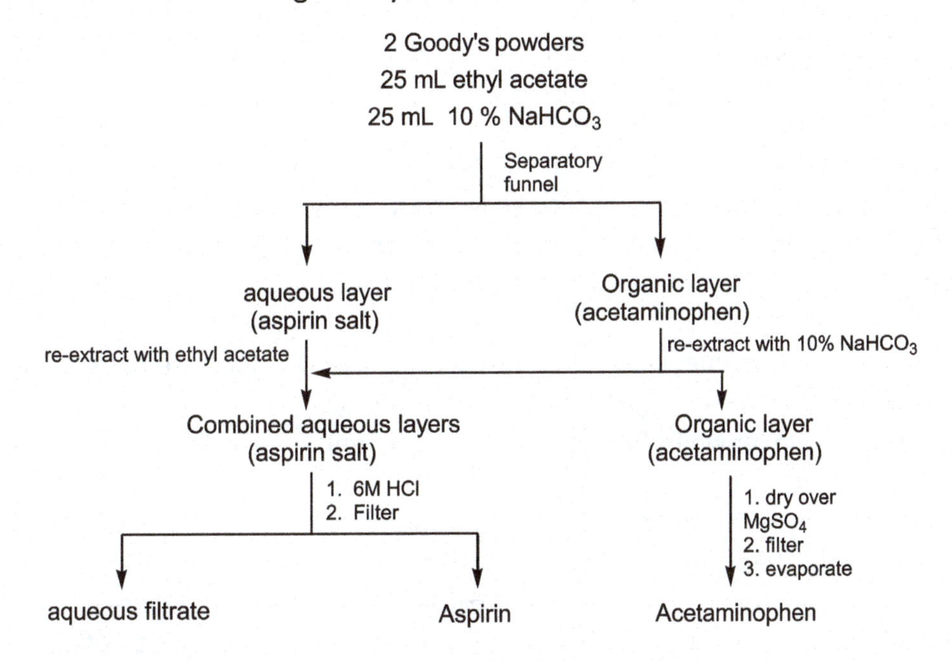

Figure 6.4 Flowchart for separation of aspirin and acetaminophen.

In the presence of aqueous sodium bicarbonate, aspirin is converted to its carboxylate salt and is extracted into the aqueous sodium bicarbonate layer, leaving acetaminophen in the organic layer. An additional extraction of the organic phase with aqueous $NaHCO_3$ removes any remaining aspirin from the organic phase. The combined aqueous layers are "back-extracted" with ethyl acetate to remove any residual acetaminophen from the aqueous layer.

The aqueous layer, containing the carboxylate salt of aspirin, is acidified with HCl. The resulting re-protonated carboxylic acid is now relatively insoluble in water and precipitates from the aqueous solution. The aspirin is filtered and dried for further analysis and purification (Experiment 7, Recrystallization, Melting Point, and HPLC Analysis of Analgesics).

The combined organic layers contain the acetaminophen. However, since water is also slightly soluble in ethyl acetate (3.3 g/ 100 g), it must be removed before further purification and analysis of the acetaminophen. Magnesium sulfate ($MgSO_4$) is used to dry the ethyl acetate solution. After filtration and evaporation of the solvent, the acetaminophen will be further purified (Experiment 7, Recrystallization, Melting Point, and HPLC Analysis of

Analgesics). After obtaining the weights of your aspirin and acetaminophen, you can calculate the percent recovery for each. You will evaluate the effectiveness of your extractions by HPLC.

Objectives

In this experiment the technique of acid–base extraction will be introduced. This technique will be used to separate aspirin from acetaminophen in Goody's Body Pain powders. The techniques of drying organic solvents and vacuum filtration will also be used. Once isolated, the purity of your aspirin and acetaminophen samples will be analyzed using HPLC analysis. Both samples will be used in a following experiment.

Experimental Procedure

Purification:

Acid–Base Extraction

- Set up a ring stand and separatory funnel. Make sure the separatory funnel is clean and does not leak. The separatory funnel is suspended on an iron ring, cushioned with a rubber flask support. Remember to close the stopcock before use. The stopcock will be in the horizontal position when closed.
- Using two 125 mL Erlenmeyer flasks from your lab drawer, label one flask "**O**" and the other flask "**A**" to represent your organic and aqueous phases.
- Using a short-stem powder funnel, pour the contents of two Goody's Powders into the separatory funnel.
- Add 25 mL of ethyl acetate and place the stopper in the funnel. Give the stopper a twist to ensure a tight fit. Shake funnel to mix.
- Add 25 mL of 10% sodium bicarbonate (10% $NaHCO_3$) to the separatory funnel. Stopper the funnel and while holding the cap in place, invert the funnel pointing the tip up and into the hood. Open the stopcock to vent any pressure that builds up or from any heat generated during the acid–base reaction that has taken place in the funnel.
- Close the stopcock. While still holding the cap, shake the funnel for ~ 10 seconds, invert the funnel and vent again. Repeat this process a few times and then place the separatory funnel in a rubber flask support on an iron ring attached to a ring stand and let the two layers separate.
- Remove the stopper and drain the lower (**aqueous**) layer through the stopcock into the "**A**" flask and set aside. Pour the top (**organic**) layer out of the top of the separatory funnel into the "**O**" flask and set aside.
- Pour the aqueous layer (**"A"**) back into the separatory funnel. Add an additional 10 mL of ethyl acetate to the separatory funnel. Agitate and vent as instructed above. Allow the layers to separate, and then drain the bottom aqueous layer from the separatory funnel back into the "**A**" flask.
- Add the first organic layer (**"O"**) back into the separatory funnel. Add an additional 10 mL of 10% $NaHCO_3$, agitate and vent as instructed above. This step will generate a *second* aqueous layer in the funnel. Drain the bottom aqueous layer into the "**A**" flask.
- Pour the top organic layer out the top of the separatory funnel back into the "**O**" flask. You should now have one flask marked "**O**," which contains your acetaminophen dissolved in

ethyl acetate, and another flask marked "**A**," which contains the aspirin salt dissolved in the aqueous base. **DO NOT DISCARD ANY SOLUTIONS!**

Recovering the Acetaminophen; Drying the Organic Phase

- Add a small portion (~1/2" on the tip of the scoopula) of anhydrous $MgSO_4$ to the "**O**" flask containing the organic layers from above (ethyl acetate solution of acetaminophen). Swirl the flask and watch how the solid settles. If it clumps and settles very quickly, add more $MgSO_4$. Repeat this process until the newly added $MgSO_4$ does not clump, but flows freely when the flask is swirled (think: snow globe).
- Set up a suction filtration apparatus (Appendix A). Place a small filter paper in the top of the Büchner funnel, and seat (or *wet*) the filter paper with ice cold ethyl acetate.
- Apply vacuum by connecting a red vacuum hose to the vacuum line. Slowly pour the solution into the center of the Büchner funnel, and filter off the drying agent under vacuum.
- Transfer the liquid filtrate from the filter flask to a *preweighed* 150 mL beaker containing two to three boiling chips. Place the beaker on **WARM** hot plate (setting of 3, ***NO HIGHER***) and heat until all of the ethyl acetate has evaporated, leaving a colorless residue.
- Reweigh the beaker to determine the mass of the recovered acetaminophen. This value is the ***actual recovery*** of acetaminophen (in grams) after the extraction. Record this value in the laboratory notebook, and in Table 6.1 (provided online). **DO NOT DISCARD SAMPLE!**
- Calculate percent recovery for acetaminophen. Record this value in the laboratory notebook, and in Table 6.1. ***PROCEED TO PRODUCT ANALYSIS.***

Recovering the Aspirin

- The carboxylate anion of aspirin will be re-protonated to regenerate aspirin, which is much less soluble in water, especially in cold water.
- Cool the aqueous layers from above (aqueous solution of the aspirin) by placing the "**A**" flask in an ice bath prepared in your 400 mL beaker.
- Add ~8 mL of 6M HCl, 1 mL at a time. As the acid is neutralizing the excess $NaHCO_3$, you will notice vigorous bubbling of CO_2, after which the re-protonated aspirin will precipitate from solution. Using pH Hydrion litmus paper, check to see whether the solution is acidic. If not, add more HCl. When acidic, add 10 drops more of the 6M HCl to ensure maximum aspirin precipitation. Allow the mixture to cool in the ice bath for ~10 minutes to maximize precipitation of aspirin.
- Set up a suction filtration apparatus (see Appendix A). Place a small *preweighed* filter paper in the top of the Büchner funnel, and seat the filter paper with ice cold deionized water.
- Slowly pour the aspirin solution into the center of the Büchner funnel, and isolate the aspirin crystals from the liquid under vacuum.
- Rinse the flask with ~5 mL of ice-cold deionized water and filter through the funnel containing the crystals. Use as much ice-cold water necessary to transfer all solid aspirin from your flask to the Büchner funnel.
- Allow the solid to dry under vacuum for several minutes. ***PROCEED TO PRODUCT ANALYSIS.***
- At the beginning of the next lab period, reweigh the aspirin sample to obtain the ***actual recovery*** of aspirin. Calculate the percent recovery of aspirin. Record these values in the laboratory notebook, and in Table 6.1 on the POST LAB Assignment (provided online).

Product Analysis:

HPLC Analysis

Acetaminophen Sample Preparation:

- Prepare an HPLC sample of your pure acetaminophen. Place a few crystals of the purified acetaminophen into an auto analyzer vial, and add HPLC solvent solution (ethyl acetate).
- Cap the vial and shake until all of the solid dissolves (samples with visible solid will be discarded). Place the sample into a vial slot and sign out on sample sheet.
- Prepare the remaining acetaminophen for storage. Dissolve the acetaminophen residue in the beaker with a small amount of **reagent** acetone and transfer this solution to a large sample vial. Use the minimal amount of acetone necessary to transfer **all** acetaminophen solution/crystals to the sample vial.
- Label this vial with your name and your instructor's name, and submit the vial to your lab instructor until the next lab period.

Aspirin Sample Preparation:

- Repeat the steps above to prepare an HPLC sample of your pure aspirin.
- Prepare the remaining aspirin for storage. Weigh a large filter paper. Using a microspatula, carefully transfer the aspirin crystals and small filter paper from the Büchner funnel to the larger filter paper. Secure the aspirin sample by folding the large filter paper and securing the edges with a small amount of tape.
- Label this filter paper with your name and your instructor's name, and submit the filter paper to your lab instructor until the next lab period.

HPLC Sample Analysis:

- When sample results are posted, use the provided standard chromatogram to identity the compounds present in each sample. This will allow you to determine the success of the separation method. Record data in Table 6.2 on the POST LAB Assignment (provided online).

SAFETY

****All experiments should be performed in a fume hood with appropriate safety glasses. Gloves will be provided by request.****

All organic solvents used in this experiment are flammable, irritants, and can be toxic if ingested or absorbed through skin. Sodium bicarbonate is toxic if ingested. Hydrochloric acid is extremely corrosive.

WASTE MANAGEMENT

Place all liquid waste in the container labeled "LIQUID WASTE—EXTRACTION". The drying agent ($MgSO_4$) and filter paper should be placed in the container labeled "SOLID WASTE".

References

Palleros, Daniel R. (2000). *Experimental Organic Chemistry*. New York: John Wiley and Sons.

Pavia, D. L., Lampman, G. M., & Kriz, G. S. (1999). *Introduction to Organic Laboratory Techniques, A Microscale Approach*, 3rd ed. New York: Saunders College Publishing.

Name:			
	Max	Score	Total Grade
Tech	10		
Pre	20		
In	20		
Post	50		

Exp. 6 Extraction of Analgesics

PRE-LAB ASSIGNMENT: *(**EACH STUDENT** will complete and submit an original copy at the beginning of the lab period. **Without a complete pre-lab assignment, you will not be allowed to perform the experiment, and will receive a zero for the lab.**)* .....*max score = 20 pts.*

1. **Objective** *(Write a brief purpose of the experiment in **complete** sentences, addressing all of the following points.)*
   - What new techniques will be introduced in this experiment?
   - What goal will be accomplished using these techniques?
   - How will the efficiency of the method and the purity of the compounds be determined?

2. **Physical Data** *(Complete the following table before coming to lab.)*

	aspirin	acetaminophen	sodium bicarbonate	ethyl acetate	acetone
structure					
bp (°C)	XXX	XXX	XXX		
mp (°C)			XXX	XXX	XXX
d (g/ml)	XXX	XXX	XXX		

3. Experimental Outline *(Give a brief description of the procedure that will be followed in this experiment in 5 lines or less.)*

1	
2	
3	
4	
5	

4. Pre-Lab Questions *(Answer the following questions prior to lab.)*

A. In an extraction procedure, when mixed with water, ethyl acetate would form the:

a. Top layer

b. Bottom layer

B. What is the purpose of adding $MgSO_4$ to the ethyl acetate layer after the extraction?

I have read and understood the experimental procedure for this experiment. I am familiar with the hazards and the required disposal procedures for this experiment.

Sign here: ____

Experiment 7

Recrystallization, Melting Point, and HPLC Analysis of Analgesics

Introduction

In Experiment 6 you isolated aspirin and acetaminophen through acid–base extraction. The aspirin was obtained as a white powder and you saved your acetaminophen as a solution in acetone. In this experiment you will purify each by recrystallization. After drying, you will analyze both by melting point and HPLC.

Purification of organic compounds is important for a variety of reasons. Impurities in drugs, like analgesics, can be very toxic and must be removed before the drug is suitable for medicinal use. It is also necessary to ensure that no impurities are present for future chemical reactions using the compound and to determine the compound's structure spectroscopically.

Recrystallization

Recrystallization is an important purification technique for solids. Recrystallization relies on the differential solubility of a compound in hot vs. cold solvents. For example, as seen from Experiment 6, aspirin is much more soluble in warm water than in cold water. To recrystallize a compound, the solid is dissolved in a **minimum** amount of hot solvent. If the solution is contaminated with particulates, it is filtered while hot. If colored impurities are present, the hot solution may be treated with decolorizing charcoal and then filtered. Finally, the hot, saturated solution is allowed to **cool slowly** so that the desired compound crystallizes slowly, leaving impurities in solution. When crystallization is complete, the crystals are isolated from the "*mother liquor*" by filtration.

It is important that the crystals form slowly, rather than precipitate out, which is the rapid formation of an amorphous solid. As a hot saturated solution cools, it becomes supersaturated. Crystal nuclei can then form, often on the walls of the container, at the surface or on a dust particle. As the solution cools slowly, molecules diffuse to the crystalline surfaces of the nuclei and join the crystal lattice. Slow cooling will result in slow crystallization and exclude impurity molecules from the growing crystal lattice. In contrast, a precipitated solid may trap impurities during the rapid formation of the solid.

An ideal solvent for recrystallization is one in which the compound is very soluble in the hot solvent but relatively insoluble in the cold solvent. The melting point of the solid must also be higher than the boiling point of the solvent so that the compound dissolves, rather than melts in the hot solvent. The solvent should be relatively nontoxic and moderately volatile so that the solvent evaporates readily (refer to Appendix J). It is also important to consider the solubility of impurities vs. the solubility of the desired compound when choosing a recrystallization solvent. Although the impurities are most likely present in much lower concentrations than the major component, the impurities should remain soluble in the cold solvent, while the desired compound crystallizes, excluding the impurities.

Choosing a good recrystallizing solvent requires some knowledge of the relative solubility of different compounds in various solvents. Compounds that contain -O-H or -N-H groups (alcohols, carboxylic acids, amines, amides) are generally soluble in polar solvents like alcohols and water. As the amount of hydrocarbon character in the compound increases, the compound's solubility in polar solvents decreases, whereas its solubility in nonpolar solvents (like diethyl ether, hexane, etc.) increases ("like dissolves like").

Often, a mixed solvent pair is chosen as the recrystallization solvent. A solvent pair consists of two miscible liquids, one of which dissolves the compound readily, while the other does not. For example, many polar organic compounds are soluble in alcohol, but are relatively insoluble in cold water. To recrystallize such a compound, it is dissolved in a small amount of hot ethanol. Hot water is then added drop by drop until the solution becomes cloudy. A few drops of ethanol are added to dissolve the precipitated solid and the saturated solution is then allowed to cool slowly so that crystallization occurs.

Occasionally the sample will not crystallize from solution, even though the solution may have been saturated with the solute at an elevated temperature. If the compound fails to crystallize out of solution, the easiest method to induce crystallization is to introduce a pure seed crystal of the desired compound. If a seed crystal is not available, scratching the insides of the flask with a glass stirring rod can induce crystallization. Freshly scratched glass has planes and angles corresponding to the crystal structure, and crystals actually grow on these spots. Another solution is to reduce the volume of solvent by evaporation. It is easy to add too much solvent, resulting in a solution that is not saturated with the solute, and therefore will likely never crystallize. Finally, by lowering the temperature of the solution, the pure crystals are more likely to form due to their reduced solubility at lower temperatures.

After the crystallization is complete, the crystals are collected by vacuum filtration using a Büchner funnel. The original flask is washed with a small amount of ice-cold recrystallizing solvent. This wash is then used to wash the crystals on the filter paper in the funnel. The crystals then need to be dried before further analysis.

Melting Point

The melting point of a pure crystalline solid is an important characteristic of the compound. It can be used as further evidence to confirm the identity of a compound or help to identify an unknown. As a crystal is heated, the molecular motion of the molecules in the solid increases until the solid eventually changes into a freely moving liquid state. The melting point is defined as the temperature at which the solid changes to a liquid at 1.0 atmosphere pressure. Just as a pure liquid has a sharp characteristic boiling point, a pure solid has a sharp characteristic melting point.

The melting point is determined by slowly heating (~1°C/min) a small sample of the solid and observing when the first droplet of liquid appears (T_i) and when the sample is finally converted to a clear liquid (T_f). Thus, a melting *range* is recorded as T_i - T_f. A pure crystalline solid will have a very sharp melting range (< 1°C). However, a solid that contains impurities will often have a ***broader*** and ***lower*** melting point range (see Figure 7.1). Thus, by comparing the experimental melting points to literature values the melting range of a compound can serve as a criterion of purity.

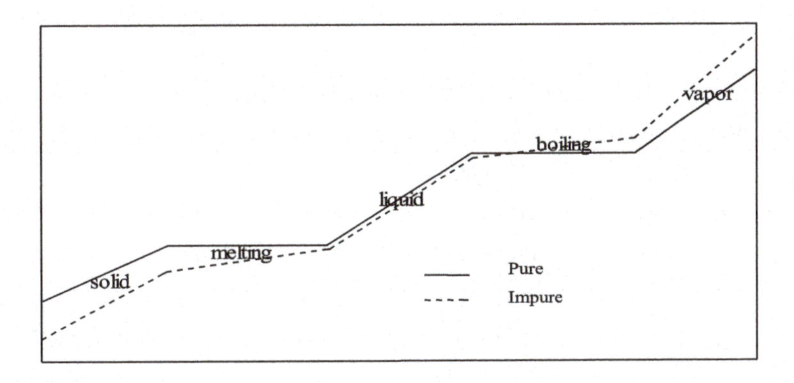

Figure 7.1 Heating curves of a pure and an impure substance.

Objectives

In this experiment the purification technique of recrystallization will be introduced. Using this technique, the aspirin and acetaminophen samples isolated during the previous experiment will be further purified. The technique of melting point analysis will also be introduced, and used to analyze the purity of the aspirin and acetaminophen samples, along with HPLC analysis.

Experimental Procedure

Recrystallization:

Before the Purification of Aspirin...

- Reweigh your aspirin crystals from the previous experiment, and record the actual recovery of aspirin in Table 7.1 on the POST LAB Assignment (provided online) and in the lab notebook. Not only is this the actual recovery of the extracted aspirin, but it is also the theoretical recovery of aspirin after the recrystallization.
- Prepare two melting point capillaries of your "*extracted aspirin sample.*" Insert the open end of a capillary tube into the sample so it will pick up a small amount of sample (a depth of 1 to 2 mm of sample in the capillary tube is desirable).
- Invert the capillary tube and tap it on the bench top to force the sample into the closed end. Set aside until the melting point analysis can be performed (see *Melting Point Analysis* below).

Purifying Aspirin

- Transfer the remaining crystals to a 50 mL Erlenmeyer flask using a powder funnel. Add 2 mL of warm ethanol. Gently heat the flask on a warm hot plate (setting 3) to dissolve all the crystals.
- Add 4 mL hot deionized water in 1 mL increments (do **not** exceed 4 mL of water). Then add two to three drops of hot ethanol to the solution to clear any cloudiness and ensure all aspirin has dissolved.
- Remove the flask from the hot plate, and allow the solution to cool to room temperature.

- Once the solution in the flask is at room temperature, place in an ice water bath for 15 minutes. If crystals do not appear, you may need to scratch the sides of the flask with a glass rod to help induce crystallization.
- Set up a suction filtration apparatus (Appendix A).
- Weigh a small filter paper. Place the small filter paper into the Büchner funnel, and wet the filter paper with ice cold deionized water. Apply the vacuum.
- Transfer the aspirin crystals to the Büchner funnel slowly, using a small amount of ice-cold deionized water to quantitatively transfer the solid. Rinse the crystals in the funnel with a few mL of ice-cold deionized water. Leave under vacuum for five minutes. ***PROCEED TO PRODUCT ANALYSIS.***
- ***At the beginning of the next lab period***, obtain your recrystallized aspirin sample from the instructor. Reweigh the sample to determine the *actual recovery* of aspirin after recrystallization. Calculate the percent recovery. Record this data in the lab notebook, and in Table 7.1 on the POST LAB Assignment (provided online).
- Prepare two melting point capillary tubes of the aspirin, and perform a melting point analysis as described below. The crystals will need to dry until the next lab period before final mass can be obtained, or a melting point analysis can be performed.

Before the Purification of Acetaminophen…

- You saved your extracted acetaminophen from Experiment 6 in a sample vial as a solution of acetaminophen dissolved in acetone. Transfer this solution to a 50 mL beaker. Rinse the sample vial with 1–2 mL of reagent acetone and transfer this rinse to the beaker.
- Heat the solution on a warm hot plate (setting 3, **NO HIGHER**) until all of the acetone has evaporated. Do ***not*** allow the acetaminophen to sit on the hot plate after all of the liquid is gone.

Purifying Acetaminophen

- Once all of the acetone has been evaporated resulting in solid acetaminophen, add hot deionized water drop wise until all of the acetaminophen just dissolves (do **not** exceed 5 mL of water). This may require additional heating in order to dissolve all of the acetaminophen solid.
- Remove the beaker from the hot plate. Allow the solution to cool slowly to room temperature.
- Once cooled, place the beaker in an ice bath for 15 minutes. If crystals do not appear, you may need to scratch the inside of the beaker with a glass rod to help induce crystallization.
- Set up a suction filtration apparatus using a CLEAN Büchner funnel.
- Weigh a small filter paper and a large filter paper. Place the small filter paper into the Büchner funnel, and wet the filter paper with ice-cold deionized water. Apply the vacuum.
- Transfer the acetaminophen crystals to the Büchner funnel slowly, using a small amount of ice-cold deionized water to quantitatively transfer the solid. Rinse the crystals in the funnel with a few mL of ice-cold deionized water.
- Leave under vacuum for five minutes. ***PROCEED TO PRODUCT ANALYSIS***.

- ***At the beginning of the next lab period***, obtain your recrystallized acetaminophen sample from the instructor. Reweigh the sample to determine the *actual recovery* of acetaminophen after recrystallization. Calculate the percent recovery. Record this data in the lab notebook, and in Table 7.1 in the POST LAB Assignment (provided online).
- Prepare two melting point capillary tubes of the acetaminophen, and perform a melting point analysis as described below.

Product Analysis:

HPLC Analysis

Acetaminophen Sample Preparation:

- Prepare an HPLC sample of your pure acetaminophen. Place a few crystals of the purified acetaminophen into an auto analyzer vial, and add HPLC solvent solution (ethyl acetate).
- Cap the vial and shake until all of the solid dissolves (samples with visible solid will be discarded). Place the sample into a vial slot and sign out on sample sheet.
- Prepare the remaining acetaminophen for storage. Weigh a large filter paper. Using a microspatula, carefully transfer the acetaminophen crystals and small filter paper from the Büchner funnel to the larger filter paper. Secure the acetaminophen sample by folding the large filter paper and securing the edges with a small amount of tape.
- Label this filter paper with your name and your instructor's name, and submit the filter paper to your lab instructor until the next lab period.

Aspirin Sample Preparation:

- Repeat the steps above to prepare an HPLC sample of your pure aspirin.
- Prepare the remaining aspirin for storage. Weigh a large filter paper. Using a microspatula, carefully transfer the aspirin crystals and small filter paper from the Büchner funnel to the larger filter paper. Secure the aspirin sample by folding the large filter paper and securing the edges with a small amount of tape.
- Label this filter paper with your name and your instructor's name, and submit the filter paper to your lab instructor until the next lab period.

HPLC Sample Analysis:

- When sample results are posted, use the provided standard chromatogram to identity the compounds present in each sample. This will allow you to determine the success of the separation method. Record data in Table 7.2 on the POST LAB Assignment (provided online).

Melting Point Analysis

- During the next lab period, you will perform melting point analyses on your ***recrystallized aspirin*** and ***recrystallized acetaminophen*** from this experiment.
- During this lab period, perform melting point analyses on a provided sample of aspirin, along with the sample of the extracted aspirin prepared at the beginning of the lab period.
- Prepare a melting point sample of the provided sample of ***impure aspirin***. You have already prepared a melting point sample of your ***extracted aspirin*** sample prior to recrystallization.

- Insert the melting point capillary tubes in the heating block closed end down (one to three tubes may be inserted at a time).
- Turn the light switch on and adjust the voltage control to a setting of 4. An accurate melting point can only be obtained if the rate of heating is no more than about 1°C/min during melting. As the temperature reaches 15–20 °C below the expected melting point, turn the dial down slightly to slow down the rate of temperature increase.
- Observe the crystals through the view lens and note the temperatures at which the first indication of liquid occurs and when the last crystal melts. Record these temperatures, separated by a dash, as the melting range of your sample in the laboratory notebook, and in Table 7.3 on the POST LAB Assignment (provided online).

SAFETY

All experiments should be performed in a fume hood with appropriate safety glasses. Gloves will be provided by request.

All organic solvents used in this experiment are flammable, irritants, and can be toxic if ingested or absorbed through skin.

WASTE MANAGEMENT

Place all liquid waste from recrystallization in container labeled "LIQUID WASTE—RECRYSTALLIZATION" in the waste hood. Used melting point capillary tubes are discarded in the broken glass container.

Name:			
	Max	Score	Total Grade
Tech	10		
Pre	20		
In	20		
Post	50		

Exp. 7 Recrystallization, Melting Point and HPLC Analysis of Analgesics

PRE-LAB ASSIGNMENT: *(**EACH STUDENT** will complete and submit an original copy at the beginning of the lab period. **Without a complete pre-lab assignment, you will not be allowed to perform the experiment, and will receive a zero for the lab.**) .....max score = 20 pts.*

1. **Objective** *(Write a brief purpose of the experiment in **complete** sentences, addressing all of the following points)*
   - What new techniques will be introduced in this experiment?
   - What goal will be accomplished using these techniques?
   - How will the efficiency of the method and the purity of the compounds be determined?

2. **Physical Data** *(Complete the following table before coming to lab.)*

	aspirin	acetaminophen	ethanol	ethyl acetate	acetone
structure					
bp (°C)	XXX	XXX			
mp (°C)			XXX	XXX	XXX
d (g/ml)	XXX	XXX			

3. <u>Experimental Outline</u> *(Give a brief description of the procedure that will be followed in this experiment in 5 lines or less.)*

1	
2	
3	
4	
5	

4. <u>Pre-Lab Questions</u> *(Answer the following questions prior to lab.)*

A. The ***ideal*** solvent for a recrystallization is one that: (*circle the correct answer*)

a. Dissolves a moderately large amount of the compound when hot.
b. Does not react with the compound.
c. Dissolves only a small amount of the compound when cool.
d. Boils at a temperature below the compound's melting point.
e. All of the above.

B. During a recrystallization, what are ***3 ways*** to induce crystallization of a sample if it at first it fails to crystallize out of solution?

I have read and understood the experimental procedure for this experiment. I am familiar with the hazards and the required disposal procedures for this experiment.

Sign here: ____

Experiment 8

Distillation and Gas Chromatography of Alkanes

Introduction

In this experiment you will explore the relationship between molecular structure, intermolecular forces, and boiling points of straight chain and branched alkanes. You will also be introduced to the techniques of simple distillation, fractional distillation, and gas chromatography (GC). You will use these techniques to separate and identify alkanes in an unknown mixture.

London Dispersion Forces and Boiling Points

As described in Experiment 1, alkanes are nonpolar molecules. London dispersion (attractive) forces are the only intermolecular attractive forces acting between alkane molecules. As the relative size and surface area of the alkane increases, the net London forces between individual molecules increase. It is the attractive forces between molecules that hold them together as either a liquid or a solid. When molecules of a liquid vaporize, the kinetic energy of individual molecules becomes greater than the attractive forces between molecules. Thus, as London forces between molecules increase, the boiling point increases; therefore, larger alkanes to have higher boiling points than smaller alkanes. Very large alkanes ($>C_{16}$) are actually waxy solids.

Alkanes can also be branched. For example, *n*-hexane has five constitutional isomers, only one of which is a straight chain alkane and four that are branched. Individual alkanes are purified by fractional distillation from petroleum. It is very difficult and expensive to obtain pure *n*-hexane, free of other isomers. In fact, hexane is often sold as "Hexanes," a mixture of hexane isomers (plus methylcyclopentane). Petroleum ether is a commonly used nonpolar solvent and is a mixture of low boiling (35–60°C) alkanes. Although branched hydrocarbons still have only London dispersion attractive forces, the *boiling points of branched alkanes are lower than those of straight chain (or normal) alkanes*. Thus, the shape of the molecule is responsible for its net intermolecular attractive forces, which in turn controls the compound's physical characteristics, such as its boiling point.

Distillation

Distillation is the most commonly used method for purification and separation of liquids. It is also used as a criterion of purity and identity of a substance, since a pure substance has a characteristic sharp boiling point (<1° range), whereas mixtures typically boil over a broader range. The boiling point is defined as the temperature at which the vapor pressure of the liquid equals the external pressure at the surface of the liquid. As a liquid is heated, the temperature of the liquid increases, as does the vapor pressure above the liquid. When the vapor pressure equals the atmospheric pressure, the liquid boils. The process of distillation involves vaporization of the liquid by boiling (in the distilling flask, see Figure 8.1A), then condensation of the vapor back into a liquid (in the condenser) and collection of the condensate (distillate) in a separate flask or container.

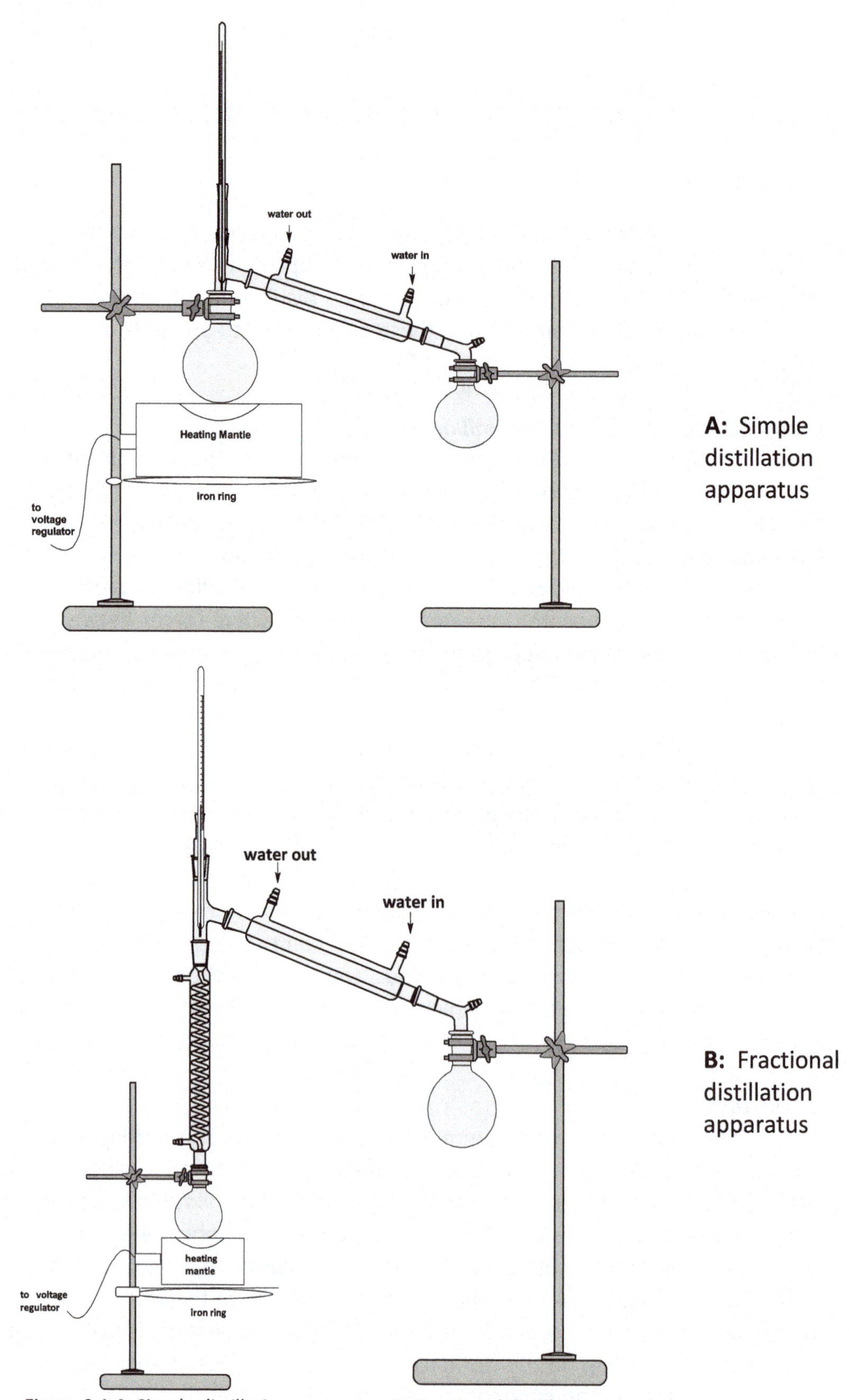

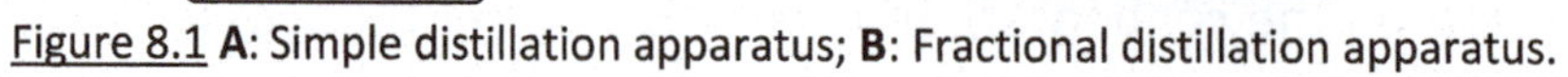
Figure 8.1 **A**: Simple distillation apparatus; **B**: Fractional distillation apparatus.

In the distillation of a pure liquid, the temperature of the distillate, measured by a thermometer placed in the path of the vapor (shown in Figure 8.1), rises rapidly to the boiling point of the liquid. When the distillation apparatus has reached equilibrium, the temperature of the distillate remains constant as the liquid distills (Figure 8.2A). As the distillation nears completion, the temperature at the thermometer again drops.

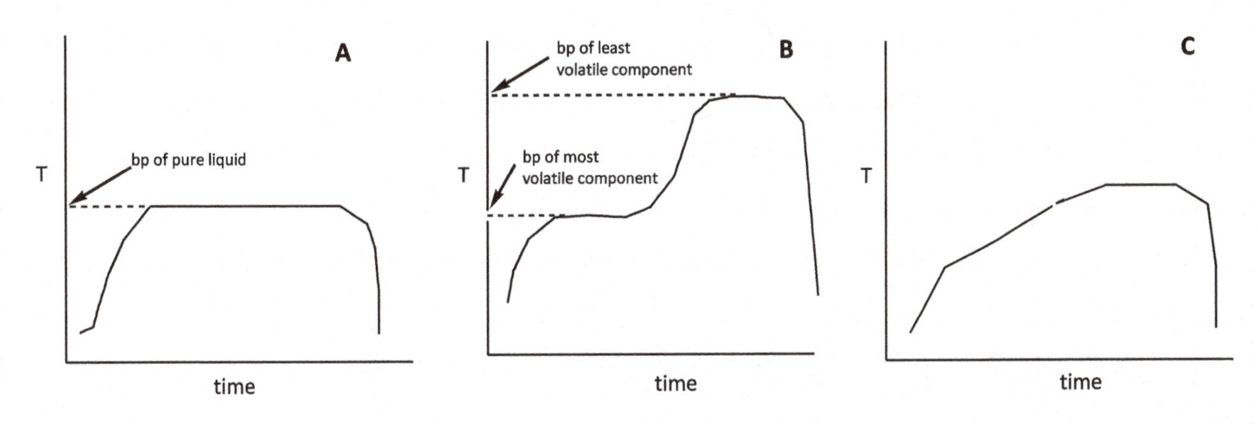

Figure 8.2 Graphical representations of temperature (T) vs. time for: (**A**) simple distillation of a pure liquid, (**B**) simple distillation of two liquids with a boiling point difference >70°C, and (**C**) simple distillation of two liquids with a boiling point difference <70°C.

Simple Distillation

The simple distillation described above is applicable when the mixture contains only one volatile component (i.e., has only nonvolatile impurities, such as distilling pure water from salt water). If the mixture contains more than one volatile component, simple distillation may be effective if the boiling points of the components differ by at least 70°C.

In a simple distillation of a mixture of two liquids (Figure 8.2B), if the boiling point difference between the two liquids is >70°C, the distillation temperature will rise to the boiling point of the lower boiling liquid and remain constant until all of that liquid has been distilled. The temperature will then rise to the boiling point of the higher boiling liquid and stay constant as the second component distills over.

However, it is often necessary to separate mixtures of liquids with boiling points less than 70°C. In a simple distillation of such a mixture (Figure 8.2C), the distillation temperature rises quickly to the boiling point of the lower boiling liquid and then rises more or less steadily during the distillation as mixtures of the two liquids distill.

A vapor–liquid phase composition diagram for a simple distillation of two liquids A and B is shown in Figure 8.3. Each evaporation and condensation step results in a vapor that is enriched with the more volatile component. Simple distillation of this mixture of 75% A and 25% B would provide (initially) 25% A and 75% B. Therefore, initial fractions collected during the distillation would be enriched in the more volatile component, B, while later fractions would be more enriched in the less volatile component, A. However, none of the fractions would contain pure A or B. Fractional distillation is necessary for purification of liquids with boiling point differences of <70°C.

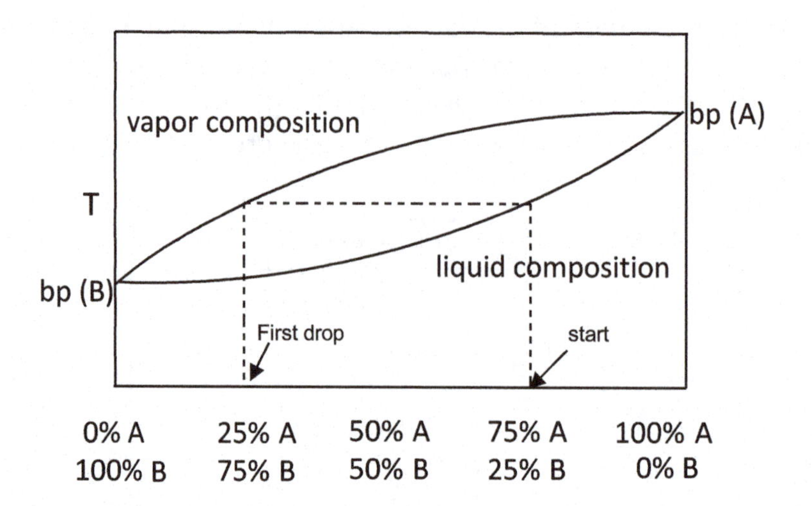

Figure 8.3 Vapor–liquid phase composition diagram for simple distillation. Simple distillation of a mixture of 75% A and 25% B would provide (initially) 25% A and 75% B.

Fractional Distillation

Better separation of two liquids with boiling point differences of less than 70°C can be achieved through fractional distillation. In fractional distillation, a fractionating column is placed between the distillation flask and the condenser (Figure 8.1B). The fractionating column provides more surface area, either through a type of column packing, such as glass beads or metal turnings, or through indentations in the wall of the column. As the vapor rises through the column, it condenses and re-vaporizes continuously, because of the increased surface area the vapor encounters. Each re-vaporization of the condensed liquid is equivalent to another simple distillation. Each of these "distillations" leads to a distillate enriched in the lower boiling component. The vapor–liquid phase diagram for a fractional distillation is shown in Figure 8.4. If a mixture of 75% A and 25% B is fractionally distilled, after a second re-vaporization and condensation, the distillate would be ~ 95% B.

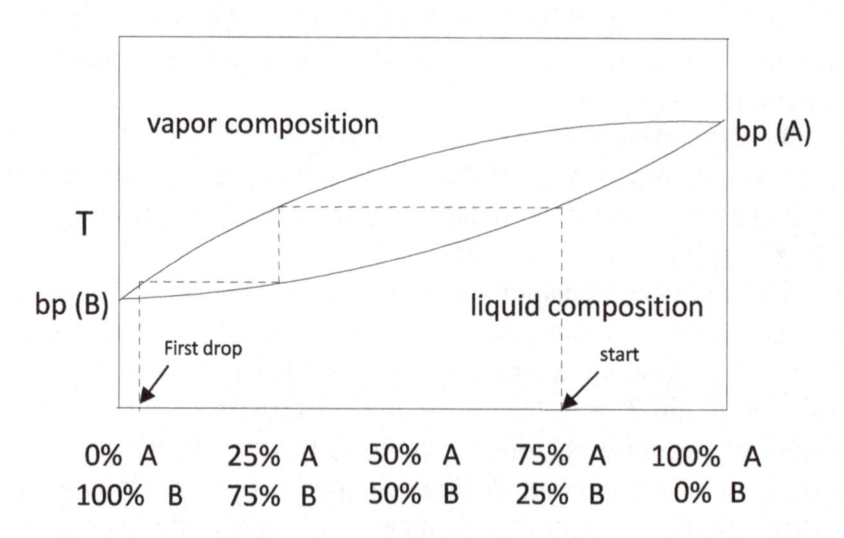

Figure 8.4 Vapor–liquid phase composition diagram for fractional distillation. Fractional distillation of a mixture of 75% A and 25% B would provide distillate, which is ~ 95% B.

Most laboratory fractionating columns can effectively separate components with a boiling point difference of 30–40°C. The temperature profile of such a fractional distillation would resemble that of the simple distillation of two components with a boiling point difference of 100°C (Figure 8.2B). In research laboratories and in industry, larger (in some cases hundreds of feet tall) and more intricately designed fractionating columns are used to separate components with boiling point differences of only a few degrees.

Gas Chromatography *(see also Appendix E)*

Gas chromatography (GC), also known as gas liquid chromatography (GLC), can be used to determine the composition of a mixture qualitatively (what is present) and quantitatively (how much of each component is present). Gas chromatography using modern instrumentation and columns requires only a trace of sample dissolved in a volatile solvent.

Instrumentation

In GC, the sample is injected (using a syringe) through a rubber septum into the *injection port*, which leads directly to the chromatographic column (Figure 8.5). The *injection port* is maintained at a very high temperature (250°C). The sample vaporizes immediately upon entering the injection port and is forced into the column by the *carrier gas*. The carrier gas (the *mobile phase* in GC) is an inert gas, usually helium, and flows through a *capillary column* at ~1 mL/min. The capillary columns we will use are very narrow (0.25 mm) and very long (30 m) quartz columns, which are coated with very thin (25 μm) *stationary phase*. The column is housed in an oven so that the temperature of the column can be varied, depending on the types of compounds to be analyzed.

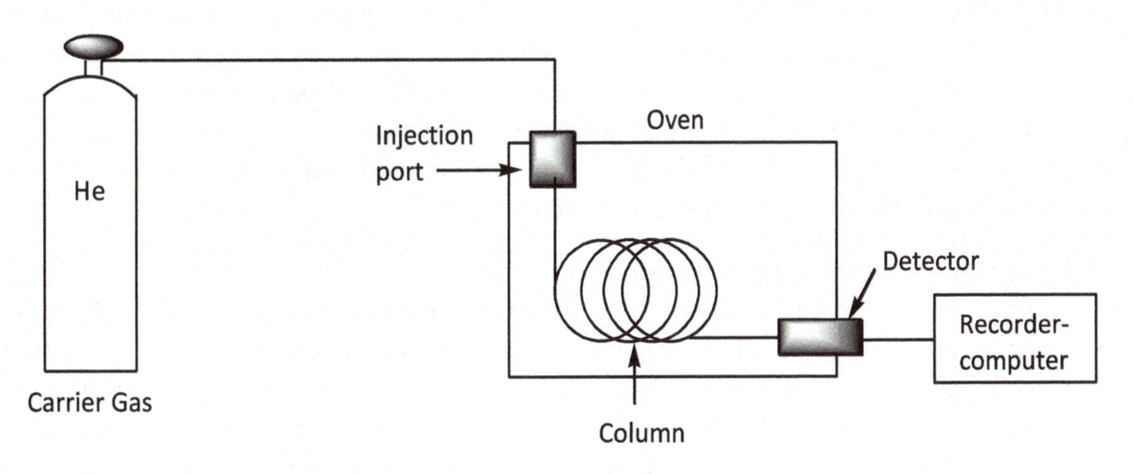

Figure 8.5 Gas chromatograph schematic.

As the sample mixture advances through the column, the components may interact with the stationary phase by any of the types of interactions between molecules: London dispersion forces or dipole–dipole forces (including hydrogen bonding). Those components that interact strongly with the stationary liquid phase are retained (slowed down, as shown for compound A in Figure 8.6), whereas those components that interact weakly with the stationary phase spend more time in the gas phase and advance more rapidly through the column (like compound B in

Figure 8.6). The result is that different substances take different amounts of time (*elution time)* to *elute* from the column and the mixture is separated into its components.

Typically, when a nonpolar packing material (such as the DB-5 capillary column of our GC) is used, *components with similar polarity elute in order of volatility*. Thus for alkanes, components will elute in order of increasing boiling points; lower boiling components will have shorter *retention times* than higher boiling components. Figure 8.7 shows the gas chromatogram of all possible unknown alkanes used in this experiment. You can see that the alkanes elute in order of boiling points. Using these standard chromatograms, a retention time can be assigned to each alkane.

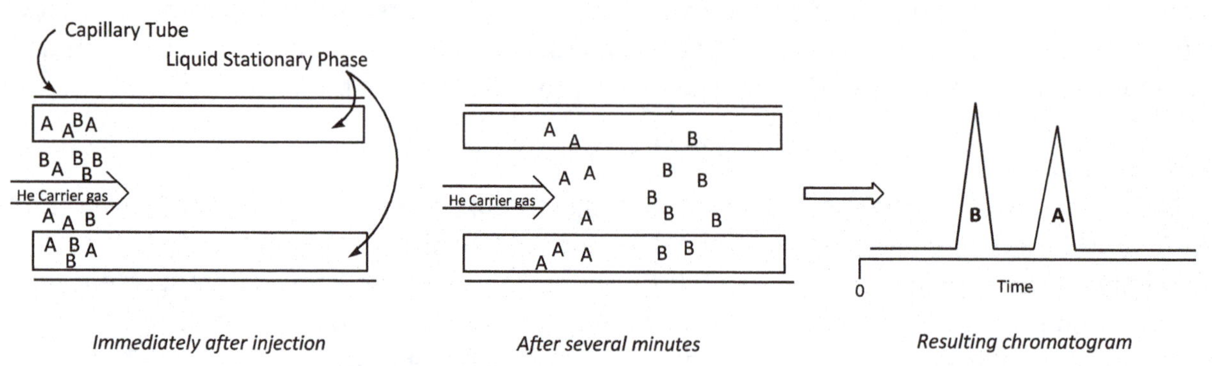

Figure 8.6 Elution of compounds A and B through a GC capillary column

Eluted components pass through a detector. There are several types of GC detectors, including thermal conductivity, flame ionization, electron capture, and mass spectrometry. The *flame ionization detector* (FID) is a very widely used detector in organic chemistry. As an organic compound passes through the flame of the FID and burns, it produces ionic intermediates and electrons. The charged species are attracted to and captured by a collector and the resulting ion current is amplified. The resulting signal is sent to the GC software where it is converted to a peak. The flame ionization detector is especially good for analysis of organic compounds because the number of ions produced (and therefore the intensity of the peak) is roughly proportional to the number of number of carbons present in the molecule.

Identifying Compounds Using Retention Times

The detectors send the electronic signals collected over time to the GC software, which generates a chromatogram. The GCs we will use in CHML 211/212 use Hewlett-Packard ChemStation software. The characteristic data given with the chromatogram are the *retention time* of each component and its area. The retention time is the time it takes from the point where the compound is injected into the instrument to the point where the compound reaches the detector. We can determine the ***identity*** of an unknown compound by comparing the ***retention time*** of the unknown to that of known compounds, called **standards**, under identical conditions. Assuming that the instrument is properly maintained and operated:

1. When comparing retention times, if the retention times are the same, the two compounds are probably identical.
2. The retention time of an unknown may be slightly different from the retention time of a standard by < 0.1 minutes (6 sec) for a variety of reasons. Usually, all retention times are offset by the similar amounts.

3. If the retention times differ by more than 0.1 min, the compounds may not be the same compound.
4. Comparison of retention times allows for identification of components of a mixture, as long as a **standard** is available. It is very important to use a standard run along with a particular sample set.

The GC chromatograms of the individual alkanes and of an *example* standard alkane mixture are shown below. The individual alkanes are dissolved in pentane prior to GC analysis.

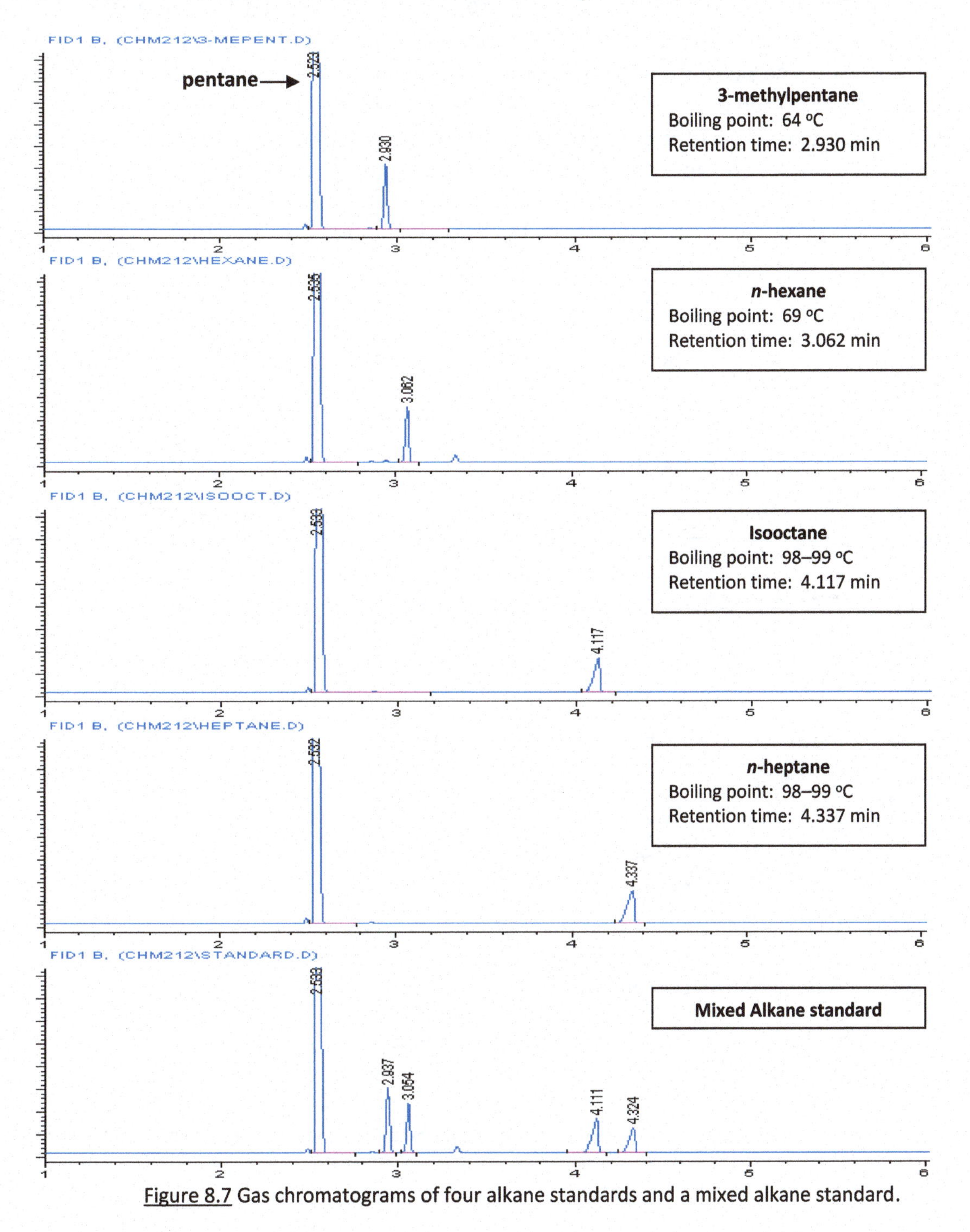

Figure 8.7 Gas chromatograms of four alkane standards and a mixed alkane standard.

Quantifying Compounds Using Area Percent

The GC software also integrates the area under each peak and generates *area percent* values for each peak of the chromatogram (Figure 8.8). The ratio of the peak areas of individual components in a mixture is not necessarily identical to their molar ratio or their weight ratio. Not only will different detectors give different relative peak areas, different compounds often give different responses with a given detector. The area percent is identical to the weight percent of the mixture if all components have identical *response factors*. [To determine the accurate percent composition of a mixture, it is necessary to analyze a standard composed of known concentrations of each component and determine *response factors* (or correction factors) for each component. The peak area of each component is multiplied by its response factor before the percent composition of the mixture is calculated.] However, for many of the mixtures we will study, the response factors of the components of the mixture are often very similar, so that using percent areas gives a good approximation of the weight percent composition of a mixture.

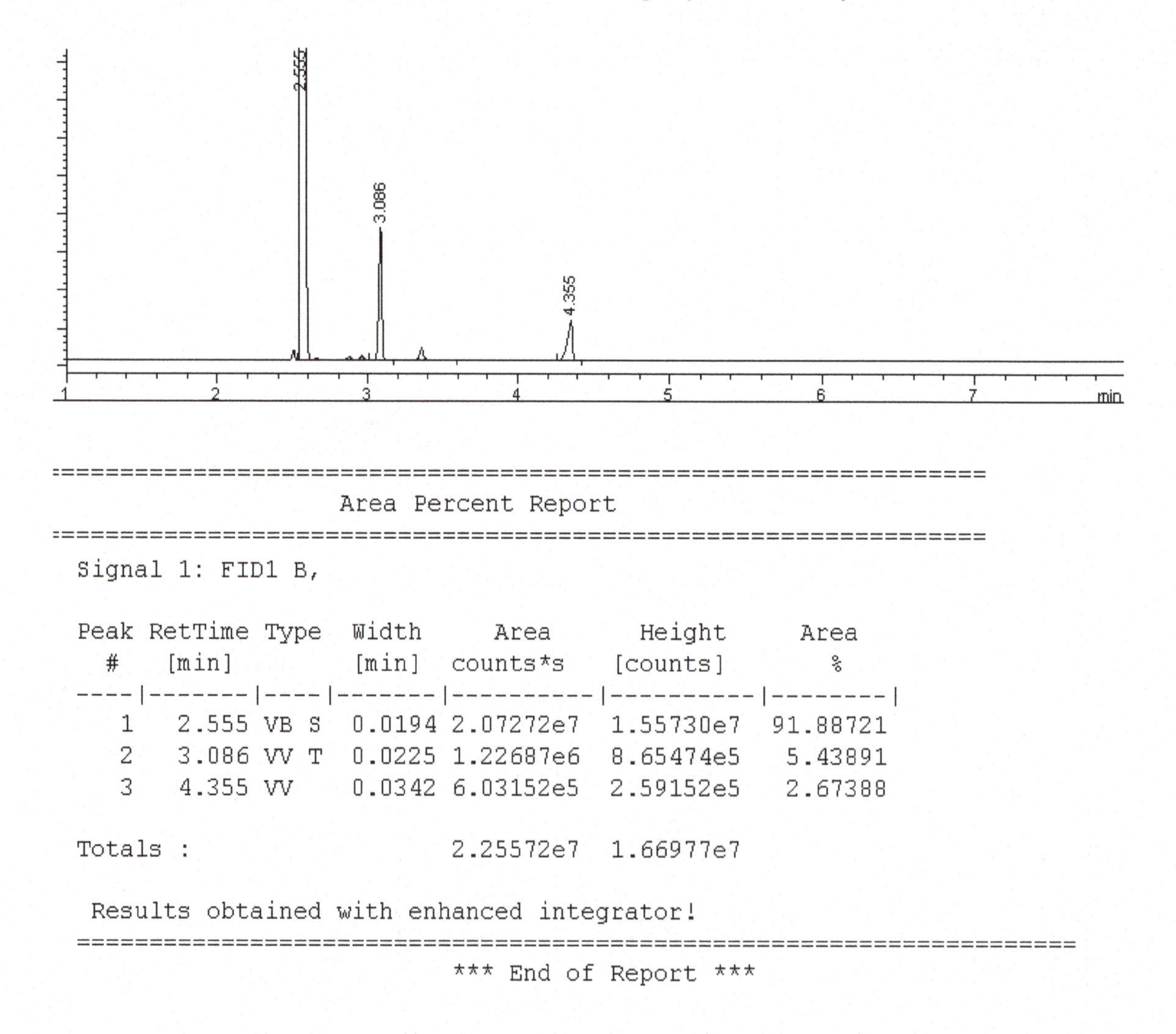

Area Percent Report

Signal 1: FID1 B,

Peak #	RetTime [min]	Type	Width [min]	Area counts*s	Height [counts]	Area %
1	2.555	VB S	0.0194	2.07272e7	1.55730e7	91.88721
2	3.086	VV T	0.0225	1.22687e6	8.65474e5	5.43891
3	4.355	VV	0.0342	6.03152e5	2.59152e5	2.67388

Totals : 2.25572e7 1.66977e7

Results obtained with enhanced integrator!

*** End of Report ***

Figure 8.8 Gas chromatogram of a fraction from the distillation of an unknown alkane mixture

Adjusted Area Percent

Most often you will dissolve your compound or mixture in a low boiling solvent for GC analysis. The relative areas of the components of interest must therefore be adjusted to exclude the large area percent of the solvent peak. In general, to calculate the *adjusted area percent,* the area of each peak of interest is divided by the sum of the areas of all peaks of interest (excluding the solvent peak) and that value is multiplied by 100%. The data in the table below are from the gas chromatogram shown in Figure 8.8. It was obtained from the analysis of a fraction from simple distillation of an unknown alkane mixture. Five drops of the distillate was dissolved in pentane. The integration shows that pentane accounts for ~91% of the total peak area. Since we are interested in the ratio of the two alkanes with retention times of 3.086 and 4.355 minutes, the area of the huge pentane peak can be ignored.

	Standard Retention Time (min)	Sample Retention Time (min)	Identification	Area %	Adjusted Area %
Peak 1	2.533	2.555	pentane	91.407	---
Peak 2	3.054	3.086	*n*-hexane	5.439	67.0
Peak 3	4.324	4.355	*n*-heptane	2.634	33.0

Thus, the ***adjusted*** areas are:

Adjusted Area % (peak at 3.086 min) = [5.439 / (5.439 + 2.634)] x 100% = 67.0%

Adjusted Area % (peak at 4.355 min) = [2.634 / (5.439 + 2.634)] x 100% = 33.0%

Objectives

In this experiment you will attempt to separate two alkanes present in an unknown mixture through simple distillation ***OR*** fractional distillation. You will identify the two alkanes in the unknown mixture using GC data from your distillation fractions. The area percent from the two fractions analyzed will allow you to determine the purity of each fraction and evaluate the effectiveness of the distillation as a separation method.

Experimental Procedure

Simple Distillation:

- Obtain 25 mL of an unknown alkane mixture (be sure to record in your notebook which unknown you are using) in a 50 mL round bottom flask. Add a 2-3 boiling chips to the flask.
- Assemble the assigned distillation apparatus shown in Figure 8.1A or 8.1B (and in Appendix A). Assemble the glassware starting with the 50 mL round bottom flask clamped securely to a ring stand. Build the apparatus one piece at a time, starting with the flask, twisting the pieces as you push them together to ensure a tight fit. Place a blue plastic keck clip on the joint between the condenser and the head adapter, and another one the joint between the condenser and the receiver adapter.

- Be sure that your thermometer is placed properly. The entire bulb should be below the lowest part of the side arm. This is necessary to obtain a correct reading of the temperature of the condensing vapor. (If the thermometer is placed too high, vapor could "sneak" by into the side arm without colliding with, imparting kinetic energy to, and thereby warming the thermometer.)
- **Have your apparatus checked by your instructor before you begin distilling.**
- After your instructor has approved your apparatus, turn the condenser water on just enough to ensure a steady gentle flow. (If it is flowing too fast, it can force the hose off and cause a water spill.) Be sure the drain hose is in the drain.
- Heating will be done using ceramic heating mantles, always plugged into a regulator, never directly into the socket. Heat slowly to achieve the best separation. Start your distillation with the power supply initially set at ~40.
- Collect four 5 mL fractions in small test tubes, leaving about 5 mL in the original distilling flask. **DO NOT distill to dryness.** Record the temperature range of the distillate collected in each of the first four fractions. The temperature range is obtained by recording the thermometer reading from the point at which the first drop enters the test tube, and again at the point at which the last drop enters the test tube before the test tube is switched. Fraction 5 will be the liquid remaining in the distillation flask.
- Record the distillation ranges for each fraction in the laboratory notebook, and in Table 8.1 on the POST LAB Assignment (provided online).

Product Analysis:

GC Analysis

- Prepare a GC sample of your second and fifth fractions from the distillation by adding **5** drops of each fraction to 1 mL of GC solvent (*n*-pentane) in a GC auto analyzer vial. Fill in the GC sample sign out sheet using your first and last name, and place your vials in the proper place in the rack (as described during lab).
- Once the GC data is returned, use this data to complete Table 8.2 on the POST LAB Assignment (provided online).

SAFETY

****All experiments should be performed in a fume hood with appropriate safety glasses. Gloves will be provided by request.****

All organic solvents used in this experiment are flammable, irritants, and can be toxic if ingested or absorbed through skin. Be sure to plug the ceramic heating mantle into the voltage regulator. Never plug the heating mantle directly into the outlet.

WASTE MANAGEMENT

Pour all liquid waste into the bottle labeled "LIQUID WASTE—DISTILLATION" in the waste hood. These mixtures will be analyzed and reused next year.

Reference

Klein, David. (2015). *Organic Chemistry*, 2nd ed. Hoboken: John Wiley and Sons.

Name:			
	Max	Score	Total Grade
Tech	10		
Pre	20		
In	20		
Post	50		

Exp. 8 Distillation and Gas Chromatography of Alkanes

PRE-LAB ASSIGNMENT: *(**EACH STUDENT** will complete and submit an original copy at the beginning of the lab period.* ***Without a complete pre-lab assignment, you will not be allowed to perform the experiment, and will receive a zero for the lab.****)* *.....max score = 20 pts.*

1. **Objective** *(Write a brief purpose of the experiment in **complete** sentences, addressing all of the following points)*
   - What is the goal of the experiment, and what new lab technique will be used to accomplish this goal?
   - What information will be used to identify the compounds in this experiment?
   - How will the purity of compounds and effectiveness of the method be determined?

2. **Physical Data** *(Complete the following table before coming to lab.)*

	n-pentane	*n*-hexane	3-methylpentane	*n*-heptane	isooctane
structure					
MW (g/mol)					
bp (°C)					
d (g/ml)					

3. Experimental Outline *(Give a brief description of the procedure that will be followed in this experiment in 5 lines or less.)*

1	
2	
3	
4	
5	

4. Pre-Lab Questions *(Answer the following questions prior to lab.)*

A. Explain how an FID detector works to detect compounds in GC analysis.

B. For identification purposes, why is it important to compare sample retention times to a standard run under ***identical*** conditions?

I have read and understood the experimental procedure for this experiment. I am familiar with the hazards and the required disposal procedures for this experiment.

Sign here: ____

Experiment 9

Conversion of an Alcohol to an Alkyl Bromide with Rearrangement

Introduction

In this experiment you will convert a secondary alcohol to isomeric alkyl halides using a S_N1 dehydration reaction with hydrobromic acid (Figure 9.1). This will demonstrate the ability to control the type of reaction that occurs by simply using a halogenated acid as opposed to an oxygenated acid, which will be used in Experiment 10. This experiment will also show the facility with which carbocation intermediates rearrange.

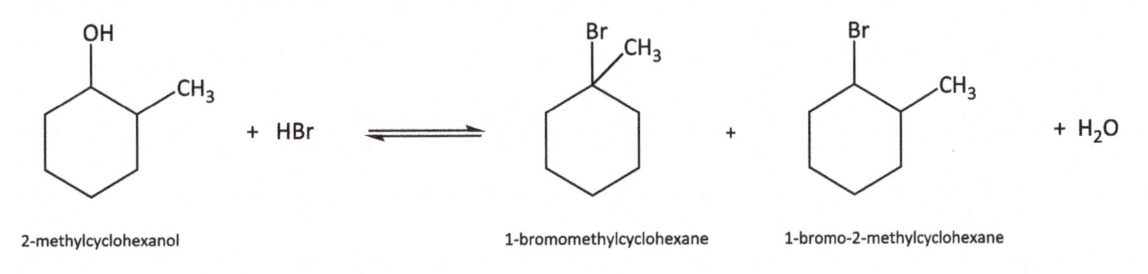

Figure 9.1 Halogenation of 2-methylcyclohexanol to give two isomeric alkyl halides.

Protonation of the alcohol group of 2-methylcyclohexanol results in a loss of water to generate a carbocation. Halides are good nucleophiles. Treating an alcohol with HBr or HCl under controlled conditions can favor conversion of the alcohol to an alkyl halide product rather than to an alkene product (Figure 9.2).

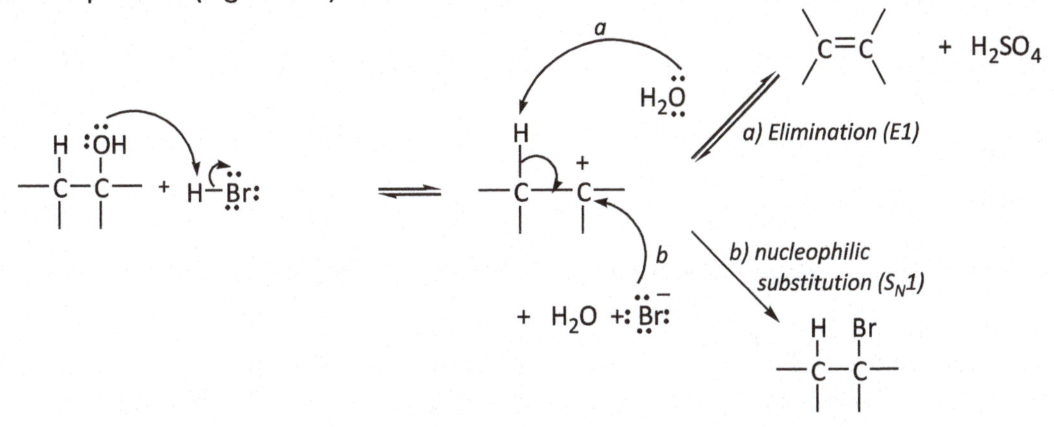

Figure 9.2 Elimination (*E1*) versus substitution (*S_N1*) of an alcohol.

Rearrangement of Carbocations

Carbocations are prone to rearrangement to a more stable carbocation whenever a low-energy pathway is available. A secondary carbocation can undergo a hydride or methide shift, rearranging to a more stable tertiary carbocation.

Protonation and subsequent loss of water from 2-methylcyclohexanol generates a secondary carbocation. Nucleophilic substitution of the intermediate secondary carbocation will give the secondary alkyl bromide (Figure 9.3a). An alternate pathway involves carbocation rearrangement to generate a more stable carbocation intermediate. The shift of a hydrogen atom and its electron pair to the adjacent cationic center, called a hydride shift, gives the more stable tertiary carbocation. Nucleophilic substitution of the tertiary carbocation intermediate generates the tertiary alkyl halide (Figure 9.3b). Thus, it is possible that more than one isomeric alkyl bromide will be produced in the reaction of 2-methylcyclohexanol with HBr. The reaction outlined in Figure 9.3 represents an example of the tendency of carbocation intermediates to rearrange and helps emphasize both the mechanism of the reaction and the limitations of reactions involving carbocations.

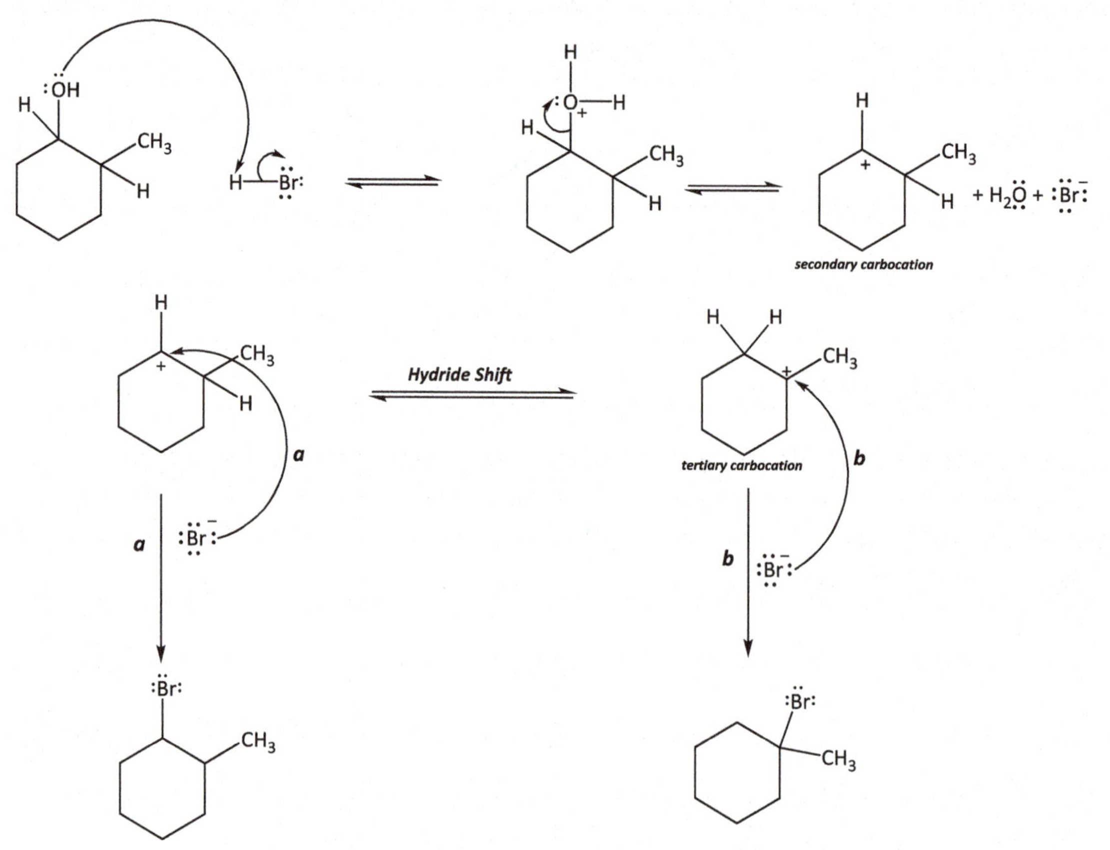

Figure 9.3 Mechanism for S_N1 substitution with rearrangement.

A major difference between the two isomeric alkyl bromide products that are obtained from the reaction mixture is that one is a secondary alkyl bromide, while the other is a tertiary alkyl bromide. Analysis of alkyl bromides by GC is difficult, since they tend to decompose upon entering the very hot injector port of the GC. However, the differing reactivity between 1°, 2°, and 3° alkyl bromides can be used to determine whether the major product is a 2° or 3° alkyl bromide.

When an alkyl halide is treated with a solution of alcoholic silver nitrate, a precipitate of silver halide forms (Figure 9.4). The alkyl halide releases the halide ion through an S_N1 reaction and the halide immediately forms an insoluble salt with the silver ion (AgBr). The rate of formation of AgBr depends on the stability of the carbocation intermediate formed during the reaction. Since 3° carbocations are more stable, they are more likely to form. Since there are more 3° carbocations present, they are more likely to collide with the $AgNO_3$, to release a Br^-, which in turn reacts with the Ag^+ to form the insoluble precipitate. The order of reactivity of alkyl halides is 3° > 2° > 1°. By cooling the silver nitrate solution before adding it to the alkyl halide, the degree of substitution of the alkyl halide can be determined by observing the relative rates of reactions.

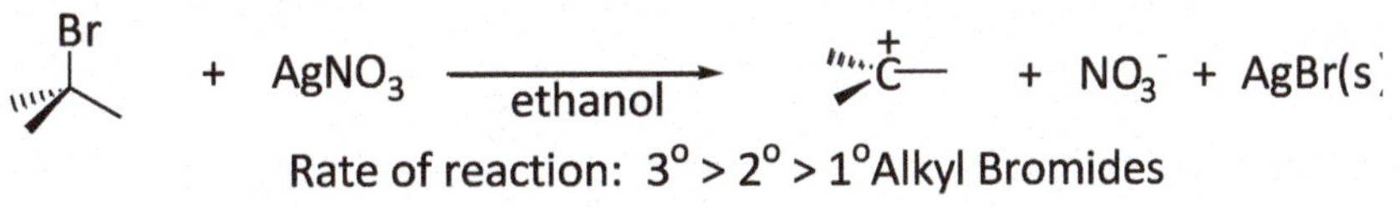

$$(CH_3)_3CBr + AgNO_3 \xrightarrow{\text{ethanol}} (CH_3)_3C^+ + NO_3^- + AgBr(s)$$

Rate of reaction: 3° > 2° > 1°Alkyl Bromides

Figure 9.4 Reaction of alkyl halide with alcoholic silver nitrate.

Functional Groups in Organic Compounds

A *functional group* is the particular atom or groups of atoms that impart reactivity to an organic compound (e.g., the C=C bond in alkenes). The functional group therefore controls how the molecule behaves or functions. They determine the organic compounds physical and chemical properties, such as melting point and reactivity. Functional groups such as the –OH of alcohols or the C=O of ketones or aldehydes also give characteristic absorptions in infrared spectroscopy, making it a valuable tool for the identification of functional groups (Table 9.1).

FUNCTIONAL GROUP	CLASSIFICATION	EXAMPLE	FUNCTIONAL GROUP	CLASSIFICATION	EXAMPLE
R—X: X = Cl, Br, or I	**Alkyl halide**	Cl:	O, R, R	**Ketone**	O
C=C	**Alkene**		O, R, H	**Aldehyde**	O, H
—C≡C—	**Alkyne**		O, R, O, H	**Carboxylic acid**	O, O, H
R—OH	**Alcohol**	OH	O, R, X:	**Acyl halide**	O, Cl:
R—O—R	**Ether**	O	O, O, R, O, R	**Anhydride**	O, O, O
R—SH	**Thiol**	SH	O, R, O, R	**Ester**	O, O
R—S—R	**Sulfide**	S	O, R, N, R, R	**Amide**	O, NH_2
	Aromatic (or arene)		R, R, N, R	**Amine**	N, H

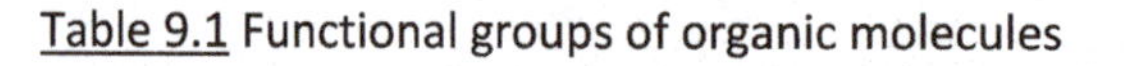

Table 9.1 Functional groups of organic molecules

Infrared Spectroscopy

Infrared (IR) spectroscopy is an extremely valuable tool for determination of functional groups Infrared radiation is a portion of the electromagnetic spectrum between the visible and microwave regions. Bonds in organic molecules bend or stretch at specific frequencies, depending on the type or the strength of the bond. When a molecule is irradiated with infrared light, energy is absorbed when the frequency of the irradiation is equal to the frequency of the bond vibration. Different types of bonds, and therefore different functional groups, absorb IR radiation at different frequencies (see Appendix I). A typical IR spectrum is a plot of transmittance of IR radiation vs. frequency of IR light in reciprocal centimeters (cm^{-1}). There are several characteristic regions and a fingerprint region (Figure 9.5). The fingerprint region, between 400 cm^{-1} to 1600 cm^{-1}, is composed of signals produced by single bonds. Each compound has a unique fingerprint region, however it contains many signals and is more difficult to analyze. Between 1600 cm^{-1} and 4000 cm^{-1}, known as the diagnostic region, generally has fewer signals, and can provide much clearer information when using IR spectroscopy to characterize a compound.

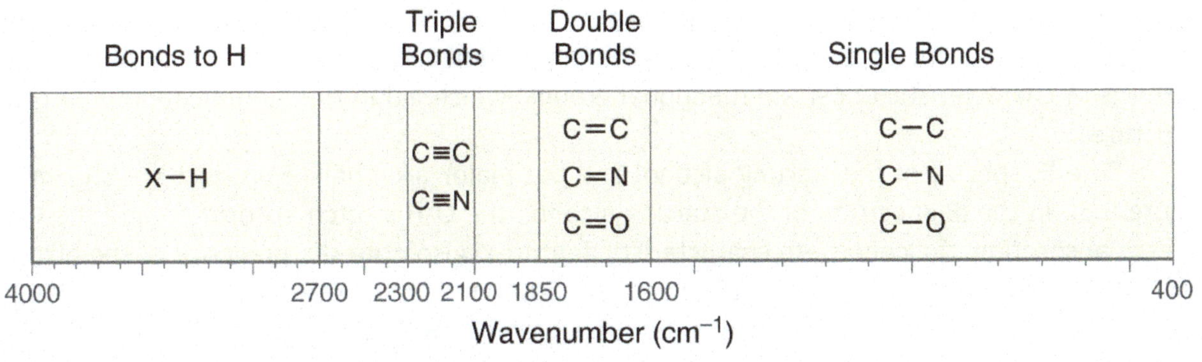

Figure 9.5 Characteristic regions in the IR spectrum.

Hydrocarbons-Alkanes, Alkenes, and Arenes

IR spectra of alkanes are usually simple with few peaks. The strongest peaks in the IR spectrum of an alkane are usually in the C-H stretching region. Peaks with multiple splittings between 2800 cm^{-1} and 3000 cm^{-1} are characteristic of sp^3 C-H stretching. Also present are several bending vibration absorptions of $-CH_2$ and $-CH_3$ groups in the fingerprint region.

Alkenes show less intense sp^2 CH stretch absorptions at around 3000–3300 cm^{-1}. The C=C stretch appears between 1600 cm^{-1} and 1700 cm^{-1}, and is also of weak to medium intensity. Aromatic rings, or arenes, show a medium to weak sp^2 CH absorption around 3030 cm^{-1}. They also show C=C bond stretching absorptions between 1450 cm^{-1} and 1600 cm^{-1}.

Alcohols and Ethers

The position of the O-H stretching absorption and its intensity depend on the extent of H-bonding. Normally, there is extensive H bonding between alcohol molecules, so the O-H stretching occurs as a strong, broad peak which appears between 3000-3700 cm^{-1}. The stretching vibration of the C-O bond in alcohols is strong, and appears between 900 cm^{-1} and 1300 cm^{-1}. Ethers do not have an O-H stretching absorption, but the C-O stretching vibration in aliphatic ethers is in the same region as alcohol C-O stretches, between 1050-1260 cm^{-1}.

Amines

The most distinctive absorptions in primary and secondary amines are due to the N-H stretching vibrations, and appear in the region of 3000–3700 cm^{-1}. Primary amines have two absorptions in this region, an asymmetric stretch and a symmetric stretch. Secondary amines have a single absorption in this region, whereas tertiary amines have none, since there is no N-H bond present.

Carbonyl Compounds—Aldehydes, Ketones, and Esters

Aldehydes, ketones, and esters show characteristic strong IR absorptions due to the C=O bond stretching. This absorption occurs in the range of 1640-1820 cm^{-1}. Esters display a strong C=O stretching absorption in the region between 1735 cm^{-1} and 1800 cm^{-1}. They also display two strong C-O stretching absorptions in the region of 1100–1300 cm^{-1}.

The correlation tables provided in Appendix I can be used to identify possible functional groups within an unknown structure. Several features can be used to distinguish between two functional groups. The presence or absence of a broad OH stretch can be used to distinguish between an alcohol and ether. The presence or absence of a C=O stretch can be used to distinguish between an ether and an ester. The correlation tables provide data on absorption patterns of characteristic of certain functional groups, which aid in the identification unknown structures.

The IR spectra of the starting alcohol and the major alkyl halide product are shown in Figure 9.6. In the IR spectrum of the starting alcohol, the O-H stretch absorption and the C-O stretch absorption of alcohols are characteristic features. Also note the presence of the alkane sp^3 C-H stretch absorptions between 2800 cm^{-1} and 3000 cm^{-1}. In the spectrum of the alkyl halide product, the sp^3 C-H stretch is present, but the broad O-H stretch observed in the alcohol spectrum is no longer present. However, there is a new absorption between 500 cm^{-1} and 700 cm^{-1}, which is due to the C-Br stretch absorption present *only* in the alkyl halide product. Correlation tables of characteristic stretching frequencies of certain functional groups are available in Appendix I.

It is important to note that the intensity of the absorptions can also be a function of the concentration of the sample. If two samples of the same compound are analyzed at different concentrations, the same characteristic absorptions will occur; however, the intensity of the absorptions may be affected.

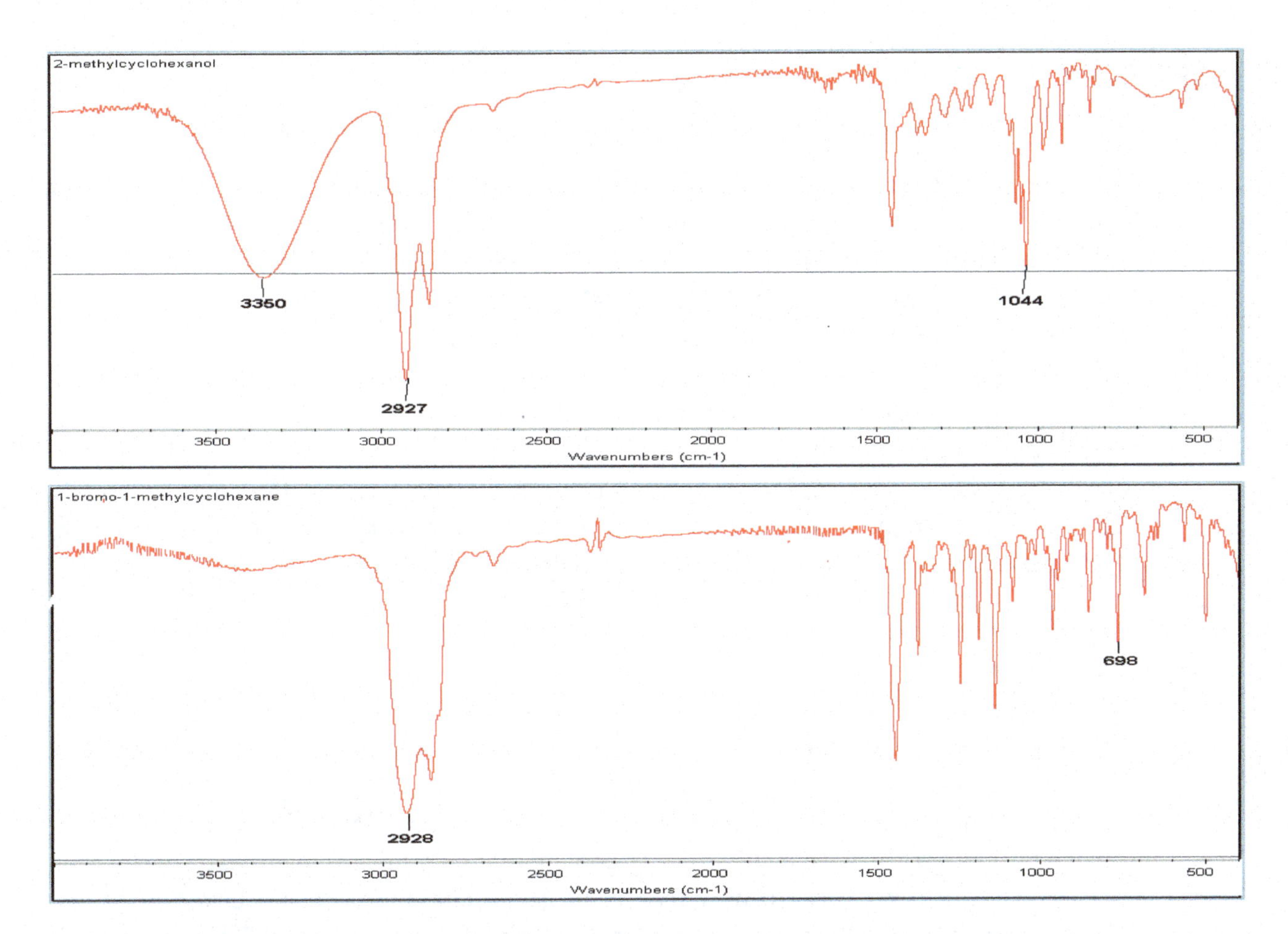

Figure 9.6 Infrared spectra of starting alcohol and major alkyl halide product.

Objectives

In this experiment you will synthesize isomeric alkyl halides from an alcohol using a S_N1 dehydration reaction. The product will be purified using a simple extraction. You will determine the product substitution of your product based on the reaction rate with $AgNO_3$. Finally, you will learn how to use IR spectroscopy to distinguish between the reactant and product using the provided IR spectra.

Experimental Procedure

Synthesis:

- Weigh ~5.0 g of 2-methylcyclohexanol directly into a 100 mL round bottom flask. Add a few boiling chips.
- Add 10 mL 48% HBr to the flask and swirl to mix.
- Clamp the reaction flask to the ring stand and set up a reflux apparatus.
- Set the VR at 40, and begin water flow. When the solvent begins to boil, condense, and drop back down into the reaction flask indicating reflux, allow the reaction to reflux for 30 minutes.
- Lower the heating mantle and cool the reaction flask in a beaker of tap water for 2 to 3 minutes, then a beaker of ice water for 2 to 3 minutes.

Purification:

- Transfer the cooled solution to a separatory funnel. Be sure to close the stopcock before adding any solution to the separatory funnel!
- Add 2.0 mL concentrated H_2SO_4 to the separatory funnel. (***WARNING: Sulfuric acid is very corrosive!)*** The sulfuric acid is used to protonate any unreacted alcohol so that it is drawn into the aqueous layer, removing it from the neutral organic alkyl halide product.
- Stir gently with a glass rod. **DO NOT INVERT AND AGITATE!** Drain off the bottom aqueous layer into a 125 mL Erlenmeyer flask labeled "Acidic Waste."
- Using a glass Pasteur pipette with rubber bulb, transfer the top layer to a pre-weighed 50 mL beaker. Cover the beaker with a watch glass except during weighing. Alkyl bromides are *lachrymators* (they make you cry).
- Reweigh the beaker and record the final mass of product. Calculate the percent yield for the reaction and observe the product appearance. Record this data in the laboratory notebook, and complete Table 9.1 on the POST LAB Assignment (provided online).

Product Analysis:

Silver Nitrate Test

- Add three drops of 1-bromobutane to a small test tube. Add three drops of 2-bromobutane to a second test tube, 2-bromo-2-methylpropane to a third test tube, and three drops of *your product* to a fourth test tube.
- Obtain 4.0 mL of the $AgNO_3$ solution in 10 mL graduated cylinder and cool the solution in an ice bath. After the $AgNO_3$ solution has cooled, quickly add ~1.0 mL to each of the four tubes. Pay close attention to how quickly (if at all) cloudiness or a precipitate forms. Record your observations in the laboratory notebook, and complete Table 9.2 on the POST LAB Assignment (provided online).

IR Analysis

- Using the provided IR spectra in Figure 9.6, assign all characteristic frequencies of the starting alcohol and product. Complete Table 9.3 on the POST LAB Assignment (provided online).

SAFETY

****All experiments should be performed in a fume hood with appropriate safety glasses. Gloves will be provided by request.****

All organic compounds used in this experiment are irritants, and can be toxic if ingested or absorbed through skin. All alkyl halide products, along with the alcohol reactant, are flammable and can be irritating if inhaled. Sulfuric acid and hydrobromic acid are ***extremely*** corrosive and can cause burns upon skin contact. Silver nitrate is corrosive, a strong oxidizer, and a possible carcinogen. Wear gloves at all times during use.

WASTE MANAGEMENT

Place the acidic solution from the extraction in the bottle labeled "LIQUID WASTE—SN1". Place your alkyl bromide product in the bottle labeled "LIQUID WASTE—SN1". Pour the contents of test tubes used in the silver nitrate test in the container labeled "SILVER NITRATE WASTE". All bottles will be located in the waste hood.

Reference

Klein, David. (2015). *Organic Chemistry*, 2nd ed. Hoboken: John Wiley and Sons.

	Max	Score	Total Grade
Tech	10		
Pre	20		
In	20		
Post	50		

Name:

Exp. 9 Conversion of an Alcohol to an Alkyl Bromide with Rearrangement

PRE-LAB ASSIGNMENT: *(**EACH STUDENT** will complete and submit an original copy at the beginning of the lab period. **Without a complete pre-lab assignment, you will not be allowed to perform the experiment, and will receive a zero for the lab.**) .....max score = 20 pts.*

1. **Objective** *(Write a brief purpose of the experiment in **complete** sentences, addressing all of the following points.)*
   - Name the *specific* reactants combined during the synthesis and name the product(s) formed.
   - What is the purification technique used in this experiment?
   - What analytical technique will be used during the experiment to identify the product substitution?
   - What type of spectral analysis will be used to characterize the reactant and product?

2. **Chemical Equation** *(Draw the chemical equation for the halogenation of 2-methylcylcohexanol using actual chemical structures.)*

3. **Physical Data** *(Complete the following table before coming to lab.)*

Compound	MW (g/mol)	bp (°C)	d (g/mL)
2-methylcyclohexanol			
48% hydrobromic acid			
sulfuric acid		X	X
1-bromobutane		X	X
2-bromobutane		X	X
2-bromo-2-methylpropane		X	X
silver nitrate		X	X
1-bromo-1-methylcyclohexane	177.08	X	X
1-bromo-2-methylcyclohexane	177.08	X	X

4. **Experimental Outline** *(Give a brief description of the procedure that will be followed in this experiment in 5 lines or less.)*

1	
2	
3	
4	
5	

5. **Pre-Lab Questions** *(Answer the following questions prior to lab.)*

A. Which of the following alcohols would generate the ***most stable*** carbocation intermediate?

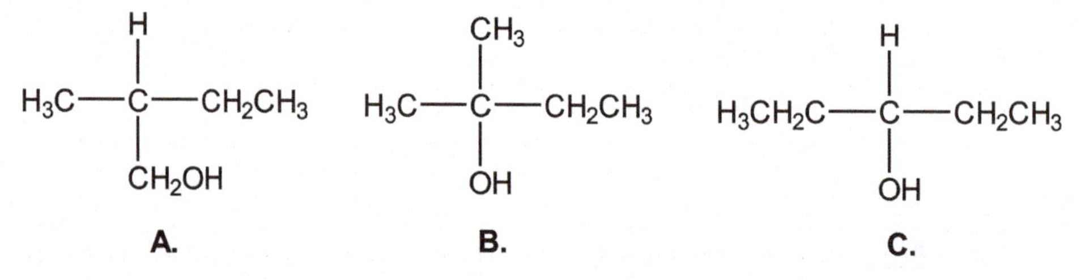

B. Why can GC analysis ***not*** be used to analyze the products synthesized in this experiment?

I have read and understood the experimental procedure for this experiment. I am familiar with the hazards and the required disposal procedures for this experiment.

Sign here: ____

Experiment 10

Acid-Catalyzed Dehydration of an Alcohol with Rearrangement

Introduction

This experiment demonstrates the process of acid-catalyzed dehydration of a secondary alcohol, 2-methylcyclohexanol, to give three isomeric alkenes (Figure 10.1). Once synthesized, GC analysis of the distilled product will be used to determine the actual ratio of isomeric alkenes formed. Like Experiment 9, this experiment shows the facility with which carbocation intermediates rearrange.

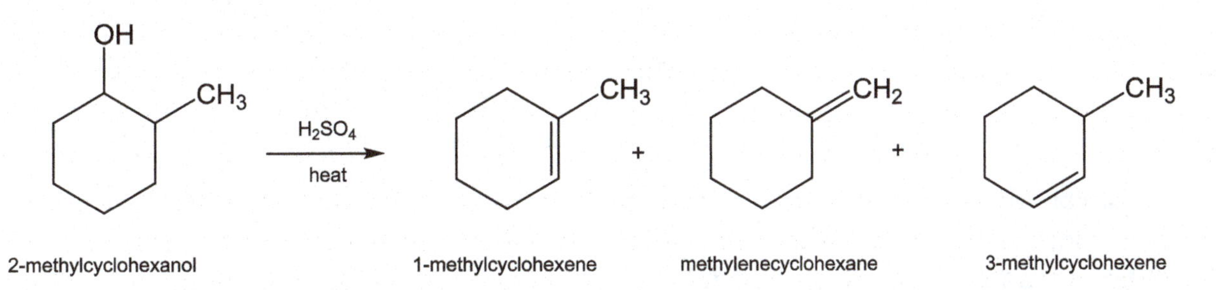

Figure 10.1 Acid-catalyzed dehydration of 2-methylcyclohexanol to give three isomeric alkenes.

Dehydration of Alcohols

Elimination reactions are fundamental reactions in organic chemistry and most important for the synthesis of alkenes. An elimination reaction requires that a leaving group, X, departs with its bonding electrons. The adjacent hydrogen atom is lost without its bonding electrons; it is lost as a proton, H^+. If the H-X lost is water, the elimination reaction is called dehydration.

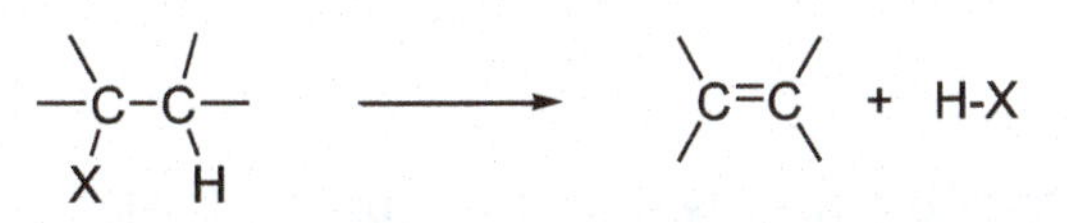

When an alcohol is heated in the presence of a strong acid, like sulfuric acid, the –OH (a poor leaving group) is protonated so that the leaving group becomes water (a good leaving group). When secondary and tertiary alcohols are heated with a strong acid, the loss of water generates the positively charged carbocation. Reaction of the carbocation with a base gives the alkene. A strong base is not necessary; in fact, water is strong enough to abstract the proton adjacent to the carbocation.

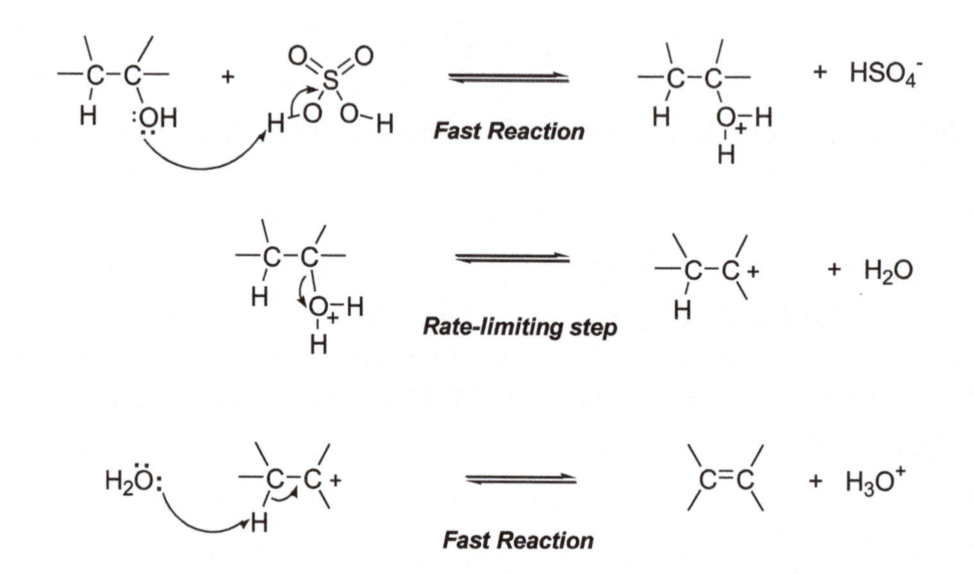

Figure 10.2 Rate-limiting step in elimination reactions (E1).

According to Zaitsev's rule, the most stable, most highly substituted alkene that arises from the most stable carbocation intermediate should be the major product. The rate of elimination of water depends on the stability of the carbocation formed. Formation of the carbocation is the most energetically unfavorable and, therefore, the slowest step of dehydration reactions (Figure 10.2). Secondary and tertiary carbocations are stable enough to be formed at a reasonable rate, but primary carbocations are so unstable that they form only at a very slow rate. H^+ abstraction is very fast compared with initial formation of the carbocation. Thus, the rate-limiting step is a unimolecular reaction, loss of water to form the carbocation. This is an example of an **E1** reaction. E1 simply means unimolecular elimination.

All of the reaction steps in E1 dehydration of alkenes are reversible. Thus, hydration of an alkene to give an alcohol is the reverse of the dehydration mechanism. Distillation of the alkene as it is formed shifts the equilibrium to favor formation of the alkene.

Rearrangement of Carbocations

Carbocations are prone to rearrangement to a more stable carbocation whenever a low-energy pathway is available. A secondary carbocation can undergo a hydride or methide shift, rearranging to a more stable tertiary carbocation (see Figure 10.3).

Protonation and subsequent loss of water from 2-methylcyclohexanol generates a secondary carbocation. Although abstraction of the adjacent proton from this intermediate (pathways *a* and *b*) would give the alkenes shown, an alternate pathway involves rearrangement to a more stable tertiary carbocation (pathway *c*). The shift of a hydrogen atom and its electron pair to the adjacent cationic center, called a hydride shift, gives the more stable tertiary carbocation. Now there are two alternate pathways (*d* and *e*) for abstraction of a neighboring proton to give the other two isomeric alkenes.

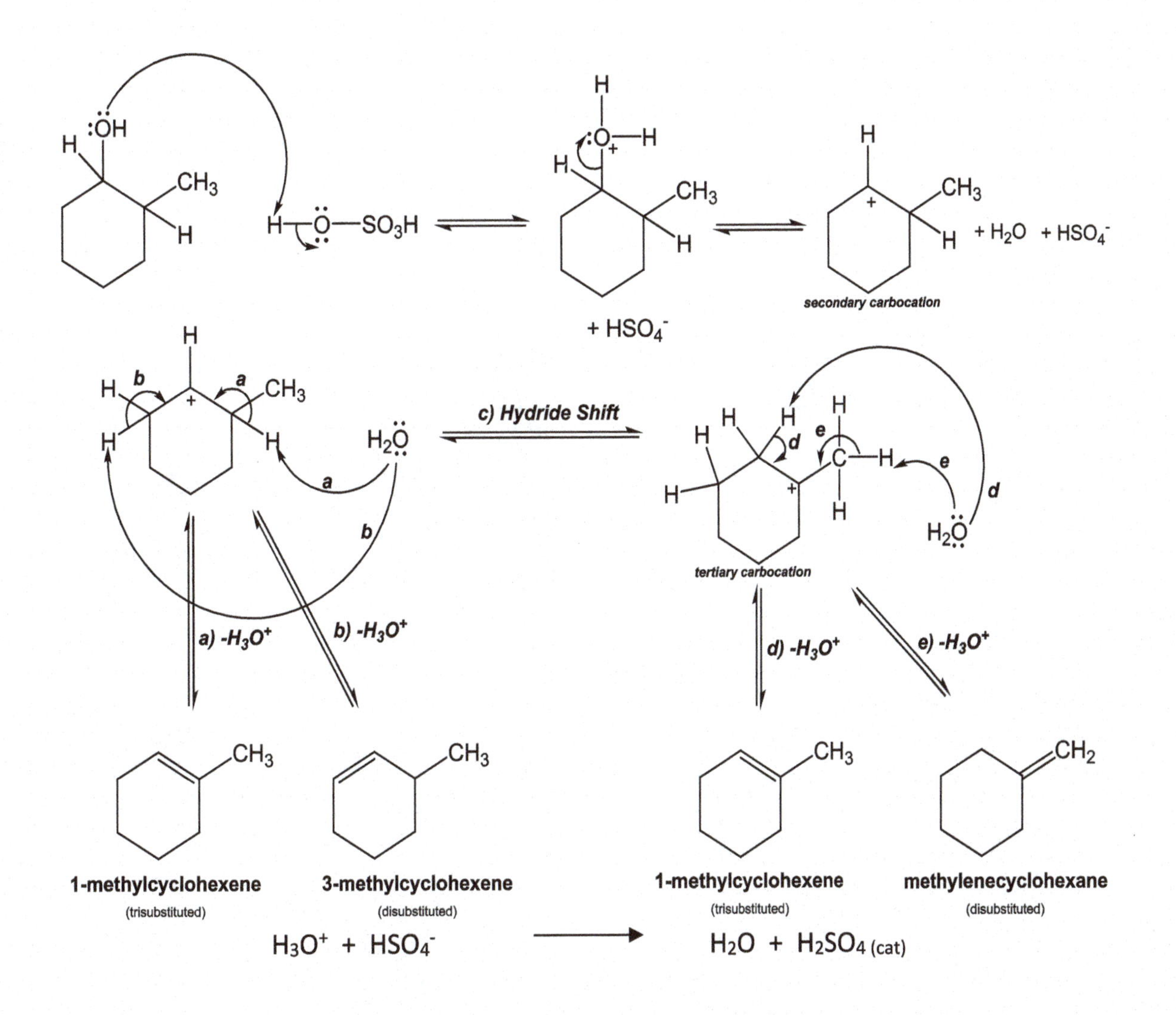

Figure 10.3 Mechanism for dehydration of 2-methylcyclohexanol.

Infrared Spectroscopy

In Figure 10.4, the IR spectra of the starting alcohol for this experiment and the resulting alkene products are shown. In the IR spectrum of the starting alcohol, the O-H stretch and the C-O stretch absorptions of alcohols are characteristic features. Also note the presence of the sp^3 C-H stretch absorptions. In the spectra of the alkene products, the sp^3 C-H stretch absorption is present, but the broad OH stretch absorption and C-O stretch absorption observed in the alcohol spectrum are no longer present. However, in the spectra of the less substituted products, there are new absorptions due to the presence of the C=C stretch, and the sp^2 C-H stretch.

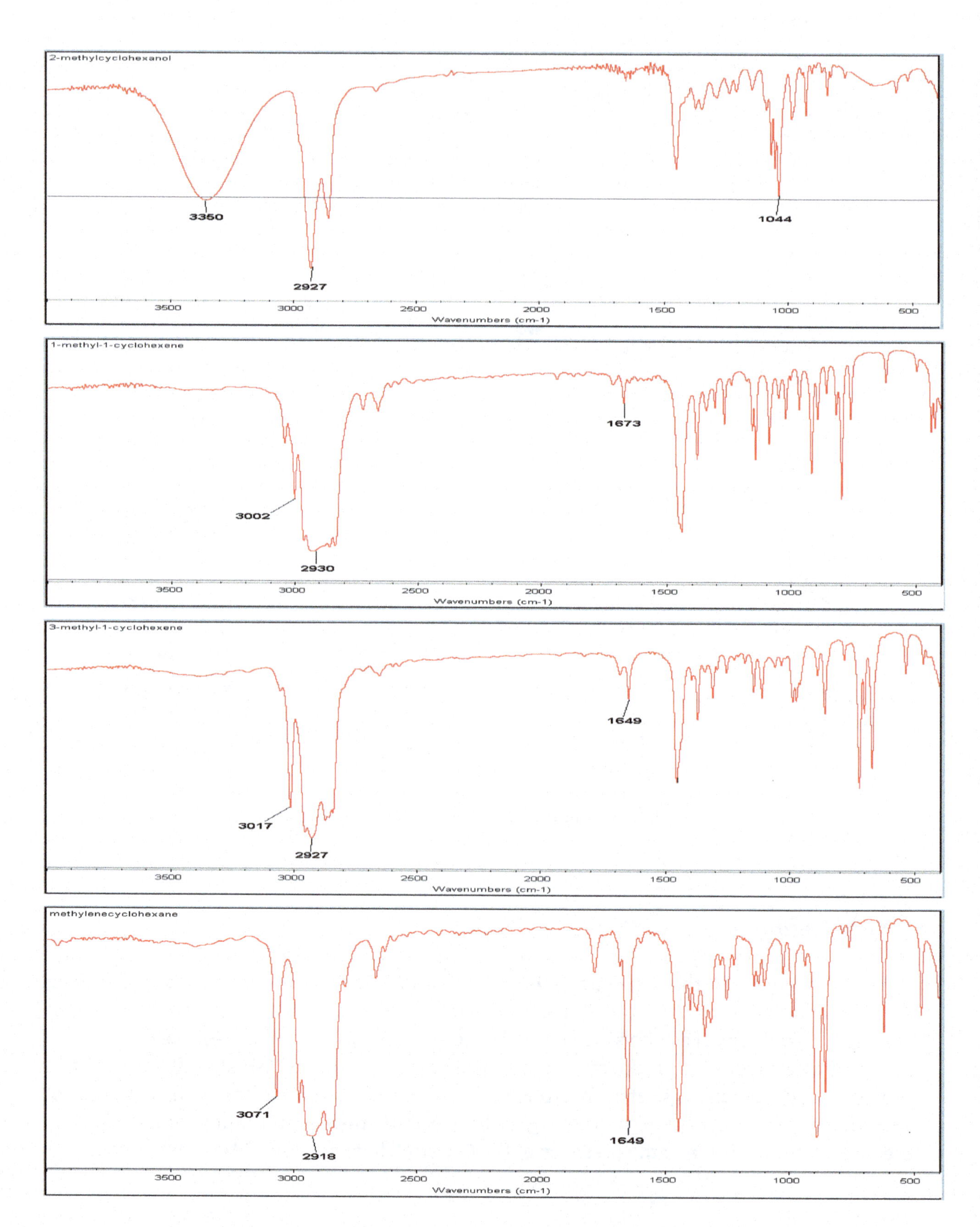

Figure 10.4 IR spectra of 2-methylcyclohexanol and dehydration products.

Objectives

In this experiment you will synthesize isomeric alkenes from a secondary alcohol using a catalytic amount of sulfuric acid. You will distill the resulting alkene mixture and analyze your distillate by Gas chromatography (GC). Results from the GC analysis (Experiment 8 and Appendix E) will enable you to identify the products, identify the major product of the synthesis, and determine the actual ratio of isomeric alkenes formed. Finally, you will use IR spectroscopy to distinguish between the reactants and products of the reaction.

Experimental Procedure

Synthesis and Purification:

- Weigh ~5.0 g of 2-methylcyclohexanol directly into a 50 mL round bottom flask. Add 10 drops of concentrated sulfuric acid and 2-3 boiling chips. (***CAUTION: Sulfuric acid is very corrosive!***)
- Clamp the 50 mL round bottom flask to a ring stand a few inches from the ring stand base, and continue to build a simple distillation apparatus one piece at a time (Experiment 8 and Appendix A). Use two blue plastic keck clips to secure the joints between the condenser and head adapter, and the condenser and receiver adapter. Place a $CaSO_4$ drying tube into the top of the head adapter, instead of a thermometer.
- Attach a *preweighed* 10 mL round bottom flask to the receiver adapter, holding in place until the flask can be securely clamped to a second ring stand.
- Connect the heating mantle to the voltage regulator, and place the heating mantle in direct contact with the reaction flask. Begin a gentle water flow, then turn the voltage regulator on 30 and allow the reaction to boil gently.
- Collect ~5 mL of the distillate. ***DO NOT ALLOW DISTILLATION TO PROCEED TO DRYNESS.***
- Lower the heating mantle away from the reaction flask and allow the flask to cool to room temperature.
- Reweigh the 10 mL receiving flask to determine the final mass of product and calculate the percent yield for this reaction. Record this data in the laboratory notebook and in Table 10.1 on the POST LAB Assignment (provided online).

Product Analysis:

GC Analysis

- Prepare a GC sample of your distillate. Place **five** drops of your distillate in an autoanalyzer vial and add 1 mL of GC solvent (methanol). Using the chromatographic results, determine the purity and identity of your product. Calculate adjusted area percent for the alkene products in the laboratory notebook. Use this data to complete Table 10.2 on the POST LAB Assignment (provided online).

IR Analysis

- Using the spectra provided in Figure 10.4, assign all characteristic frequencies of the reactant and products. Complete Table 10.3 on the POST LAB Assignment (provided online).

SAFETY

All experiments should be performed in a fume hood with appropriate safety glasses. Gloves will be provided by request.

All organic solvents used in this experiment are flammable, irritants, and can be toxic if ingested or absorbed through skin. All alkene products, along with the alcohol reactant, are flammable and can be irritating if inhaled. Sulfuric acid is ***extremely*** corrosive and can cause burns upon skin contact.

WASTE MANAGEMENT

Add a small amount of water to the reaction flask residue (mainly sulfuric acid) (CAUTION: SPLATTERING MAY OCCUR!), and transfer to the bottle labeled, "LIQUID WASTE—E1" in the waste hood. After your alkene mixture has been weighed and a GC sample has been submitted, pour it into the container labeled "LIQUID WASTE—E1".

References

Klein, David. (2015). *Organic Chemistry*, 2nd ed. Hoboken: John Wiley and Sons.

Name:			
	Max	Score	Total Grade
Tech	10		
Pre	20		
In	20		
Post	50		

Exp. 10 Acid-Catalyzed Dehydration of an Alcohol with Rearrangement

PRE-LAB ASSIGNMENT: *(**EACH STUDENT** will submit an original copy at the beginning of the lab period. **Without a complete pre-lab assignment, you will not be allowed to perform the experiment, and will receive a zero for the lab.**) .....max score = 20 pts.*

1. **Objective** *(Write a brief purpose of the experiment in **complete** sentences, addressing all of the following points.)*
   - Name the *specific* compounds combined during the synthesis and name the product(s) formed.
   - What is the purification technique used in this experiment?
   - What analytical technique will be used during the experiment to identify and quantify products?
   - What type of spectral analysis will be used to characterize the reactant and products?

2. **Chemical Equation** *(Draw the chemical equation for the dehydration of 2-methylcylcohexanol using actual chemical structures.)*

3. **Physical Data** *(Complete the following table before coming to lab.)*

Compound	MW (g/mol)	bp (°C)	d (g/mL)
2-methylcyclohexanol			
sulfuric acid	X	X	
1-methylcyclohexene			0.810
3-methylcyclohexene			0.800
methylenecyclohexane			0.810
methanol	X		

4. <u>Experimental Outline</u> *(Give a brief description of the procedure that will be followed in this experiment in 5 lines or less.)*

1	
2	
3	
4	
5	

5. <u>Pre-Lab Questions</u> *(Answer the following questions prior to lab.)*

A. The rate of dehydration depends on the **stability** of the carbocation intermediate formed during the reaction.

a. True
b. False

B. A **tertiary** carbocation will typically rearrange to a **secondary** carbocation.

a. True
b. False

I have read and understood the experimental procedure for this experiment. I am familiar with the hazards and the required disposal procedures for this experiment.

Sign here: ____

Experiment 11

Bromination of Stilbene—Green Synthesis

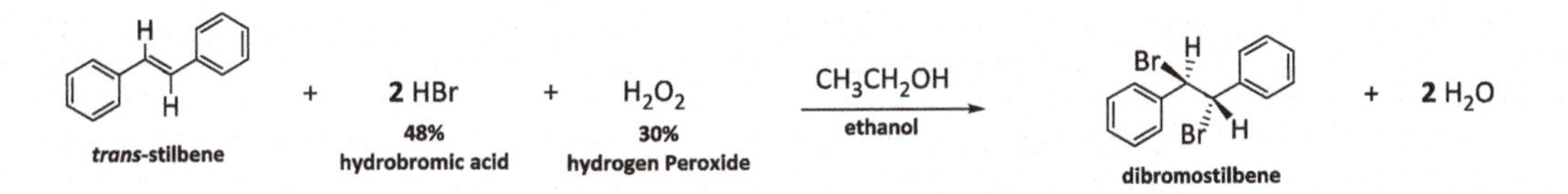

Figure 11.1 Bromination of *trans*-stilbene using hydrobromic acid and hydrogen peroxide.

Introduction

In this experiment an alkene will be brominated using HBr and hydrogen peroxide (Figure 11.1). A typical bromination of alkene using elemental bromine (Br_2) has been introduced in the textbook. The general mechanism for this reaction is shown in Figure 11.2.

Typically, bromination of an alkene is performed in a chlorinated solvent, such as carbon tetrachloride (CCl_4) or dichloromethane (CH_2Cl_2). Both of these chlorinated solvents are considered to be carcinogenic. Additionally, elemental bromine is volatile and highly corrosive. It can cause severe burns upon contact with the skin and is very irritating upon inhalation. An alternate source of elemental bromine comes from the oxidation of hydrobromic acid with hydrogen peroxide. Instead of a chlorinated solvent, ethanol can be used as the solvent in this synthesis. Although HBr and hydrogen peroxide must be handled with care, they provide an improvement over elemental bromine. This alternative bromination method is an example of a **"green chemistry"** synthesis, an increasingly important area of chemistry.

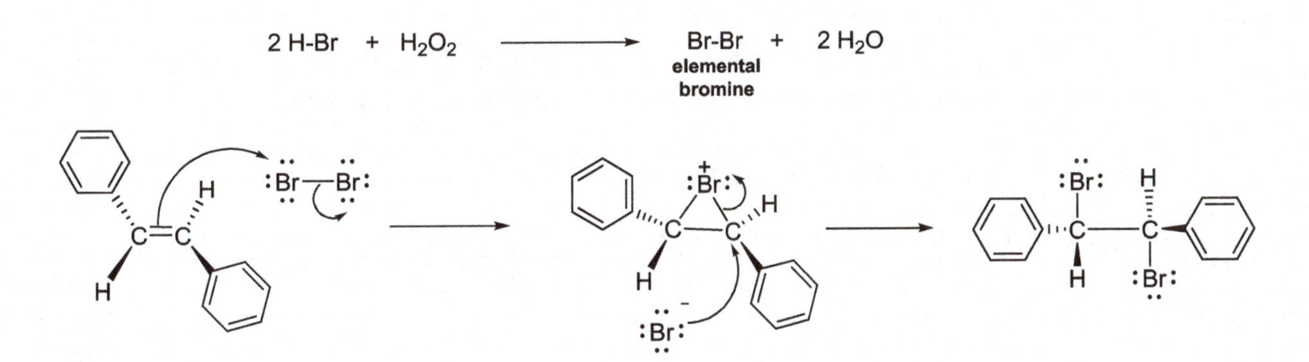

Figure 11.2 Mechanism of bromination of *trans*-stilbene with elemental bromine.

Green Chemistry

A fundamental challenge in chemistry today is to find creative ways to minimize human exposure to, and the environmental impact of, harmful chemicals while still enhancing people's lives and scientific progress. Green chemistry—the design of chemical products and processes that reduce or eliminate the use and generation of hazardous substances—is the most fundamental approach to this challenge. Adoptions of green chemistry technologies offer environmental improvement, enhanced safety, and economic benefit. To make products benign by design, chemists must consider the hazard posed by a substance, along with its other chemical and physical properties, and select products and processes that minimize harm. Green chemistry principles focus on using safer starting materials for efficient chemical reactions, using renewable resources, conserving energy, finding safer solvents, and reusing or recycling product and waste whenever possible.

In addition to using safer starting materials and processes, the efficiency of a chemical reaction must be considered when designing a synthetic process. A "perfect" chemical reaction would be one that is completely ***selective***, affording only the desired product, highly ***efficient***, incorporating all atoms of the starting materials and reagents in the product, and entirely safe, using only ***nonhazardous*** (to the chemist and the environment) starting materials, reagents, and products. Ideally, this reaction would require no solvent or energy inputs (see Appendix J).

Theoretical and Percent Yield

Generally, chemists quantify the efficiency of a reaction by reporting the "*percent yield*," which is defined as the percentage of the theoretically possible amount of product obtained. For synthesis of dibromostilbene, there are three reactants. To determine the theoretical yield, it is first necessary to calculate the moles of product possible based on each of the three reagents used. The reagent which produces the least amount of product is the *limiting reagent* and determines the amount of product that can be produced in the reaction.

To calculate the moles of each reagent, you must know their molecular weights. In this experiment, you are starting with 0.50 g of stilbene so that:

For stilbene:

$$0.50\ \text{g stilbene} \times \frac{1\ \text{mol stilbene}}{180\ \text{g}} \times \frac{1\ \text{mol dibromostilbene}}{1\ \text{mol stilbene}} = 2.78 \times 10^{-3}\ \text{mol product formed}$$

(based on stilbene used)

To calculate the moles of H_2O_2 and HBr, you must take into consideration the volume used, the densities, and the percent compositions of the H_2O_2 and HBr.

If you used 0.8 mL of **30% H_2O_2** and 1.2 mL of **48% HBr**:

For 30 % H_2O_2:

$$0.8\ \text{mL}\ H_2O_2 \times \frac{1.11\ \text{g}}{\text{mL}} \times \frac{1\ \text{mol}}{34\ \text{g}} \times \frac{1\ \text{mol dibromostilbene}}{1\ \text{mol}\ H_2O_2} = 2.61 \times 10^{-2}\ \text{mol} \times \underline{0.30} = 7.84 \times 10^{-3}\ \text{mol product formed}$$

(based on 30% H_2O_2 used)

For 48% HBr:

$$1.2\ \text{mL HBr} \times \frac{1.49\ \text{g}}{\text{mL}} \times \frac{1\ \text{mol}}{81\ \text{g}} \times \frac{1\ \text{mol dibromostilbene}}{2\ \text{mol HBr}} = 1.10 \times 10^{-2}\ \text{mol} \times \underline{0.48} = 5.30 \times 10^{-3}\ \text{mol product formed}$$

(based on 48% HBr used)

Based on the calculations above, the reagent that produces the least amount of product is the stilbene; therefore, the stilbene is the limiting reagent that determines how much product can be made. The final step of a theoretical yield calculation is to multiply the number of moles of product by the molecular weight of the product, because the ***theoretical yield is always reported in grams***. Once the experiment is complete and the actual weight of the product is obtained, the percent yield can be determined by simply dividing the actual weight of the product obtained by the theoretical yield of the product, calculated based on the limiting reagent. For additional assistance with limiting reagent and theoretical yield calculations, see Appendix D.

Objectives

In this experiment we will brominate an alkene using green chemistry conditions. The product will be purified by recrystallization, and the pure product isolated by vacuum filtration. Finally, the degree of purity of the product will be determined using TLC analysis.

Experimental Procedure

Synthesis:

- Weigh ~0.50 g of stilbene into a 100 mL round bottom flask. Clamp the flask to a ring stand.
- Add 10 mL of ethanol, and three boiling chips to the flask.
- Set up the reflux apparatus by placing a water-cooled reflux condenser in the top of the flask (Appendix A). Start the water flow through the condenser, and begin heating the reaction with a heating mantle (set the voltage regulator on 20).
- Heat the mixture under reflux until the majority of the solid has dissolved.
- Lift the condenser just long enough to slowly add 1.5 mL of 48% hydrobromic acid, then lower back onto the flask. Some of the stilbene may precipitate, but continue to heat at reflux for another 10 minutes until the majority of the solid has dissolved. After 10 minutes, go on to the next step even if a small amount of the solid remains.
- Lift the condenser just long enough to slowly add 1.0 mL of 30% hydrogen peroxide, then lower back onto the flask (**CAUTION**! *30% hydrogen peroxide is a very strong oxidizer and will burn your skin if you get any on yourself. One person per group should handle the H_2O_2 using gloves).* The initially colorless mixture will change in color to golden yellow as the Br_2 forms in the reaction flask.
- Lower the hood sash ***completely*** and allow the reflux to continue until the yellow color fades and the mixture becomes cloudy white (10–20 minutes), indicating that the Br_2 has been consumed.

Purification:

- After synthesis, lower the heating mantle and allow the flask to cool to room temperature in a room temperature water bath.
- Add ~3 mL of 1M NaOH to neutralize any excess HBr.
- Cool the reaction mixture in an ice bath for 5 minutes to maximize crystallization of the product.
- Set up a suction filtration apparatus (Appendix A). Remember to weigh a small filter paper, and seat the filter paper with cold ethanol prior to use.
- Collect the solid by vacuum filtration. Rinse the round bottom flask with 3 mL of ice-cold ethanol and use this rinse to wash the crystals in the Büchner funnel.

- Continue to draw air through your product for an additional 5 minutes to help dry the product.
- Transfer the crystals to a preweighed watch glass and place them in the warm oven to dry for 10 minutes.
- Determine the mass of your dried product and calculate the percent yield for your synthesis. If the final mass of your product is higher than the theoretical yield, place back into the oven to dry for an additional 10 minutes.
- Record all data in the laboratory notebook, and complete Table 11.1 on the POST LAB Assignment (provided online). ***PROCEED TO PRODUCT ANALYSIS.***

Product Analysis:

TLC Analysis:

- Remove a few crystals of your product for TLC analysis. Place the crystals in a small test tube and dissolve in 1 mL of reagent acetone.
- Perform a TLC analysis with your sample from above. Apply to the TLC plate alongside the provided standards of the stilbene starting material and the expected product.
- Check the TLC plate under a UV lamp to ensure that the compounds are concentrated enough to detect after development. Apply additional compound to the origin line if necessary.
- Develop the plate in 9:1 hexane/ether.
- Visualize the plate under the UV lamp. Measure and record TLC R_f values of all spots. Sketch the TLC plate in your laboratory notebook, and complete Table 11.2 on the POST LAB Assignment (provided online).

SAFETY

****All experiments should be performed in a fume hood with appropriate safety glasses. Gloves will be provided by request.****

Hydrobromic acid is corrosive. Wear gloves when working with it. Hydrogen peroxide is a strong oxidizing agent and may damage clothing and skin. Ethanol, hexane, acetone, and diethyl ether are flammable.

WASTE MANAGEMENT

Place all liquid waste from recrystallization and TLC analysis in the container labeled "LIQUID WASTE—BROMINATION". Place solid product, TLC plates, filter papers, and pipets in the container labeled "SOLID ORGANIC WASTE". All used TLC capillaries should be placed in the broken glass container.

References

Klein, David. (2015). *Organic Chemistry*, 2nd ed. Hoboken: John Wiley and Sons.

Kirchhoff, M.,Ryan, M. (2002) *Greener Approaches to Undergraduate Chemistry Experiments*. Washington. American Chemical Society.

Doxsee, Kenneth M., Hutchinson, James E. (2003). *Green Organic Chemistry, Strategies, Tools and Laboratory Experiments*. Belmont: Thompson/Brooks and Cole.

Zubrick, James W. (1997). *The Organic Chem Lab Survival Manual, A Student's Guide to Techniques*. New York: John Wiley and Sons, pp. 255–268.

Name:			
	Max	Score	Total Grade
Tech	10		
Pre	20		
In	20		
Post	50		

Exp. 11 Bromination of Stilbene—Green Synthesis

PRE-LAB ASSIGNMENT: *(**EACH STUDENT** will complete and submit an original copy at the beginning of the lab period. **Without a complete pre-lab assignment, you will not be allowed to perform the experiment, and will receive a zero for the lab.**) .....max score = 20 pts.*

1. **Objective** (*Write a brief purpose of the experiment in **complete** sentences, addressing all of the following points.*)
   - Name the *specific* reactants combined during the synthesis and name the product formed.
   - What is the purification technique used in this experiment?
   - What analytical technique will be used during the experiment to determine purity?

2. **Chemical Equation** (*Draw the chemical equation for the synthesis of dibromostilbene using actual chemical structures.*)

3. **Physical Data** (*Complete the following table before coming to lab.*)

Compound	MW (g/mol)	mp (°C)	d (g/mL)
trans-stilbene			X
dibromostilbene			X
48% hydrobromic acid		X	
30% hydrogen peroxide		X	
ethanol	X	X	

4. Experimental Outline *(Give a brief description of the procedure that will be followed in this experiment in 5 lines or less.)*

1	
2	
3	
4	
5	

5. Pre-Lab Questions *(Answer the following questions prior to lab.)*

A. List ***3*** characteristics of a "perfect" chemical reaction based on green chemistry principles.

a.

b.

c.

B. What is meant by generating the bromine using the "*in situ*" approach?

I have read and understood the experimental procedure for this experiment. I am familiar with the hazards and the required disposal procedures for this experiment.

Sign here: ______________________________

Experiment 12

Synthesis of Diphenylacetylene

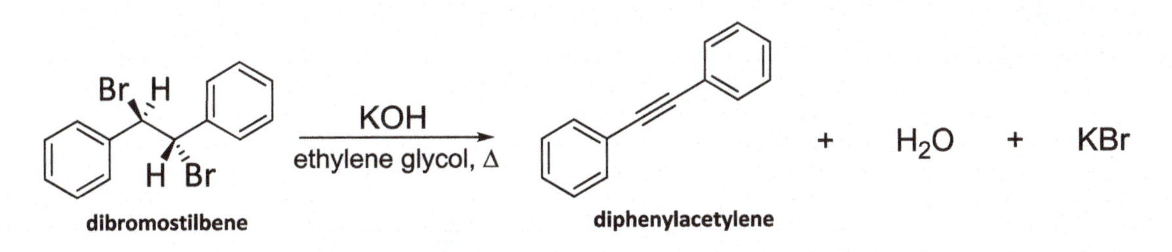

Figure 12.1 Double elimination of dibromostilbene.

Introduction

In the previous experiment, dibromostilbene was prepared by bromination of *trans*-stilbene using Green Chemistry conditions. In this experiment, the dibromostilbene prepared in the previous lab will be used as the reactant in the route to synthesize an alkyne product (Figure 12.1).

Alkynes are commonly prepared by using the standard method of brominating the double bond of an alkene to form a 1,2 dihalide, followed by the heating of this vicinal dihalide with a strong base at high temperatures. This will result in the double elimination of two equivalents of hydrogen bromide to form the desired alkyne. This reaction is referred to as a didehydrohalogenation, and will be used to convert the dibromostilbene to diphenylacetylene (Figure 12.2).

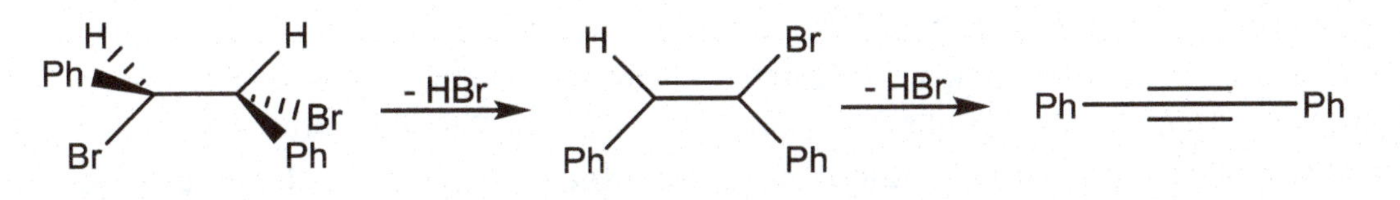

Figure 12.2 Didehydrobromination of dibromostilbene to form diphenylacetylene.

Didehydrohalogenation is a bimolecular elimination, or **E2** elimination, which occurs in a concerted fashion. No intermediates are isolated during this elimination. Recall that a typical dehydrohalogenation is favored when the five atoms involved in the E2 elimination reaction are in the same plane, known as an *anti*-periplanar conformation (Figure 12.3)

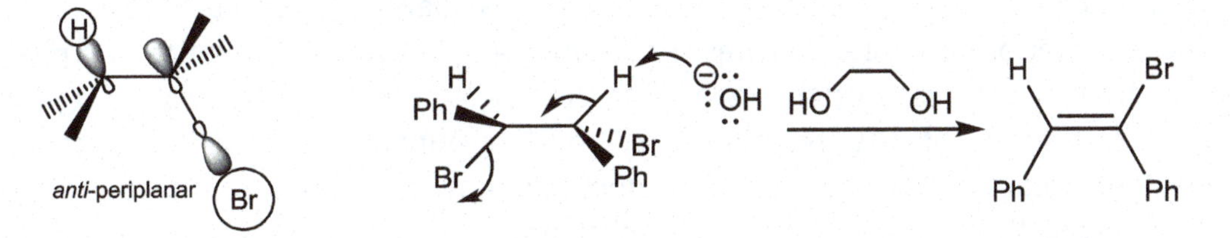

Figure 12.3 Mechanism for the dehydrobromination of dibromostilbene.

The most favored orbital interaction occurs when the atom geometry is *anti*-periplanar, as is the case with the initial elimination step. However, this is not always the case. In the second elimination step, the intermediate exists in the less favorable "*syn*-periplanar" conformation, thus the orbital overlap is reduced. For this reason, the activation energy is much higher. In order to achieve this amount of energy, higher temperatures must be used to carry out the final elimination step to yield the desired alkyne (Figure 12.4).

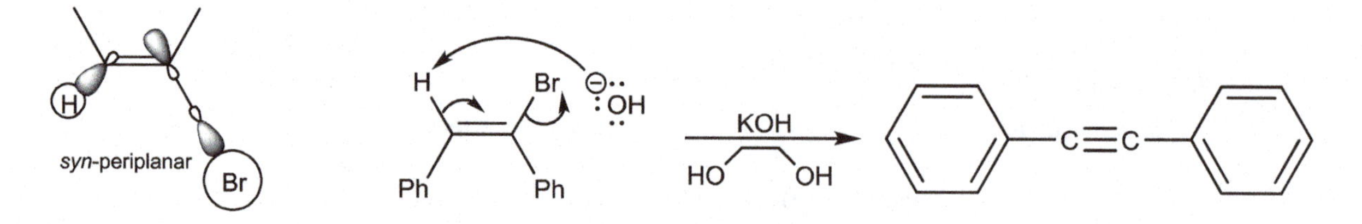

Figure 12.4 Mechanism for the dehydrobromination of (*E*)-1-bromo-1, 2-diphenylethylene.

For this reaction potassium hydroxide (KOH) will be used as the strong base, and ethylene glycol will be used as the reaction solvent. Ethylene glycol has a high boiling point (196-198°C) as compared to typical organic solvents. The high boiling solvent is required for the second elimination step, as it requires much more energy than the first.

Mass Spectrometry

Mass spectrometry can be useful to determine the size and formula of a compound. Recall that when a compound is analyzed using a mass spectrometer, the molecules are bombarded by high energy electrons. This electron bombardment converts some of the molecules to ions by knocking off a single electron. Once the molecules are converted to ions, they are accelerated in an electric field, where they are separated according to their mass-to-charge ratio (*m/z*). These ions are then detected and recorded to product a mass spectrum. The mass spectrum is simply a plot of ion abundance vs. m/z ratio.

The simple removal of a single electron from a molecule yields an ion whose weight is equivalent to the molecular weight of the original molecule. This is called the molecular ion peak ($M^{+\bullet}$). If the $M^{+\bullet}$ peak can be identified on the mass spectrum, it is possible to use the spectrum to determine the molecular weight of an unknown compound.

Once ionized, the molecule becomes unstable and will continue to break into smaller fragments. The intensity of each fragment is dependent on the stability of the fragment. The most abundant ion formed gives rise to the tallest peak in the mass spectrum, called the base peak. The relative abundances of all other peaks in the spectrum are reported as percentages of the base peak.

Mass spectrometry can be useful during organic synthesis. If mass spectra of the reactant(s) and product are available, and the $M^{+\bullet}$ is identified for each, it is possible to determine whether or not a reaction was successful. Simply by comparing the $M^{+\bullet}$ peak of the product to that of the reactant(s), along with the use of other characteristic fragmentation patterns, the success of the reaction can be determined. The mass spectra of the dibromostilbene reactant and the desired product (diphenylacetylene) are shown in Figure 12.5.

One interesting feature in the mass spectrum of alkyl bromides is the presence of an $(M+2)^{+\bullet}$ peak that is the same height as the $M^{+\bullet}$ peak, since bromine has two equally abundant isotopes, ^{79}Br and ^{81}Br. Unfortunately, it's not always possible to take advantage of this characteristic pattern since the intensity of the $M^{+\bullet}$ peak is generally very weak in dialkyl bromides, especially if there is increased branching at the α position. However, in the mass spectrum of the reactant alkyl halide, the fragment formed due to the simple loss of a single bromide ion is very evident, appearing at m/z = 259. There is also a fragment of equal intensity at m/z = 261, which represents the heavier bromine isotope. This particular fragmentation pattern is not observed in the product alkyne spectrum, indicating that there is no bromine present in the product (Figure 12.5).

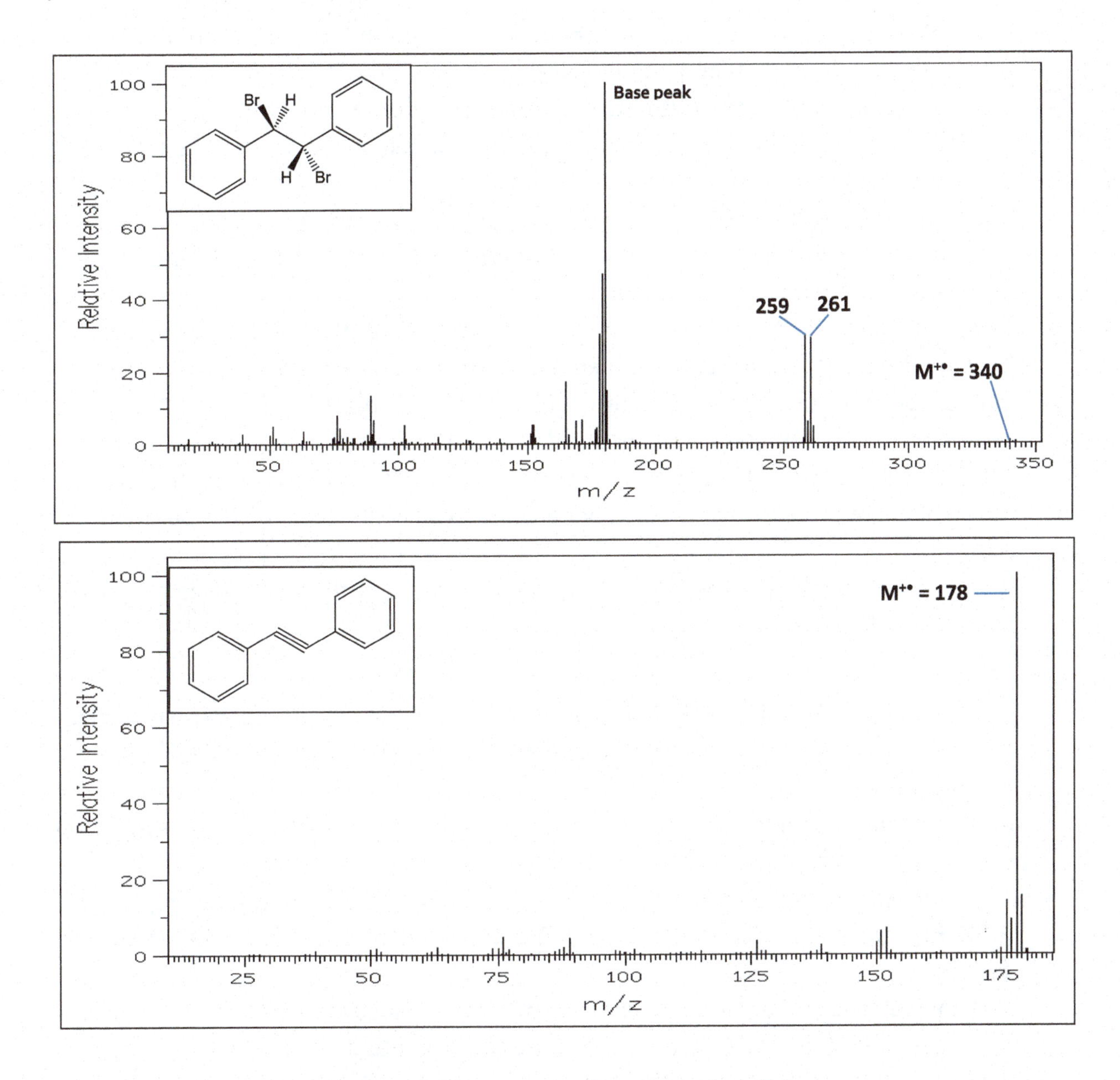

Figure 12.5 Mass spectra of reactant and product.

Objectives:

In this experiment, a double elimination of HBr will be performed on dibromostilbene, prepared in the previous experiment, to yield an alkyne. The product will be purified by recrystallization, then analyzed by TLC and melting point analysis. Finally, the reactant and product will be characterized using mass spectrometry.

Experimental

Synthesis:

- Place ~0.50 g of the dibromostilbene from the previous lab into a 25 mL round bottom flask. Add 2-3 boiling chips and clamp this reaction flask to a ring stand.
- Fill the heating mantle (connected to a voltage regulator) ½ full with sand. Lower the reaction flask into the sand bath, ensuring that the bottom half of the flask is completely covered with sand.
- Add ~ 0.40 g of crushed KOH to the reaction flask using a powder funnel. Using 4 mL of ethylene glycol, rinse any remaining solid from the funnel to the flask. Stir the contents of the vial gently with a microspatula.
- Place a condenser in the top of the reaction flask. Since the ethylene glycol is such a high boiling solvent, water will not be needed to cool the condenser walls.
- Turn the voltage regulator on a setting of 60, and bring the reaction to reflux. Once reflux begins, continue heating the flask for an additional 10 minutes.
- After 10 minutes, lower the heating mantle from the flask and allow the contents to cool to room temperature (this may take several minutes since the reaction temperature was so high).
- Once cooled, remove the condenser, and add 8 mL water to the flask. Place the flask in an ice water bath for 10 minutes.

Purification:

- Isolate the crude solid by suction filtration. Be sure to seat the filter paper with a small amount of cold water. Use as much ice cold water as necessary to transfer the solid to the Büchner funnel.
- After 5 minutes under vacuum, transfer the crude solid to a clean 25 mL Erlenmeyer flask containing 2 boiling chips.
- Add 2 mL ethanol to the flask, and place the flask on a WARM hotplate (setting @ 3, NO HIGHER!). Once the solid dissolves, add 10 drops of water to the flask until cloudiness persists.
- Add an additional 1 mL ethanol to the flask and allow the solid to dissolve. Remove the flask from the hotplate and allow to cool to room temperature slowly, then place the flask in an ice water bath for 15 minutes.
- Isolate the purified solid by suction filtration. Be sure to seat the *WEIGHED* filter paper with a small amount of cold ethanol. Leave the solid under vacuum for 10 minutes.

- Transfer the solid/filter paper to a preweighed watch glass. Reweigh to obtain final product mass. Calculate percent yield and record data in Table 12.1 on the post lab assignment (provided online).
- ***PROCEED TO PRODUCT ANALYSIS***.

Product Analysis:

Melting Point Analysis:

- Prepare a melting point sample of the product sample.
- Begin heating the sample on a MelTemp setting of 3. When the temperature reaches 50°C, lower the dial by one tic mark. Monitor the sample closely to visualize the melting range. Record T_i-T_f in Table 12.1 on the post lab assignment (provided online).

TLC Analysis:

- Prepare a TLC sample of the product by transferring a small amount of the solid to a small test tube. Add 1 mL reagent acetone to dissolve the solid.
- Prepare a TLC plate with three lanes. Apply the sample solution to the TLC plate alongside the provided standards of the dibromostilbene reactant and the diphenyl acetylene desired product.
- Develop the TLC plate in 100% hexane. Visualize the plate under UV light. Circle the reactant and product spots and calculate the R_f values.
- Record data in Table 12.2 on the post lab assignment (provided online).

Mass Spectral Analysis:

- Using the provided spectra, identify the molecular ion peak of the dibromostilbene and the diphenyl acetylene. Complete Table 12.3 on the post lab assignment (provided online).

SAFETY

****All experiments should be performed in a fume hood with appropriate safety glasses. Gloves will be provided by request.****

Potassium hydroxide is corrosive. Ethylene glycol is hazardous if ingested, inhaled, or makes contact with skin or eyes. Chronic exposure could lead to carcinogenic, mutagenic, and teratogenic effects. Ethanol, hexane, and acetone are highly flammable.

WASTE MANAGEMENT

Place all liquid waste from filtration, recrystallization and TLC analysis in the container labeled "LIQUID WASTE—ALKYNES". Place solid product in the container labeled "SOLID ORGANIC WASTE—ALKYNES". All TLC plates, filter papers, and pipets can be placed in the yellow trashcan. Used TLC capillaries and melting point capillaries should be placed in the broken glass container.

References

Mass Spectra are from: http://riodb01.ibase.aist.go.jp/sdbs/cgibin/cre_index.cgi?lang=eng.
Klein, David. (2015). *Organic Chemistry*, 2nd ed. Hoboken: John Wiley and Sons.
Adapted from: Williamson, Kenneth L., Minard, Robert D., Masters, Katherine M. (2007) *Macroscale and Microscale Organic Experiments*, 5th ed. Boston: Houghton Mifflin Company.

Name:			
	Max	Score	Total Grade
Tech	10		
Pre	20		
In	20		
Post	50		

Exp. 12 Synthesis of Diphenylacetylene

PRE-LAB ASSIGNMENT: *(**EACH STUDENT** will complete and submit an original copy at the beginning of the lab period. **Without a complete pre-lab assignment, you will not be allowed to perform the experiment, and will receive a zero for the lab.**)* .....*max score = 20 pts.*

1. **Objective** *(Write a brief purpose of the experiment in **complete** sentences, addressing all of the following points.)*
   - Name the *specific* reactants combined during the synthesis and name the product formed.
   - What is the purification technique used in this experiment?
   - What analytical technique(s) will be used during the experiment to determine purity?
   - What type of spectral analysis will be used to characterize the reactant and product?

2. **Chemical Equation** *(Draw the chemical equation for the synthesis of diphenylacetylene using actual chemical structures.)*

3. **Physical Data** *(Complete the following table before coming to lab.)*

Compound	MW (g/mol)	mp (°C)	bp (°C)	d (g/mL)
dibromostilbene			X	X
ethylene glycol	X	X		
hexane	X	X		
ethanol	X	X		
acetone	X	X		
diphenylacetylene			X	X

4. <u>Experimental Outline</u> *(Give a brief description of the procedure that will be followed in this experiment in 5 lines or less.)*

1	
2	
3	
4	
5	

5. <u>Pre-Lab Questions</u> *(Answer the following questions prior to lab.)*

A. Why must ethylene glycol be used as the reaction solvent for the synthesis of diphenylacetylene, as opposed to a typical organic solvent such as methanol or ethanol?

B. What is the difference between a ***molecular ion peak*** and a ***base peak*** in mass spectrometry?

I have read and understood the experimental procedure for this experiment. I am familiar with the hazards and the required disposal procedures for this experiment.

Sign here: ____

Experiment 13

Base-Promoted Elimination of HBr from an Alkyl Halide

Introduction

In this experiment you will synthesize an isomeric mixture of alkenes from base promoted elimination of HBr from a secondary alkyl halide, 2-bromoheptane (Figure 13.1). Varying the size of the base used for this dehydrohalogenation reaction will demonstrate the effect of base size on the product distribution (Zaitsev vs. Hofmann). Half of the students in each lab section will use sodium methoxide/methanol and the other half will use potassium *t*-butoxide/*t*-butanol. Distillation of the resulting isomeric mixtures of alkenes will allow for GC analysis. The relative area percent of the isomeric alkenes will be compared for the two bases. IR analysis will also be performed to characterize reactants and products.

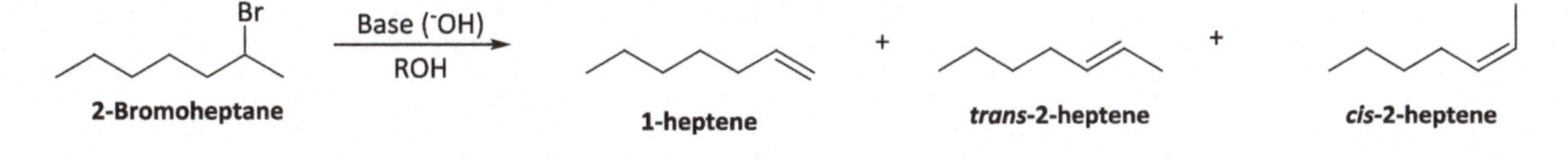

Figure 13.1 Dehydrohalogenation of 2-bromoheptane.

Dehydrohalogenation

Dehydrohalogenation is an elimination reaction that is used synthetically to produce alkenes from alkyl halides. Dehydrohalogenation reactions almost always yield an isomeric mixture of alkenes when starting with unsymmetrical alkyl halides. However, the major product can be predicted using the Zaitsev rule. In general, base-promoted elimination of HX proceeds in the direction that yields the most highly substituted alkene. In cases where *cis* or *trans* alkenes can be formed, the reaction shows stereoselectivity and the more stable *trans* isomer is the major product, as shown below for 2-bromoheptane (Figure 13.2).

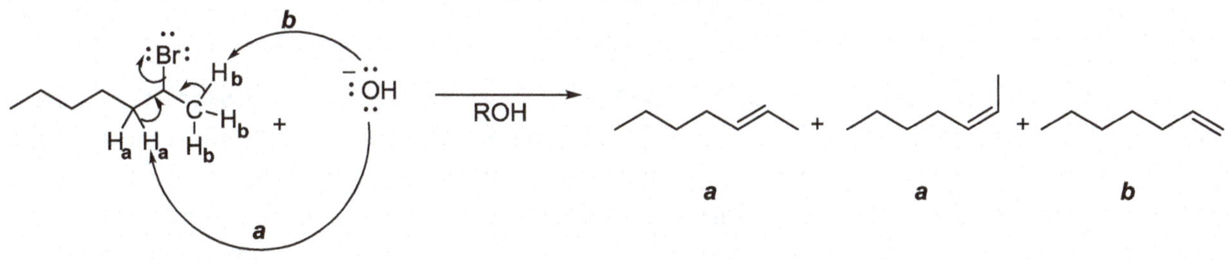

Figure13.2 E2 dehydrohalogenation of 2-bromoheptane.

The elimination shown in Figure 13.2 proceeds through an **E2** mechanism (bimolecular elimination). E2 conditions require a relatively high concentration of strong base in a relatively nonpolar solvent (as compared with water), such as sodium methoxide in methanol or potassium butoxide in *t*-butanol. The base must abstract the hydrogen (without its bonding electrons) adjacent to the halogen. A new π bond is formed with concurrent release of the halogen. No intermediates have been isolated from E2 reactions and no rearrangements occur under E2 conditions (unlike the E1 reaction we saw in Experiment 10).

Experimental evidence indicates that the five atoms involved in the E2 elimination reaction must lie in the same plane, in an *anti-periplanar* conformation, as shown in Figure 13.3 for elimination of HBr from 2-bromobutane. A small base like sodium methoxide can easily abstract the internal H, adjacent to the bromine (route 13.3*a*), to give the thermodynamically more stable 2-butene (= Zaitsev elimination). Steric strain between gauche methyl groups in structure II also helps to explain why the *trans* isomer is favored (route 13.3*b*). If a big base, like potassium *t*-butoxide, is used, it will more easily abstract the less hindered hydrogen (route 13.3*c*) of the terminal carbon to yield the terminal alkene (= Hofmann elimination).

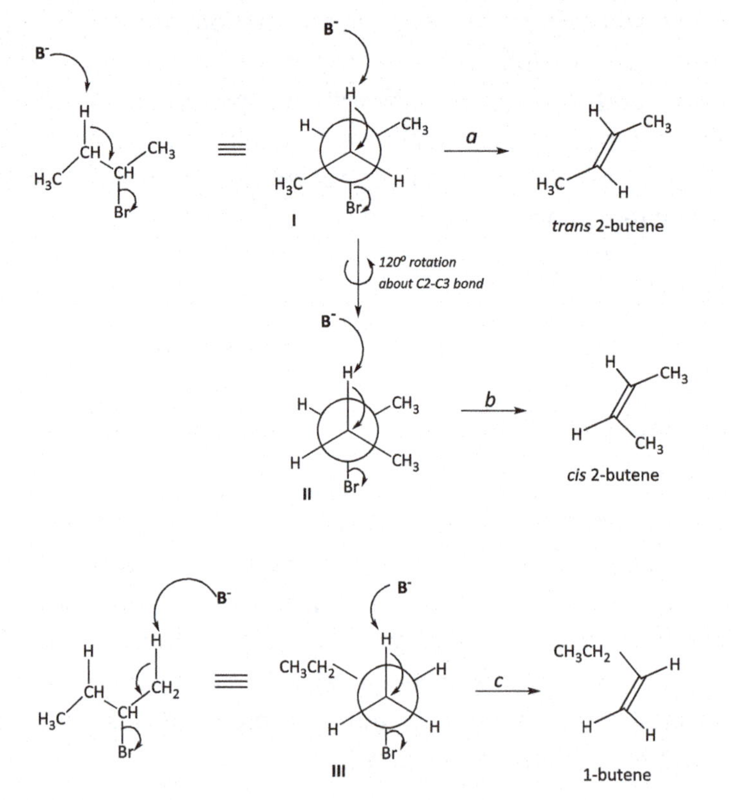

Figure 13.3 Zaitsev vs. Hofmann elimination of HBr from 2-bromobutane.

The reaction conditions for this dehydrohalogenation requires that no water is present in the reaction mixture. Exposure of the sodium methoxide or potassium *t*-butoxide to the air while transferring reagents should be minimized since the air contains water vapor. We will attach a calcium sulfate filled drying tube to the reflux condenser to ensure that no water enters the reaction flask while heating. This is especially important if Hofmann elimination, using a big base, is desired. Water will react with sodium methoxide to produce methanol and NaOH, and with potassium *t*-butoxide to give *t*-butanol and KOH (Figure 13.4). Think about the size of KOH vs. potassium *t*-butoxide and how that will affect the product distribution.

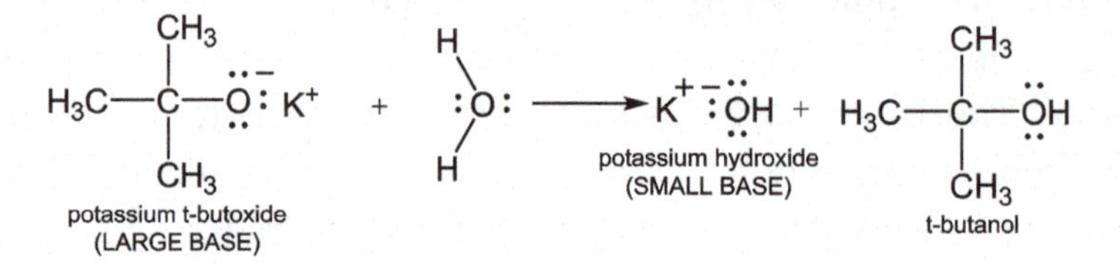

Figure 13.4 Reaction between large base and water.

IR Analysis

In Figure 13.5, the IR spectra for the starting alkyl halide and the resulting alkene products are shown. In the IR spectrum of the starting alkyl bromide, the C-Br stretch absorption is a characteristic feature. In the spectrum of the alkene products, an sp^2 C-H stretch absorption and a C=C stretch absorption appear; however, the C-Br stretch absorption observed in the alkyl halide spectrum is no longer present. Notice that in the IR spectrum of *trans*-2-butene, no C=C stretch is observed. Due to the symmetry of the molecule, the C=C bond does not absorb IR radiation. All spectra have sp^3 C-H stretch absorptions. Refer to Experiment 9 for IR correlation tables, along with Appendix I.

Objectives

In this experiment you will synthesize isomeric alkenes from an E2 base-promoted elimination of HBr from 2-bromoheptane. You will purify the products through a simple distillation. GC analysis will be used to identify and quantify the products in an effort to understand how base size affects product distribution. Finally, IR spectroscopy will be used to distinguish between reactants and products.

Experimental Procedure

Synthesis:

- Weigh 1.00 g of 2-bromoheptane directly into a clean, ***dry*** 25 mL distilling flask. Add a few boiling chips.
- Add 3.0 mL of either methanol or *t*-butanol (depending on base size assigned by instructor).
- If you used methanol, then weigh out 0.30 g of sodium methoxide on a small piece of weigh paper. If you used *t*-butanol, then weigh out 0.60 g of potassium *t*-butoxide on a small piece of weigh paper.
- Add the solid base to the reaction flask using a short-stem funnel.
- Apply a small amount of stopcock grease on the ground glass joint of the condenser (VERY IMPORTANT)!
- Place the condenser in the top of the reaction flask. Using a thermometer adapter, place a $CaSO_4$ drying tube in the top of the condenser.
- Heat the reaction mixture using a heating mantle (VR setting = 30) and reflux it for 60 minutes.

Purification:

- Cool slightly with a water bath, and rearrange the apparatus for simple distillation.
- Heat the 25 mL round bottom flask using a heating mantle (VR setting = 30), collecting the distillate in a preweighed 10 mL round bottom flask.
- Be sure to record the distillation range, and collect only the first 2-3 mL of the distillate.
- Reweigh the flask to determine product yield. Calculate percent yield for the reaction.
- ***PROCEED TO PRODUCT ANALYSIS.***

Product Analysis:

GC Analysis

- Prepare a GC sample of your distillate by placing five drops of your sample in an autoanalyzer vial and adding 1 mL of GC solvent. Using the chromatographic results, determine the identity of your products.
- Calculate the adjusted area percent for the alkene products. Record this data, along with the data from the opposing alcohol–base pair in Table 13.1 on the POST LAB Assignment (provided online).

IR Analysis

- Using the provided spectra, assign all characteristic absorptions of the reactant and products. Complete Table 13.2 on the POST LAB Assignment (provided online).

SAFETY

****All experiments should be performed in a fume hood with appropriate safety glasses. Gloves will be provided by request.****

All organic solvents used and the reaction products in this experiment are flammable, irritants, and can be toxic if ingested or absorbed through skin. Sodium methoxide and potassium *t*-butoxide are extremely corrosive. Gloves should be worn during use.

WASTE MANAGEMENT

Clean the reaction flask by adding a small amount of water to the cooled flask to dissolve the solid salts. Pour this and all liquid waste into the container labeled "LIQUID WASTE—E2".

Reference

Klein, David. (2015). *Organic Chemistry*, 2nd ed. Hoboken: John Wiley and Sons.

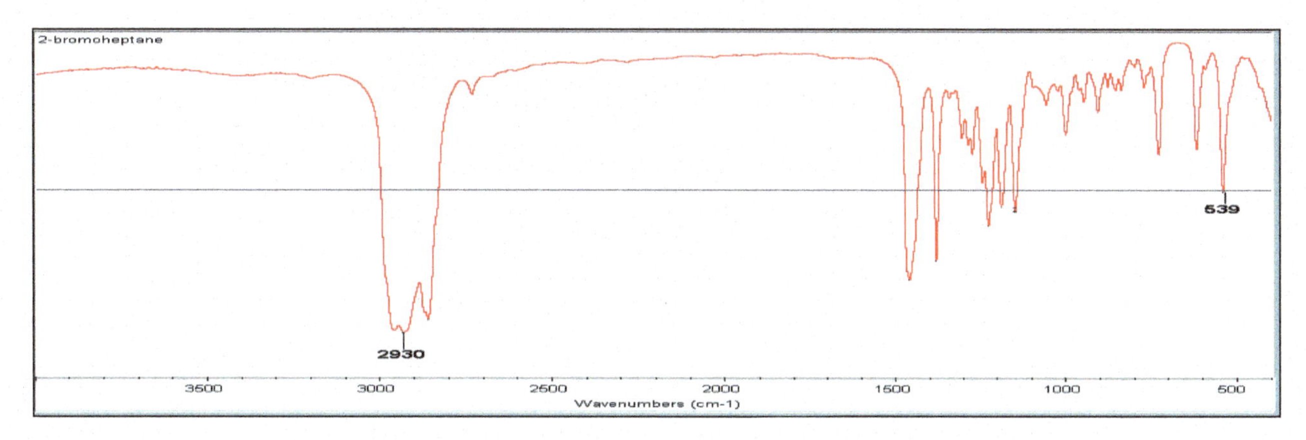

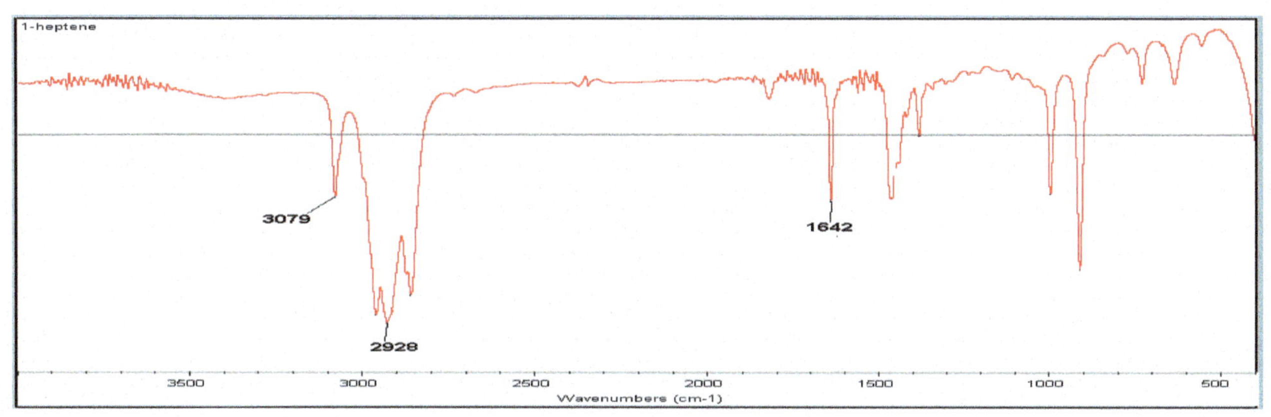

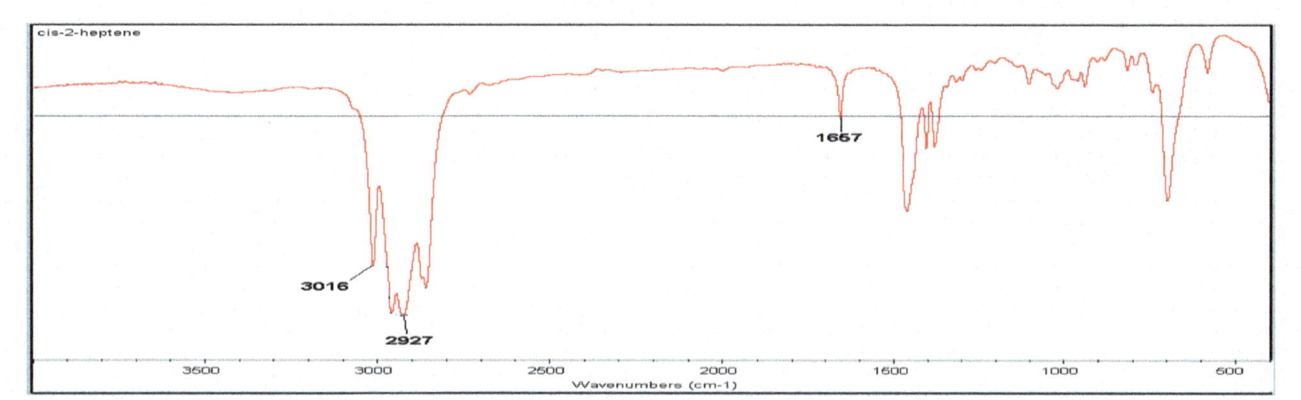

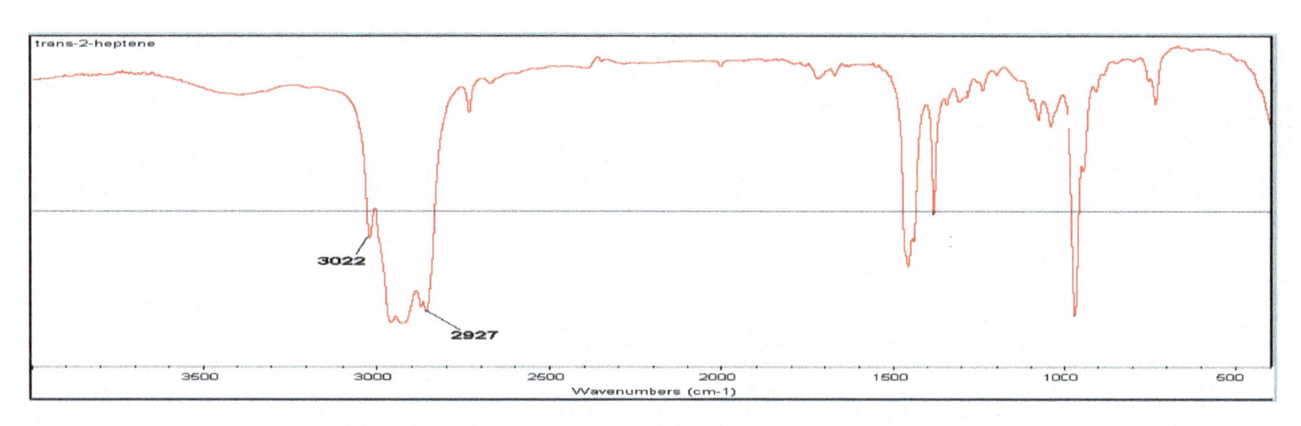

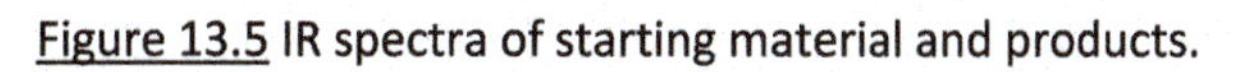
Figure 13.5 IR spectra of starting material and products.

Name:			
	Max	Score	Total Grade
Tech	10		
Pre	20		
In	20		
Post	50		

Exp. 13 Base-Promoted Elimination of HBr from an Alkyl Halide

PRE-LAB ASSIGNMENT: *(**EACH STUDENT** will complete and submit an original copy at the beginning of the lab period.* ***Without a complete pre-lab assignment, you will not be allowed to perform the experiment, and will receive a zero for the lab.***) *.....max score = 20 pts.*

1. Objective *(Write a brief purpose of the experiment in **complete** sentences, addressing all of the following points.)*

- Name the *specific* reactants combined during the synthesis and name the product(s) formed.
- What is the purification technique used in this experiment?
- What analytical technique will be used during the experiment to identify and quantify products?
- What type of spectral analysis will be used to characterize reactant and products?

2. Chemical Equation *(Draw the chemical equation for the dehydrohalogenation of 2-bromoheptane using actual chemical structures.)*

3. Physical Data *(Complete the following table before coming to lab.)*

Compound	MW (g/mol)	bp (°C)	d (g/mL)
2-bromoheptane			X
sodium methoxide		X	X
potassium *t*-butoxide		X	X
methanol	X		
t-butanol	X		
cis-2-heptene			X
trans-2-heptene			X
1-heptene			X

4. Experimental Outline *(Give a brief description of the procedure that will be followed in this experiment in 5 lines or less.)*

1	
2	
3	
4	
5	

5. Pre Lab Questions *(Answer the following questions prior to lab.)*

A. The ***PURIFICATION*** technique used in today's experiment is:

a. recrystallization
b. column chromatography
c. acid-base extraction
d. simple distillation

B. Draw the chemical structures of ***sodium methoxide*** and ***potassium t-butoxide***. Label which base is the small base used in today's experiment, and which is the large base used.

I have read and understood the experimental procedure for this experiment. I am familiar with the hazards and the required disposal procedures for this experiment.

Sign here: ____

Experiment 14

Calculation, Chromatographic, and Spectral Applications

Introduction

Several laboratory calculations and applications are introduced during the first semester organic laboratory course that you will be expected to recall throughout the second semester course. In this experiment laboratory calculations and analytical techniques introduced in the first semester organic laboratory course will be reviewed. Methods will be introduced in order to determine the efficiency of chemical reactions, in an effort to evaluate the environmental impact of syntheses performed in laboratory. Finally, spectroscopic methods will be discussed in an effort to introduce the use of spectroscopic analysis to differentiate between reactants and products in a chemical reaction.

Determination of Limiting Reagent

The limiting reagent is the reagent present in the smallest molar ratio. In a typical synthesis that requires two reagents to react to form one product, one of the two reagents typically limits the amount of product that can be formed (see Appendix D). The limiting reagent is the reagent that dictates the amount of the product that can be formed during the synthesis. In order to determine which reagent is the limiting reagent in a synthesis, you must first convert the amount of each reagent used (in grams or mL) to moles of reagent using their molecular weights (and density if the reagent is a liquid). Next, take into account the stoichiometric ratio of the reactants and products. How many moles of product can be formed from one mole of the reagent? Consideration of the stoichiometric ratio should leave the units in moles of product. Whichever reagent generates the lowest number of moles of product is identified as the limiting reagent. ***The unit for limiting reagent is always number of moles of product.***

Theoretical Yield Calculation

Once the limiting reagent has been identified, the next consideration is to determine the total amount of product that can be generated based on that limiting reagent. The calculation of limiting reagent leaves the value in the units of moles of product. In order to determine the theoretical yield, simply multiply the moles of product by the molecular weight of the product. ***The unit for theoretical yield is always number of grams of product.***

Percent Yield Calculation

After the synthesis, product isolation, and purification of the product, the actual weight of product produced during the synthesis is obtained. In order to determine the success of your synthesis, the percent yield is calculated by simply dividing the actual product yield (g) by the theoretical yield (g), and multiplying by 100 to convert it to a percentage. All of the aforementioned calculations are reviewed in Appendix D in the back of your laboratory manual.

Green Chemistry

A fundamental challenge in chemistry today is to find creative ways to minimize human exposure to and the environmental impact of harmful chemicals while still enhancing people's lives and scientific progress. Green chemistry—the design of chemical products and processes that reduce or eliminate the use and generation of hazardous substances—is the most fundamental approach to this challenge. Adoption of green chemistry technologies offers environmental improvement, enhanced safety, and economic benefit. To make products benign by design, chemists must consider the hazard posed by a substance, along with its other chemical and physical properties, and select products and processes that minimize harm. Green chemistry principles focus on using safer starting materials for efficient chemical reactions, using renewable resources, conserving energy, finding safer solvents, and reusing or recycling product and waste whenever possible.

A "perfect" chemical reaction would be one that is completely ***selective***, affording only the desired product, highly ***efficient***, incorporating all atoms of the starting materials and reagents in the product, and entirely safe, using only ***nonhazardous*** (to the chemist and the environment) starting materials, reagents, and products. Ideally, this reaction would require no solvent or energy inputs (see Appendix K).

The 12 Principles of Green Chemistry

1. ***Prevention***: It is better to prevent waste than to treat or clean up waste after it is formed.
2. ***Atom Economy***: Synthetic methods should be designed to maximize the incorporation of all materials used in the process into the final product.
3. ***Less Hazardous Chemical Syntheses***: Wherever practicable, synthetic methodologies should be designed to use and generate substances that possess little or no toxicity to human health and the environment.
4. ***Designing Safer Chemicals***: Chemical products should be designed to preserve efficacy of function while reducing toxicity.
5. ***Safer Solvents and Auxiliaries***: The use of auxiliary substances (e.g., solvents, separation agents, etc.) should be made unnecessary wherever possible, and innocuous when used.
6. ***Design for Energy Efficiency***: Energy requirements should be recognized for their environmental and economic impacts and should be minimized. Synthetic methods should be conducted at ambient temperature and pressure.
7. ***Use of Renewable Feedstocks***: A raw material or feedstock should be renewable rather than depleting, wherever technically and economically practicable.
8. ***Reduce Derivatives***: Unnecessary derivatization (blocking group, protection/deprotection, and temporary modification of physical/chemical processes) should be avoided whenever possible.
9. ***Catalysis***: Catalytic reagents (as selective as possible) are superior to stoichiometric reagents.
10. ***Design for Degradation***: Chemical products should be designed so that at the end of their function they do not persist in the environment and break down into innocuous degradation products.
11. ***Real-Time Analysis for Pollution Prevention***: Analytical methodologies need to be further developed to allow for real-time, in-process monitoring and control prior to the formation of hazardous substances.
12. ***Inherently Safer Chemistry for Accident Prevention***: Substances and the form of a substance used in a chemical process should be chosen so as to minimize the potential for chemical accidents, including releases, explosions, and fires.

In addition to using safer starting materials and processes, the efficiency of a chemical reaction must be considered when designing a synthetic process. Several methods exist to determine the efficiency of a chemical reaction, including chemical yield, atom economy, experimental atom economy, and calculation of the "$E_{product}$" of a reaction.

Green Chemistry Calculations

Green chemistry calculations are used to determine how "environmentally friendly" your choice of reagents, solvents, and experimental conditions were. The green chemistry calculations include atom economy, experimental atom economy, and "$E_{product}$." Green chemistry calculations are only based on the type and amount of reagents used, never the amount or type of solvents or catalysts used.

Atom Economy

Atom economy is based on the efficiency of reactant atoms converted to product. Ultimately, in an efficient synthesis, all reactant atoms would appear as part of the product structure. Any atoms present in the reactants that do not appear in the product structure were converted to side products or waste, which has to be disposed of. Sometimes the side products and waste generated is harmful to the experimenter, as well as the environment, so experiments should be designed to minimize the generation of waste and unnecessary side products if at all possible.

Atom economy is based on ***which*** reactants were selected to form the desired product. It assumes that the reactants were used in equivalent amounts, and no excesses were used. The atom economy is expresses as a percentage. ***The closer the atom economy is to 100%, the more efficient the reaction***. The general equation used to calculate the atom economy is shown below:

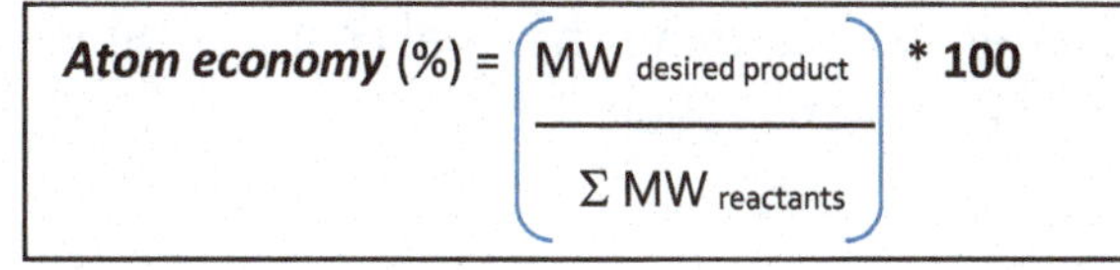

$$\textbf{\textit{Atom economy}}\ (\%) = \left(\frac{MW_{\text{desired product}}}{\Sigma\, MW_{\text{reactants}}}\right) * \mathbf{100}$$

Experimental Atom Economy

Another measure of efficiency of a reaction is based on the ***amount*** of each reactant used, not which reactants were used to form the product. This is a more precise measure of efficiency than the atom economy, since it takes into account the mass of each reactant used. This is important to consider since an excess of one reactant is commonly used in order to drive the reaction to completion, and this is not considered in the typical atom economy calculation. The experimental atom economy is expressed as a percentage. ***The closer the experimental atom economy is to the calculated atom economy, the more efficient the reaction***, because some efficiency is lost when an excess of a reagent is used, therefore being transformed into waste. The general equation used to calculate the experimental atom economy is shown below:

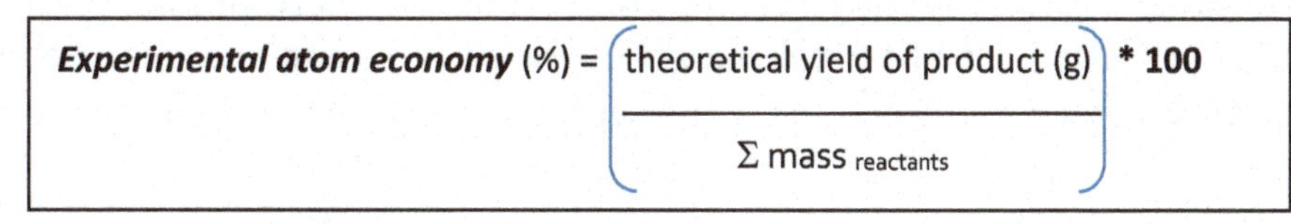

$$\textbf{\textit{Experimental atom economy}}\ (\%) = \left(\frac{\text{theoretical yield of product (g)}}{\Sigma\, \text{mass}_{\text{reactants}}}\right) * \mathbf{100}$$

"$E_{product}$"

The ultimate measure of efficiency is the calculation of the "$E_{product}$". The "$E_{product}$" value considers not only the conditions used during the synthesis, but also the amount of product that resulted under those conditions. The higher the "$E_{product}$" value, the more efficient the reaction. The general equation used to calculate the "$E_{product}$" is shown below:

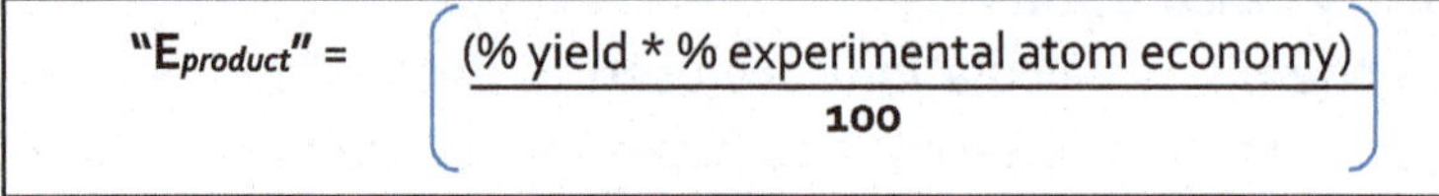

Cost Analysis

One thing typically not considered by the student is the cost required to perform a synthesis, or how cost efficient a synthesis is under typical conditions. In second semester organic chemistry experiments, this will now be introduced to give an idea of the expense of performing such experiments, and how marketable the product may be in comparison to other manufacturers.

Cost Per Synthesis

In order to determine the sale price of a bottle of your product, you must first consider how much it costs to purchase any reagents, solvents, or catalysts required to perform the synthesis. A typical synthesis requires a small amount of each reagent, solvent, or catalyst. But a manufacturing company sells these chemicals in bulk quantities. ***In order to calculate the cost of your synthesis, you must first determine the cost of the amount of each chemical used***. This is accomplished by taking the amount of the compound used during the synthesis and multiplying it by the cost of a stock bottle of that compound. Once the cost of each individual compound used is determined (including all reactants, catalysts, and solvents), the costs are added together to determine the final cost per synthesis. The general equation for this type of calculation is shown below:

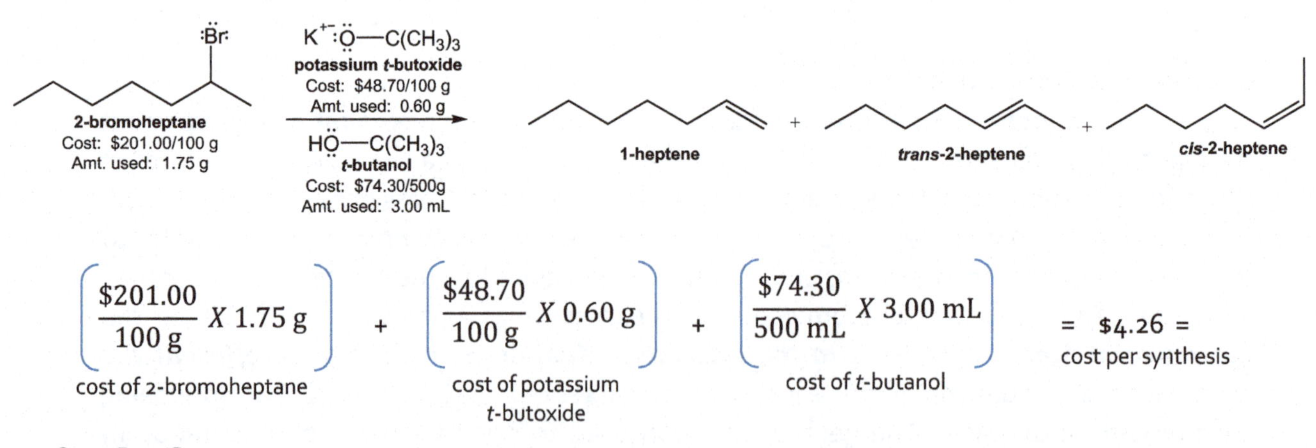

Cost Per Gram

Once the cost of all materials has been determined (cost per synthesis), the cost per gram of your product can be calculated by simply dividing the cost per synthesis by the amount of product generated during the synthesis, as shown below:

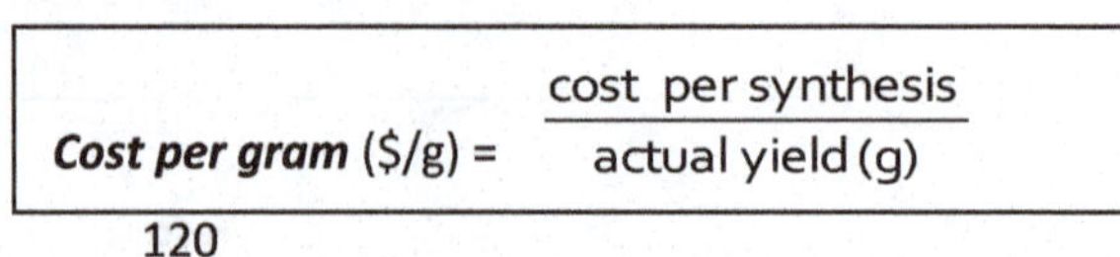

Cost Per Bottle

Once the cost per gram has been determined, the cost per bottle can be calculated by simply multiplying the cost per gram by the bottle size desired. This will allow the marketability of the product to be evaluated, as compared to cost of the same compound available from other manufacturers, such as Sigma-Aldrich, Alfa-Aesar, and Acros Organics, to name a few. These reactants are generally sold in 100 g bottles or 500 mL bottles.

> ***Cost per bottle*** (\$/g) = cost per gram x desired bottle size
> *(*Can also be calculated in mL)*

Chromatographic and Spectral Applications

Several of the analytical techniques introduced in first semester organic chemistry laboratory will be used repeatedly during the second semester and it is the responsibility of the student to review these techniques. TLC and HPLC chromatography were introduced in Experiment 5, and will be used to evaluate the success and purity of compounds in several of the second semester experiments. These techniques are also discussed in Appendices E and F in the back of the laboratory manual. IR spectroscopy was introduced in Experiment 9 and will be used to characterize reactants and products, and to distinguish between them. IR spectroscopy is discussed in first semester organic chemistry lecture, and a correlation table of the common infrared absorption frequencies of common bonds in organic compounds is available in Appendix I in the back of the laboratory manual. Students will be expected to review this material and be able to use spectral data to answer questions.

Infrared Spectroscopy

Infrared (IR) spectroscopy is an extremely valuable tool for determination of functional groups. Infrared radiation is a portion of the electromagnetic spectrum between the visible and microwave regions. Bonds in organic molecules bend or stretch at specific frequencies, depending on the type or the strength of the bond. When a molecule is irradiated with infrared light, energy is absorbed when the frequency of the irradiation is equal to the frequency of the bond vibration. Different types of bonds, and therefore different functional groups, absorb IR radiation at different frequencies (see Appendix I). A typical IR spectrum is a plot of transmittance of IR radiation vs. frequency of IR light in reciprocal centimeters (cm^{-1}). Refer to Experiment 9 and the textbook for further review of theory and important terminology.

NMR Spectroscopy

Nuclear magnetic resonance (NMR) spectroscopy has become one of the most important tools for structure elucidation in organic chemistry. By evaluating chemical shifts, integrations, and splitting patterns of NMR signals for a compound, the connectivity of hydrogens and carbons can be determined, therefore, providing a "map" of the carbon-hydrogen framework of an organic compound. A typical NMR spectrum is a plot of the effective field strength felt by the nuclei (intensity) vs. the intensity of absorption of rf energy (chemical shift). Refer to the textbook for further review of theory and important terminology.

Degree of Unsaturation

The molecular formula can provide valuable information about the structural formula of a compound. The index of hydrogen deficiency, sometimes called the degree of unsaturation, is the determination of how many hydrogen atoms need to be added to a structure in order to obtain the corresponding saturated acyclic structure. It can tell us how many π bonds and/or rings a molecule contains. If given the molecular formula, $C_cH_hN_nO_oX_x$, the degree of unsaturation is two times the number of carbons, plus two, minus the number of equivalent hydrogens. Essentially, there are two hydrogen atoms lost per π bond or ring. The following equation can be used:

$$\text{Degree of Unsaturation} = \frac{[(2c + 2) - (h + x - n)]}{2}$$

A compound with 1° of unsaturation must have one double bond or one ring, but not both. A compound with 2° of unsaturation could have a triple bond, or it could have two double bonds, or two rings, or one of each. Aromatic rings contain one ring and three double bonds, thus having 4° of unsaturation. With this information, the IR and NMR spectra can be used to determine the presence or absence of these types of functional groups.

Using ^{13}C-NMR Spectroscopy for Structure Elucidation

One of the most complex tasks for students is to learn how to propose a structure for a compound given the molecular formula and spectral data. One thing to consider is the amount of information provided by the NMR spectrum of a compound. Carbon NMR allows us to predict what kinds of carbons are present in a molecule, or the functionality of the carbons, via the chemical shift. It can also tell us about the number of unique carbons, or symmetry in the molecule, via the number of signals in the spectrum. Finally, it may tell us about the presence or absence of non-protonated carbons, based on the size of the signals, although this is not always useful. A small chart with typical ^{13}C-NMR chemical shift values is shown in Figure 14.1.

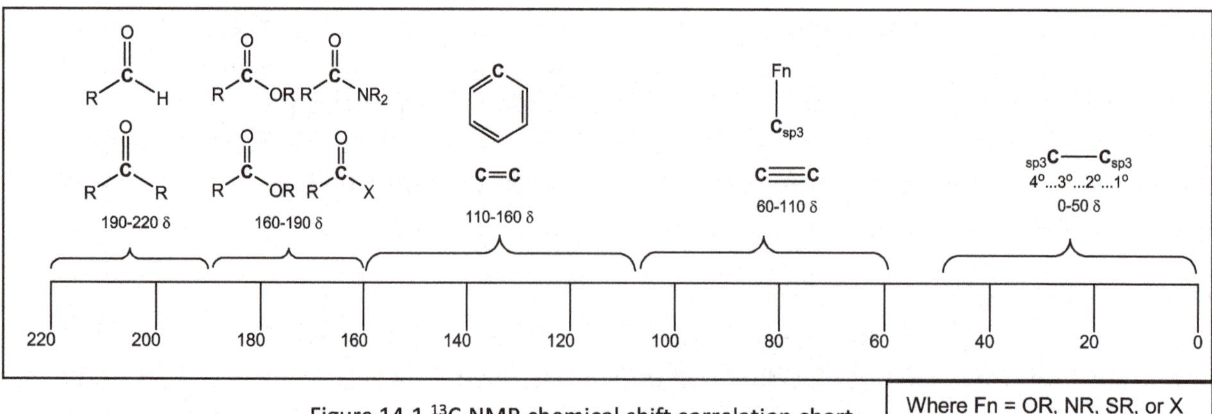

Figure 14.1 ^{13}C-NMR chemical shift correlation chart.

Where Fn = OR, NR, SR, or X

Using ¹H-NMR Spectroscopy for Structure Elucidation

Proton NMR can also be very useful when proposing the structure of a compound given the molecular formula and spectral data. One thing to consider is the amount of information provided by the NMR spectrum of a compound. Proton NMR allows us to predict what kinds of protons are present in a molecule, or the functionality of the protons, via the chemical shift. It can also tell us about the number of unique hydrogens, or symmetry in the molecule, via the number of signals in the spectrum. It can tell us the number of protons of each type per signal via the integration. Finally, it can tell us the number of neighboring protons per signal via the splitting patterns of the signals. A small chart with typical ¹H-NMR chemical shift values is shown in Figure 14.2. Full spectral correlation tables can be found in Appendix I.

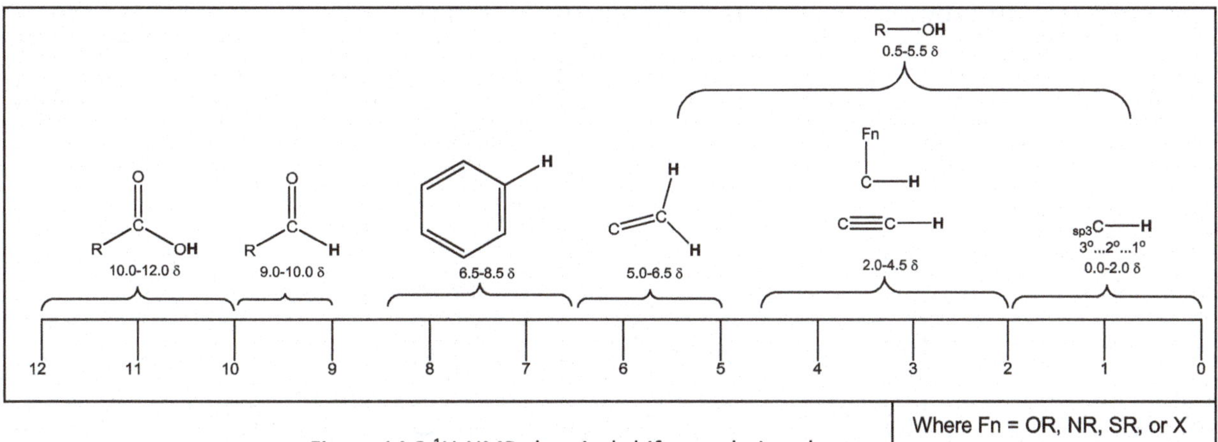

Where Fn = OR, NR, SR, or X

Figure 14.2 ¹H-NMR chemical shift correlation chart.

When given a molecular formula (MF) and a spectrum and asked to determine the structure of the molecule, first calculate the degree of unsaturation. For $C_4H_{10}O$, the DU = 0°. This typically indicates that there are no double bonds or rings present. This is important information, since it indicates that the oxygen is not part of a carbonyl group. First, look at the number of signals in the spectrum and determine if symmetry exists within the molecule. Now, propose possible structures based on the MF and the DU. If there are several possibilities, analyze each and try to predict what the spectrum would look like. Start with the number of signals predicted for each structure. This will help to eliminate some possibilities.

Using Mass Spectrometry (MS) for Structure Elucidation

Mass Spectrometry (MS) measures the mass of an ion. Therefore, if a molecule can be ionized, its molecular weight can be determined. Structural information can also be obtained from MS since charged molecules are unstable and often fragment following simple and predictable chemical pathways to generate the most stable products, which depend on structure and functional groups present.

In one of the most common types of mass spectrometers, an electron impact MS, a high-energy electron bombards the molecule and knocks out an electron, converting it to a radical cation. It is considered a radical because the molecule now has an odd number of electrons, with one unpaired electron. The positive ions are then accelerated in a vacuum through a magnetic field and are sorted on the basis of mass-to-charge ratio, m/z, where m is the mass and z is the charge. Most small molecules will carry only a single charge (z = +1). The output of the mass spectrometer shows a plot of the abundance of the ions vs. the mass-to-charge ratio (m/z). The most intense peak in the spectrum is referred to as the ***base peak*** and all others are reported as a percent of its intensity (Figure 14.3) The highest molecular weight peak observed in a spectrum will typically represent the parent molecule, minus one electron, and is called the ***molecular ion peak*** ($M^{+\cdot}$). Knowing the molecular weight, it is possible to propose molecular formulas. For example, a molecular ion of m/z = 70 could have a formula of C_5H_{10}, C_4H_6O, $C_3H_2O_2$ or $C_3H_6N_2$. A High Resolution Mass Spectrometer (HR-MS) measures the molecular mass with such high accuracy that the correct molecular formula, of the four possibilities above, can be distinguished. Another way to choose the correct molecular formula is by analyzing the rest of the mass spectrum.

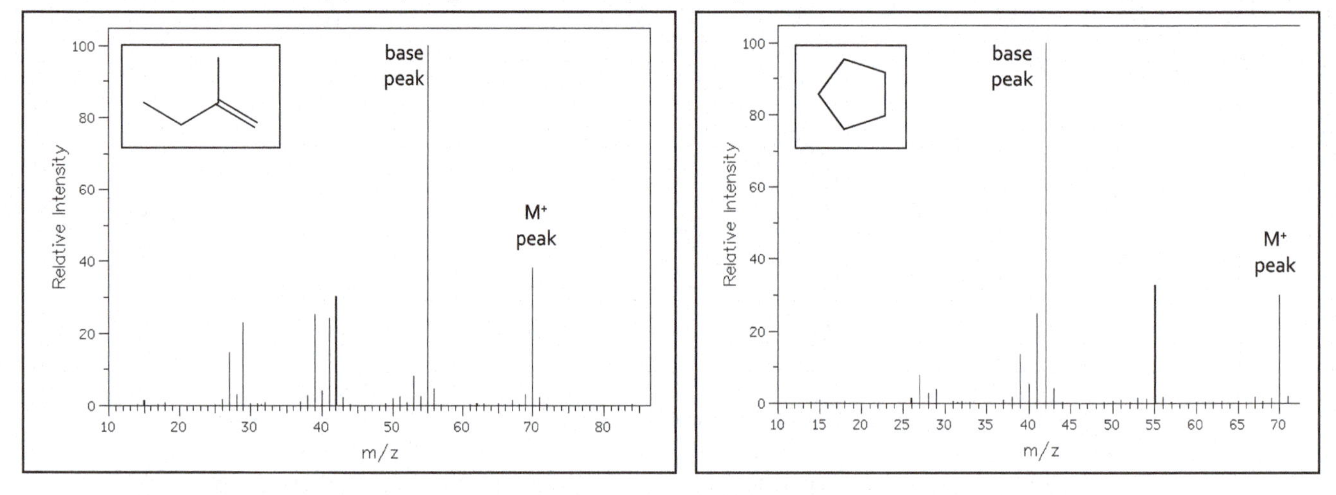

Figure 14.3 Mass spectra of 2-methyl-1-butene and cyclopentane, both C_5H_{10}.

As a radical cation, the molecular ion is usually quite unstable and will fragment into smaller pieces, cations, radicals and radical-cations. Most mass spectra are complex but fragmentations follow simple and predictable chemical pathways, producing the most stable products. Carbocation stability is as follows:

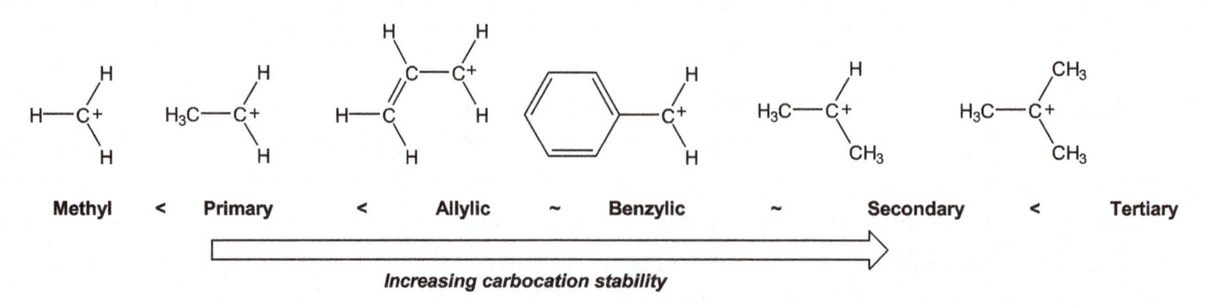

The resulting m/z is equivalent to the mass of the fragment formed, from which the fragment chemical formulas can be proposed. For example in the fragmentation of propane, upon initial ionization, an electron is lost from somewhere within the molecule. Since it is impossible to know from which sigma bond the electron is lost, brackets are placed around the structure and the symbol "•+" is put at the top right edge of the bracket to indicate that the structure is a radical ion. In propane, if an electron from a C-C bond is lost, a mechanism can be (using single pronged arrows to show movement of a single electron) written showing loss of the methyl radical (a neutral loss) leaving the ethyl cation, $C_2H_5^+$, which would be detected by the mass analyzer as m/z = 29.

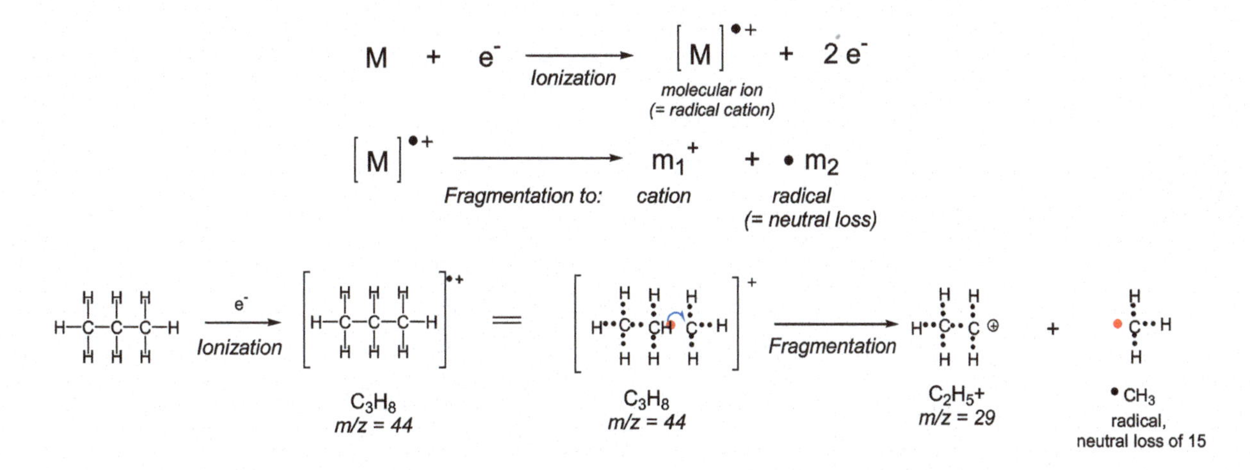

The mass spectra of two isomers of C_6H_{14}, 2-methylpentane and 3-methylpentane are shown in Figure 14.4. In both cases the molecular ion, m/z = 86, is quite small indicating that this radical cation is not very stable. Fragmentation generates cations and radicals, and the relative stability of both cations and radicals is:

tertiary > secondary > allylic > primary > CH_3

Because of this, the fragmentation of the two compounds would be predicted to be different, since the mass of the most stable tertiary cations would be different, and this is evident in the spectra shown. For 2-methylpentane, the base peak is generated by loss of the propyl radical (neutral loss of 43 = C_3H_7), generating the secondary carbocation at m/z = 43, = C_3H_7. In 3-methyl-2-pentane, the base peak is at m/z = 57 which results from loss of 29 mass units, or neutral loss of an ethyl radical, C_2H_5. The resulting secondary carbocation C_4H_9 has m/z = 57.

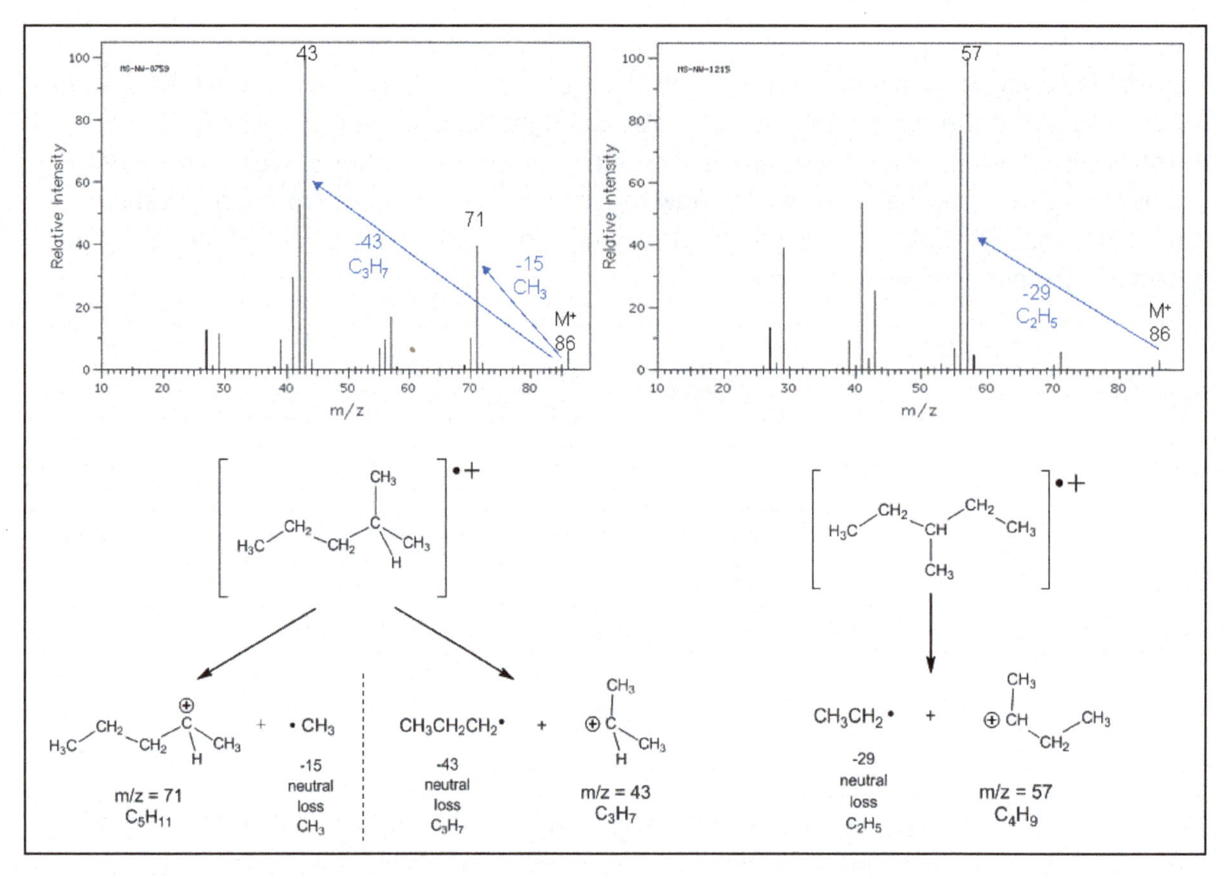

Figure 13.4 Mass spectra and fragmentation of 2-methylpentane and 3-methylpentane.

Most functional groups have characteristic fragmentation patterns generating the most stable cation and radical fragments. A brief table of some common fragments is shown below (Table 14.1).

Commonly Lost Fragments		Common Stable Ions	
m-15	$\cdot CH_3$	m/z = 43	$CH_3\overset{+}{C}\equiv O$
m-17	$\cdot OH$		
m-28	$H_2C{=}CH_2$		
m-29	$\cdot CH_2CH_3$ ***OR*** $\cdot CHO$		
m-31	$\cdot OCH_3$	m/z = 91	$C_6H_5\overset{\bullet +}{C}H_2$
m-43	$CH_3\dot{C}{=}O$		
m-45	$\cdot OCH_2CH_3$		
m-91	$C_6H_5\dot{C}H_2$		

Table 14.1 Common mass spectral fragments.

Alkenes fragment to generate resonance stabilized allylic cations (Figures 14.5 and 14.6). Alcohols fragment through alpha cleavage which generates a resonance stabilized oxygen containing cation (Figures 14.5 and 14.6). Carbonyl containing compounds most readily fragment on either side of the carbonyl, again, generating resonance stabilized carbocations. All of the fragmentations described above are from cleavage of a single bond and generate fragments with odd masses (this will be opposite for compounds with odd numbers of nitrogen atoms). Often, the most stable cation results from two bond cleavages and generate even mass cations, as for dehydration of alcohols. For carbonyl containing compounds a very common fragment is the McLafferty ion which results from rearrangement and cleavage of 2 bonds, as shown for ethyl benzoate. You will learn more about this fragmentation in future carbonyl chapters.

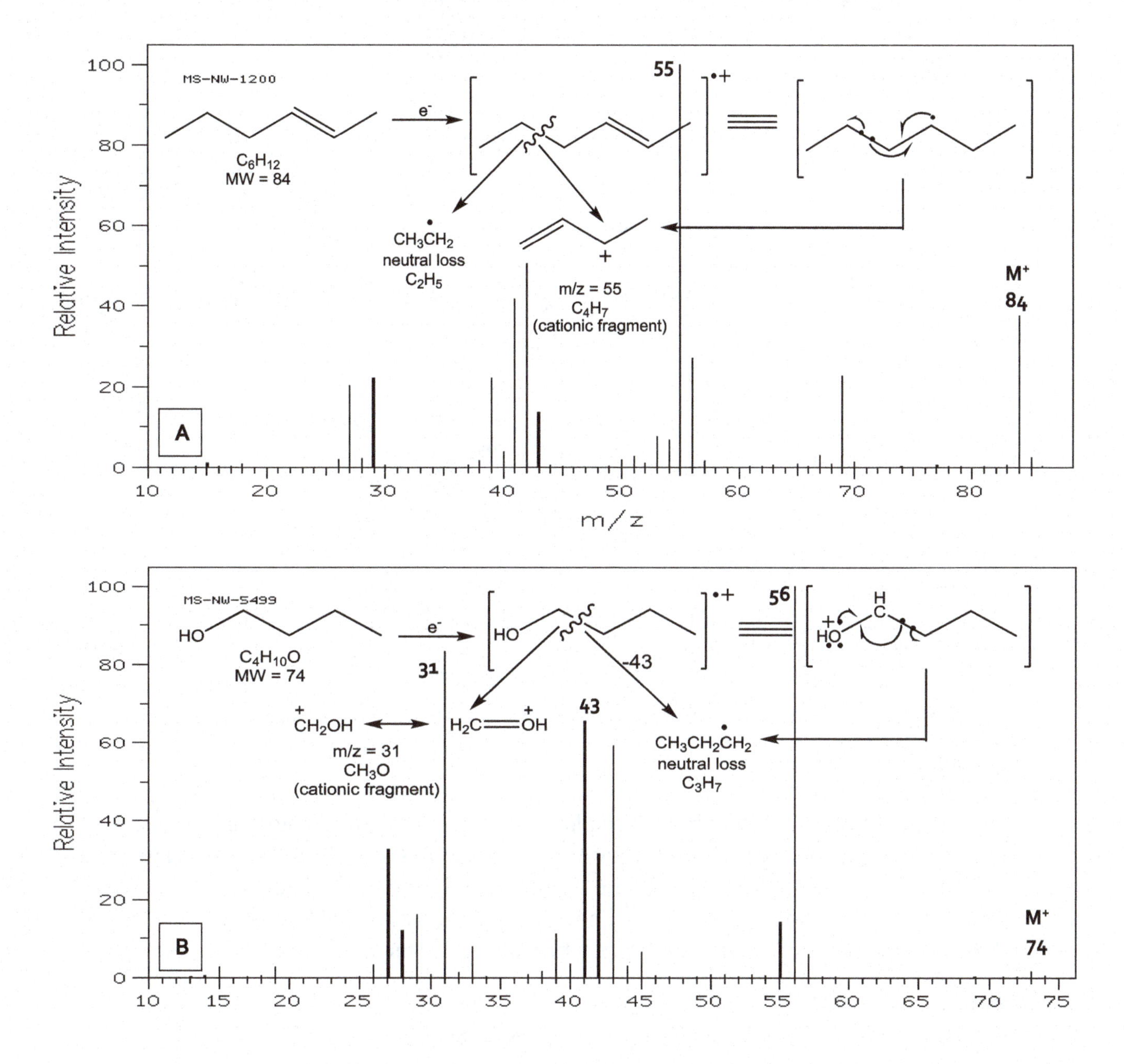

Figure 14.5 Mass spectra and fragmentation of 2-hexene (A) and 1-butanol (B).

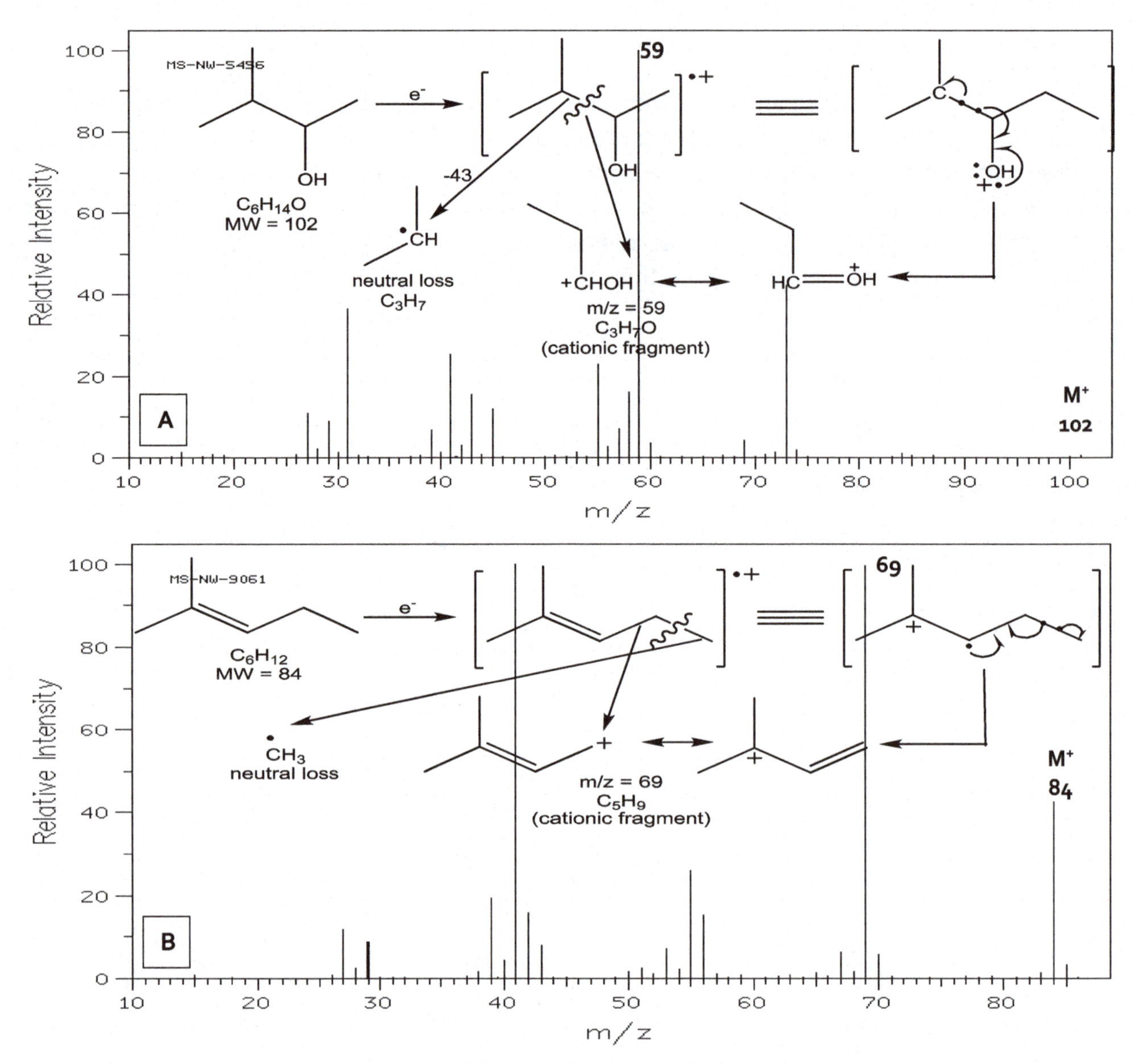

Figure 14.6 Mass spectra and fragmentation of 2-methyl-3-pentanol (A) and 2-methyl-2-pentene (B).

Objectives

In this experiment, commonly used laboratory calculations, chromatographic, and IR spectral techniques will be reviewed. New calculations will be introduced, which will be used to determine the efficiency of synthetic methods and overall marketability of products generated in the laboratory. Finally, two new spectral techniques will be introduced for use in future experiments to determine the structure of the product generated in the laboratory.

References

Klein, David. (2015). *Organic Chemistry*, 2nd ed. Hoboken: John Wiley and Sons.

Mass and IR Spectra are from:

http://riodb01.ibase.aist.go.jp/sdbs/cgibin/cre_index.cgi?lang=eng.

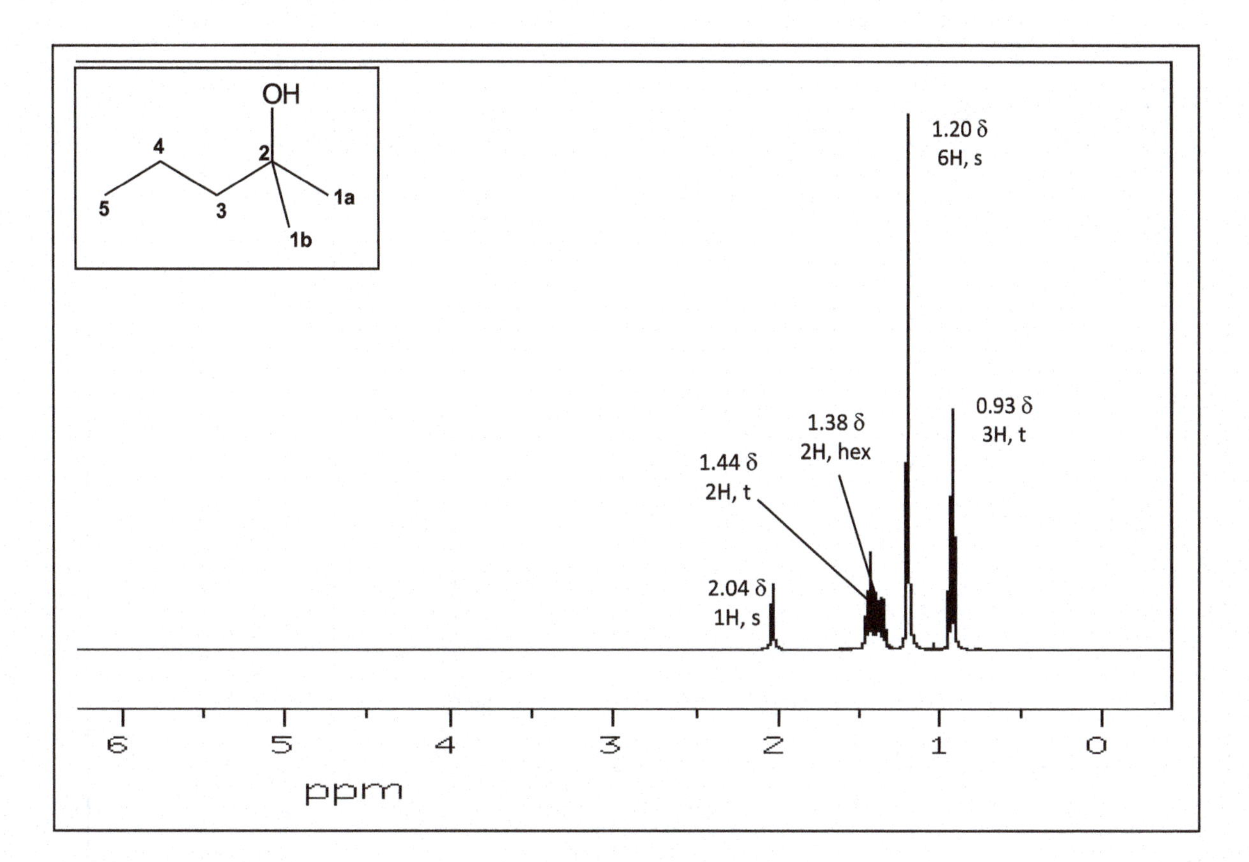

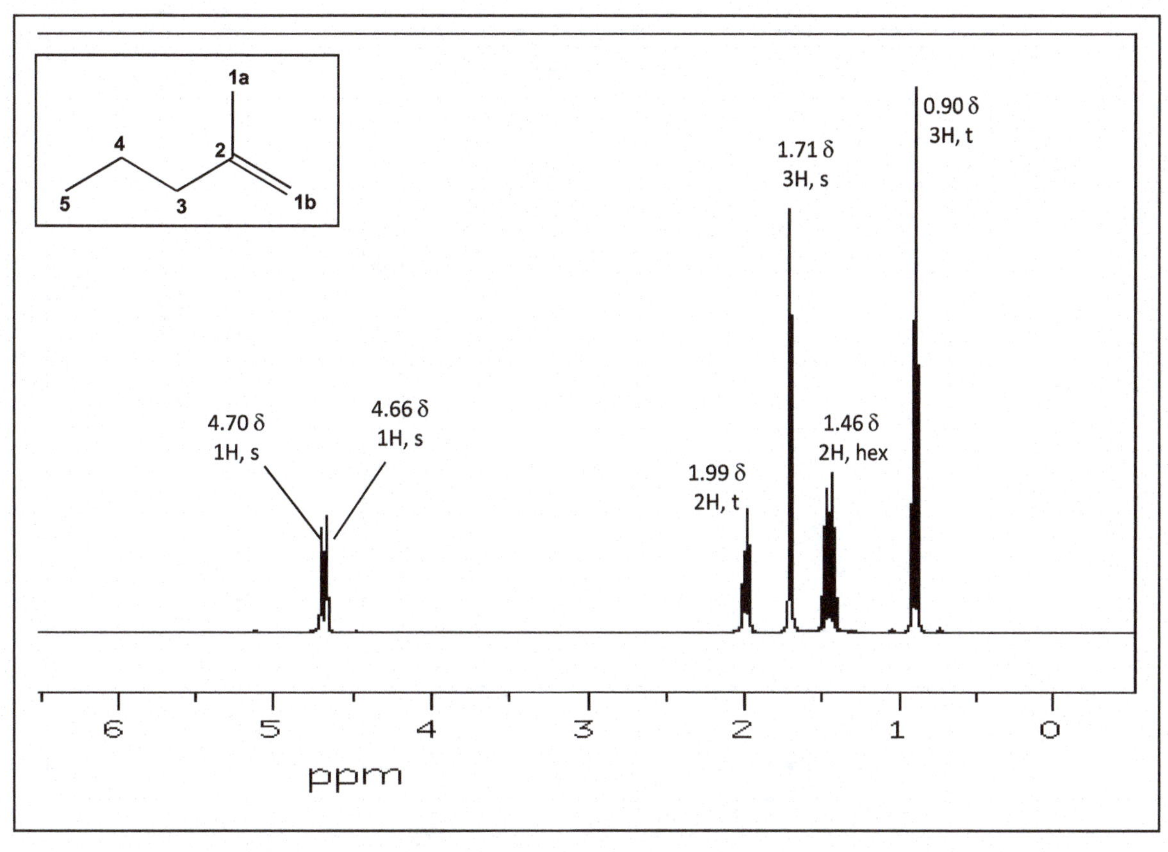

Figure 14.7 ^{1}H NMR spectra of 2-methyl-2-pentanol and 2-methyl-1-pentene.

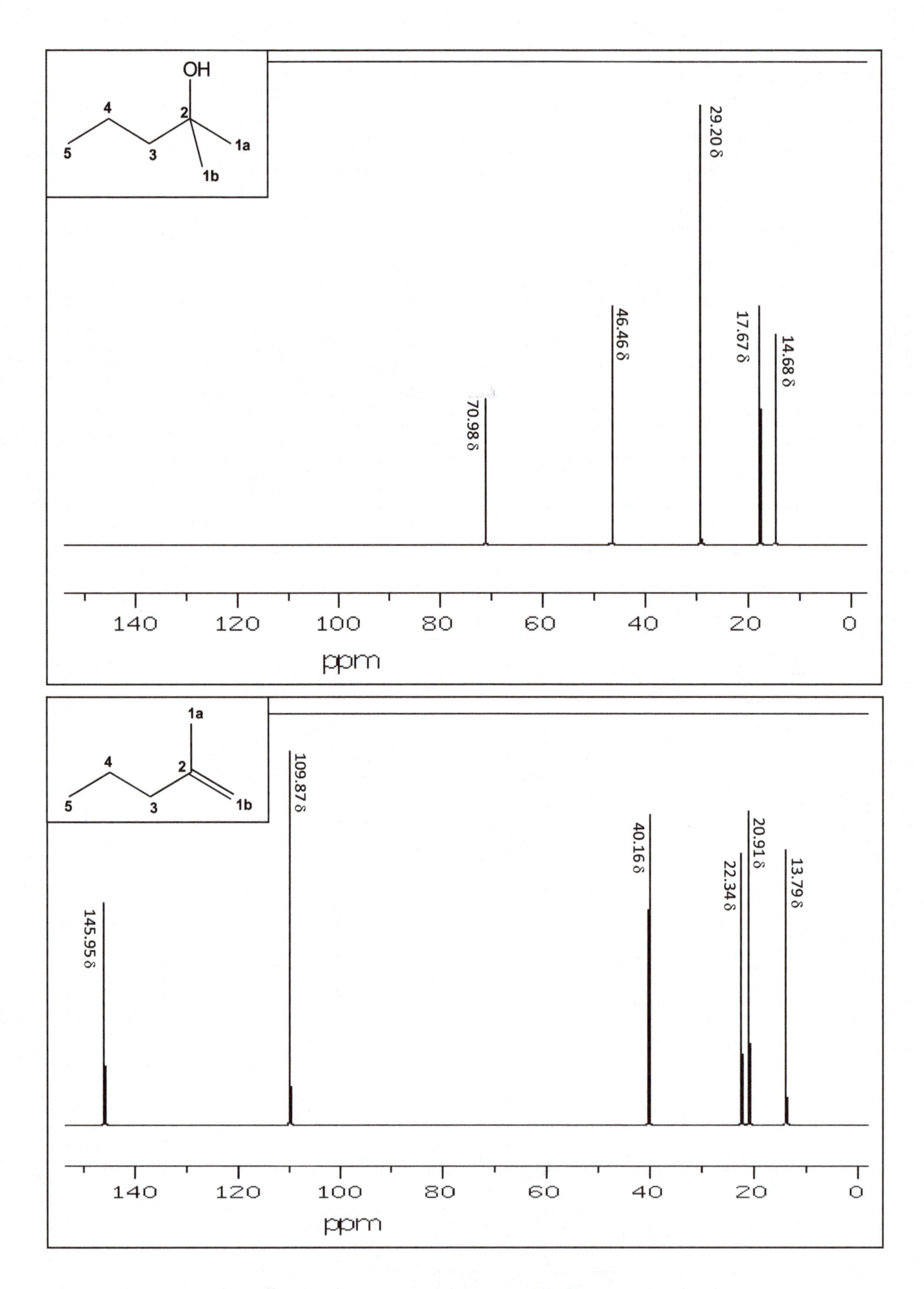

Figure 14.8 ^{13}C NMR spectra of 2-methyl-2-pentanol and 2-methyl-1-pentene.

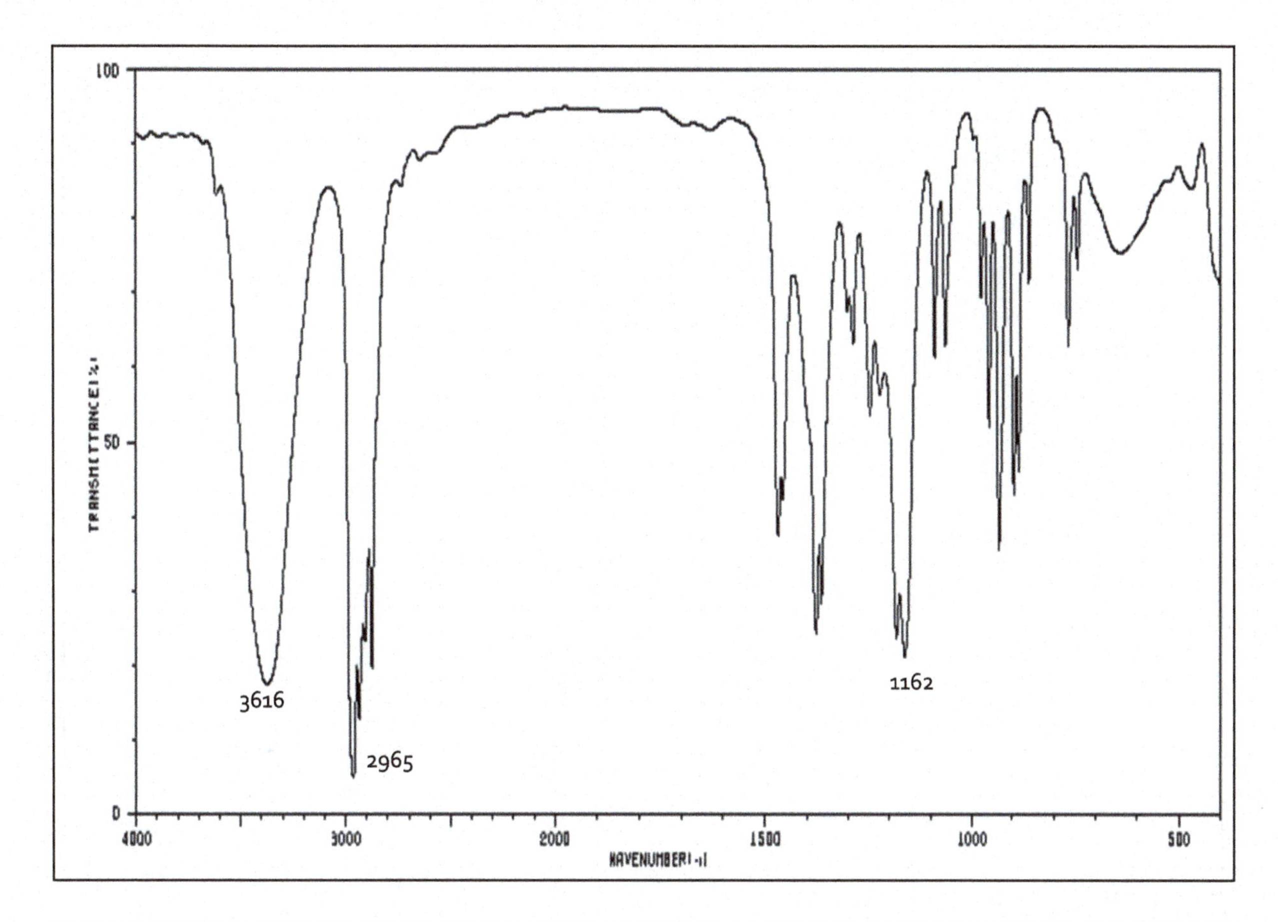

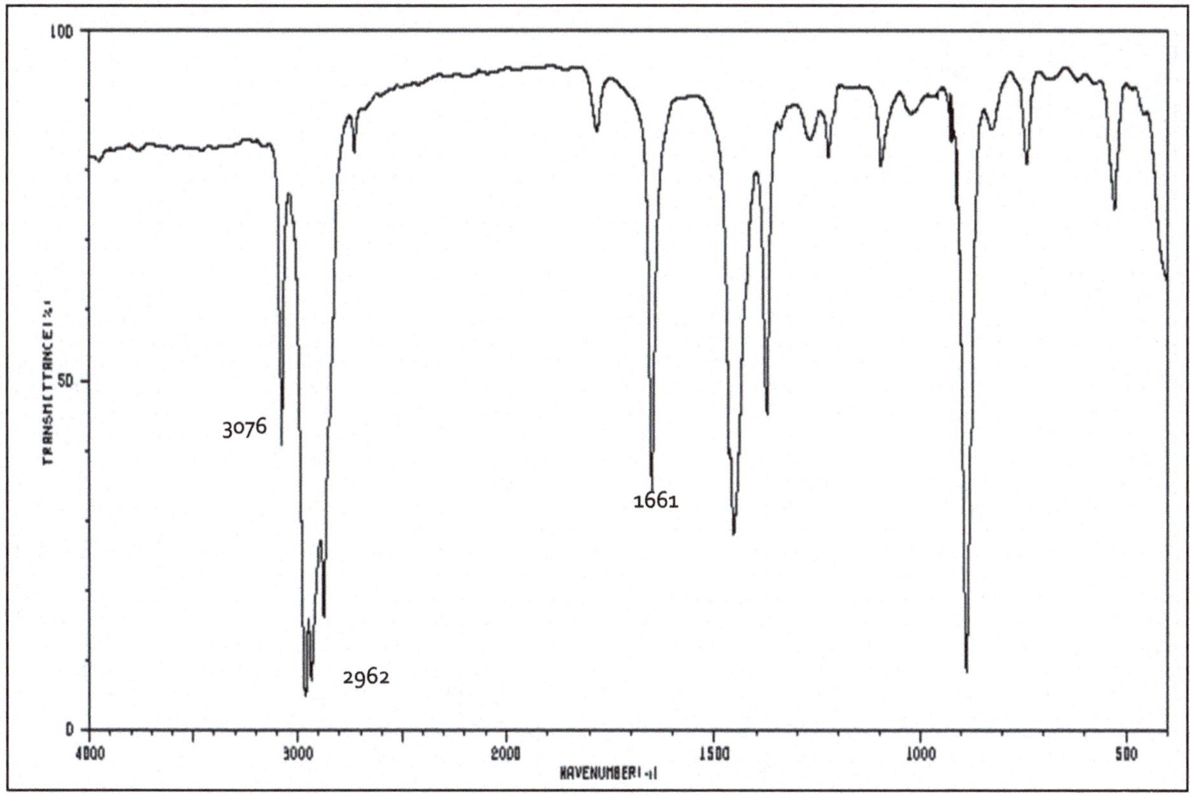

Figure 14.9 IR spectra of 2-methyl-2-pentanol and 2-methyl-1-pentene.

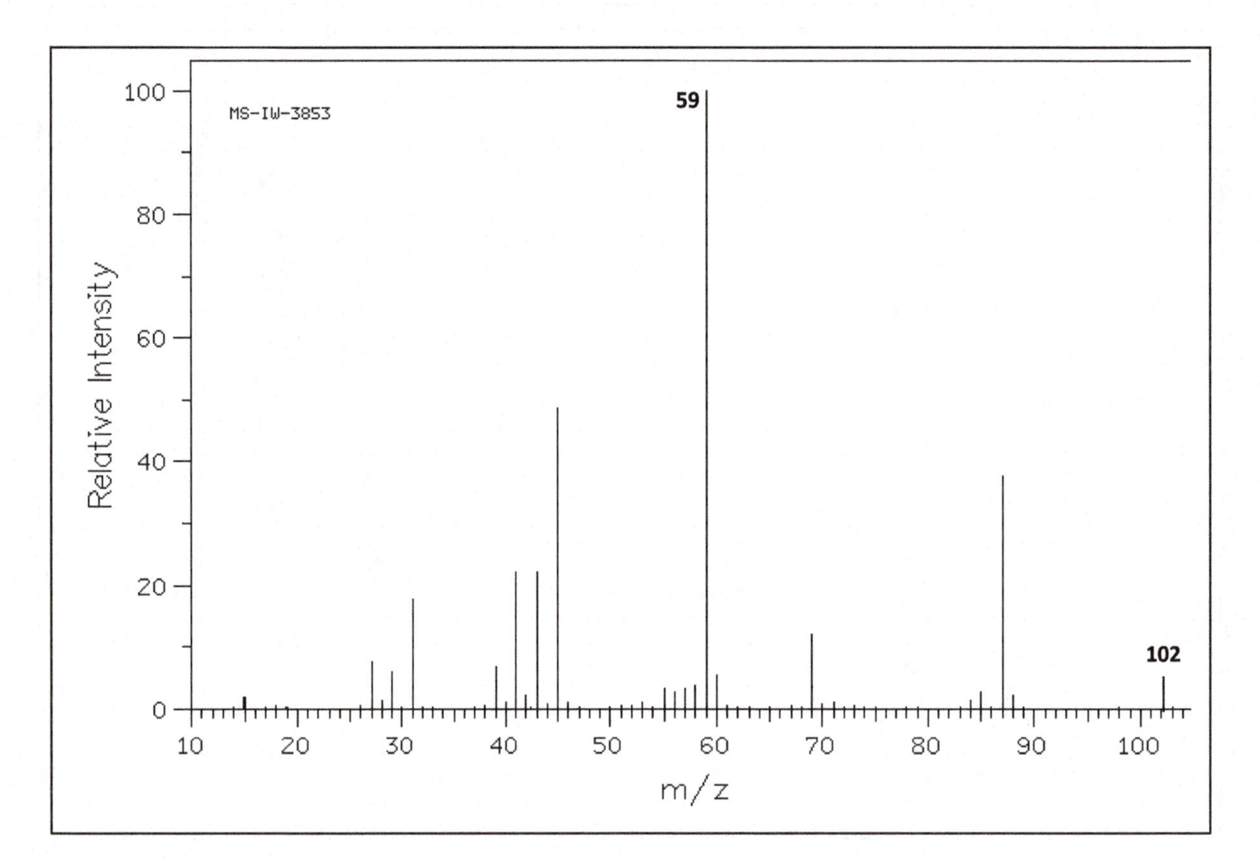

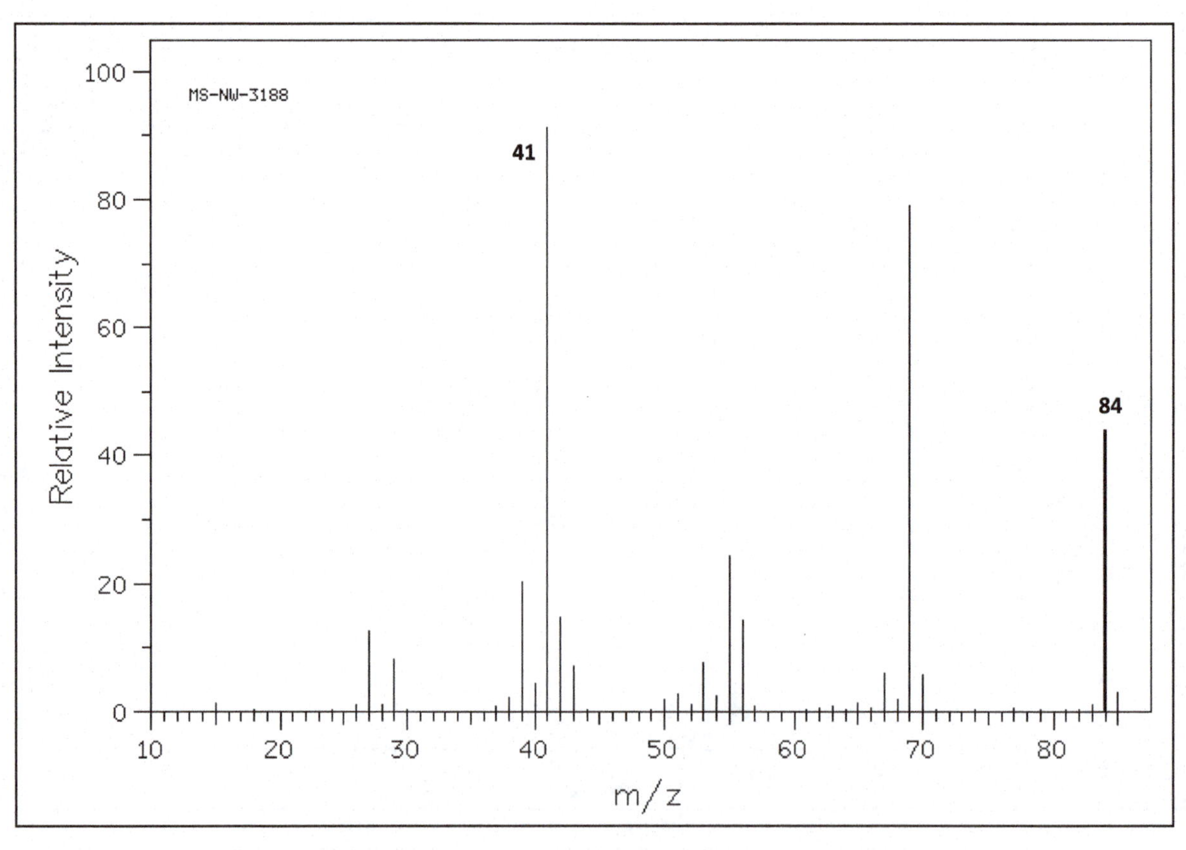

Figure 14.10 Mass spectra of 2-methyl-2-pentanol and 2-methyl-1-pentene.

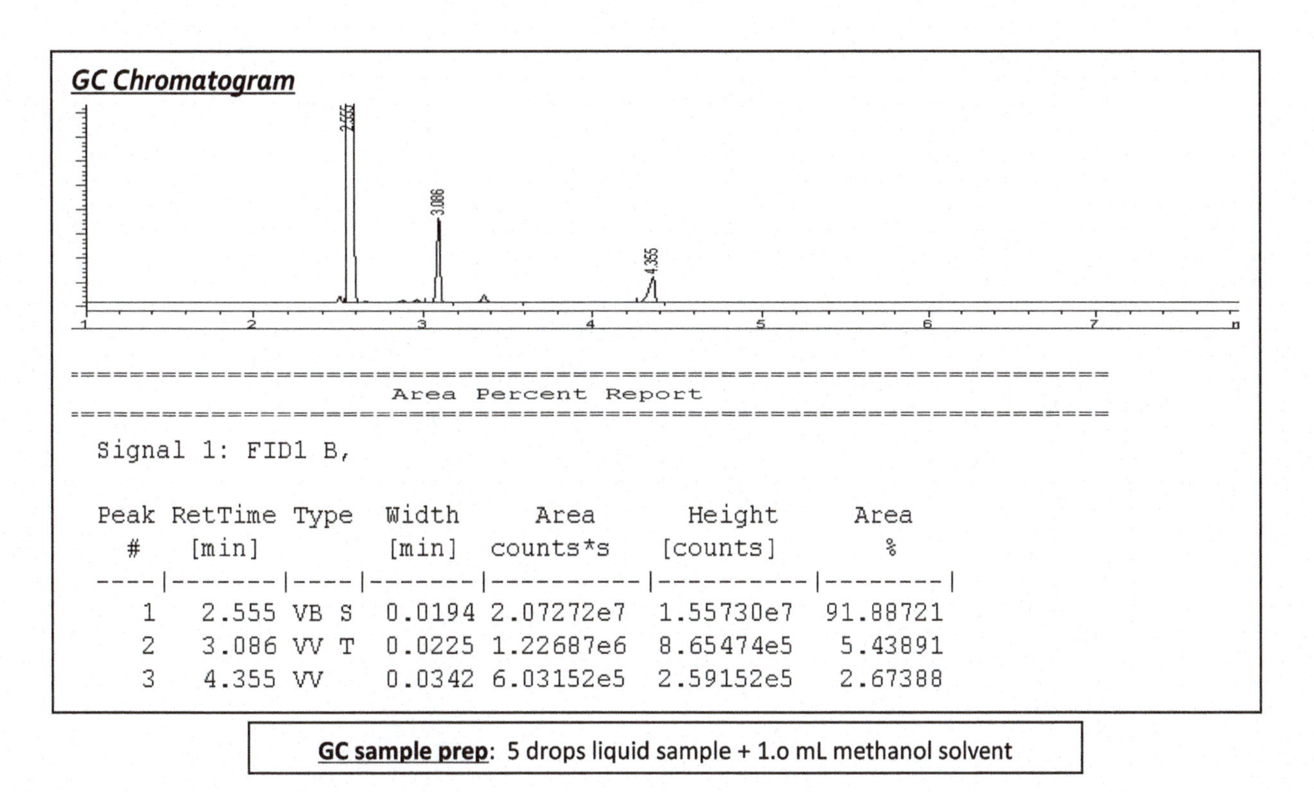

Area Percent Report

Signal 1: FID1 B,

Peak #	RetTime [min]	Type	Width [min]	Area counts*s	Height [counts]	Area %
1	2.555	VB S	0.0194	2.07272e7	1.55730e7	91.88721
2	3.086	VV T	0.0225	1.22687e6	8.65474e5	5.43891
3	4.355	VV	0.0342	6.03152e5	2.59152e5	2.67388

GC sample prep: 5 drops liquid sample + 1.o mL methanol solvent

Figure 14.11 GC Chromatogram for dehydration of 2-methyl-2-pentanol.

Experiment 15

The Grignard Reaction

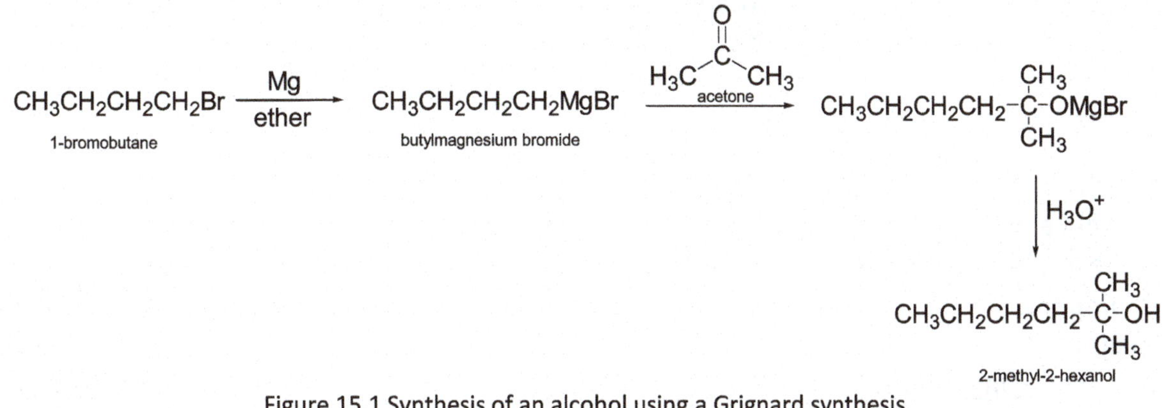

Figure 15.1 Synthesis of an alcohol using a Grignard synthesis.

Introduction

Organomagnesium compounds (Grignard reagents) are valuable intermediates in organic synthesis. These reagents are usually prepared by the reaction of an organic halide and magnesium in an ether solvent. The solvent acts as a Lewis base, complexing with the Grignard reagent, thus stabilizing it.

The initial reaction between the alkyl halide and magnesium requires a clean metal surface (free of the oxide film that forms when the metal is exposed to air), and is sometimes slow to get started. Alkyl iodides are more reactive than bromides, which are in turn more reactive than chlorides. Alkyl halides are more reactive than the corresponding aryl halides. When difficulty is encountered, especially with less reactive halides, a number of techniques have been found effective to initiate the reaction. These methods include grinding the magnesium in the presence of the halide to provide a clean active surface, the addition of a small crystal of iodine (believed either to produce a small amount of the more reactive iodide or to react with the magnesium thus exposing a fresh active surface), or the addition of a small amount of 1,2-dibromoethane which reacts rapidly with magnesium forming ethene gas and exposing fresh magnesium.

Grignard reagents are synthetically valuable because the carbon bound to magnesium is nucleophilic. This allows for a number of reactions, the most important of which is the addition to carbonyl compounds to produce alcohols (Fig. 15.1).

Grignard reagents are also strong bases. Therefore, great care must be taken in their preparation to avoid contact with water, alcohols, amines or any other compounds containing active hydrogens (N-H or O-H). The presence of any of these kinds of compounds will destroy the reagent immediately as it forms and will inhibit the initiation reaction due to the coating of salts formed on the magnesium surface (Figure 15.2).

Figure 15.2 Reaction which occurs between butylmagnesium bromide and water.

Objectives

In this experiment you will synthesize a tertiary alcohol using a Grignard synthesis between 1-bromobutane and acetone. The product will be isolated using an extraction technique, and the purity of the product will be determined using GC analysis. Finally, IR and NMR spectroscopy will be used to distinguish between reactants and products.

Experimental Procedure

Synthesis:

- Obtain a 50 mL round bottom flask from the oven in the lab. Quickly place a $CaSO_4$ drying tube into the top of the reaction flask using a thermometer adapter.
- Clamp the reaction flask to a ring stand. Support the flask using an iron ring with a wire gauze pad.
- Lift the drying tube and add 1.0 g of dried magnesium turnings. Quickly replace the drying tube.
- Return to the oven for a condenser, a Claisen adapter, a separatory funnel, and a ground glass stopper. Assemble the reflux with addition apparatus (Appendix A) and set apparatus aside. Place the glass stopper in the separatory funnel while not in use.
- Leaving the reaction flask clamped to the ring stand, remove the drying tube from the flask and quickly add 2 mL of dry ether and 1 mL of 1-bromobutane to the reaction flask.
- Using your glass stirring rod, gently scratch the surface of the magnesium turnings, and stir to induce the reaction. This may take several minutes. You may notice the appearance of a pale yellow color, followed by the evolution of heat. When the reaction has started, it will appear to boil.
- As soon as the reaction begins, quickly place the top portion of the apparatus onto the reaction flask, and place the drying tube in the top of the reflux condenser.
- Be sure to close the stopcock of the separatory funnel. Combine 4 mL of 1-bromobutane and 15 mL of ether in the separatory funnel. Remove the glass stopper, open the stopcock ***SLOWLY***, and add the mixture drop wise at a rate that maintains a steady reflux.
- Swirl the reaction flask occasionally by holding one hand on the apparatus, and gently rocking the ring stand with the other.
- After all of the 1-bromobutane/ether solution has been added, close the stopcock of the separatory funnel.
- Leaving the apparatus assembled, cool the reaction flask to room temperature using a tap water bath prepared in a small beaker.
- When the reaction has stopped boiling, the Grignard reagent should be formed. A black residue from impurities in the magnesium will be noticeable.
- While the reaction mixture is cooling proceed to add 3 mL of acetone and 8 mL of dry ether to the separatory funnel. Remove the glass stopper and open the stopcock slowly to add the mixture drop wise. Swirl the reaction flask occasionally by gently rocking the ring stand.
- Once all of the acetone/dry ether solution has added, the reaction is complete. Cool the reaction flask to room temperature using a tap water bath prepared in a small beaker. Carefully remove the top portion of the apparatus and set aside.

- Using a plastic pipet, transfer **ONLY** the liquid to a 125 mL Erlenmeyer flask, leaving any unreacted magnesium and solid salts behind. Add a small ice cube directly to the Erlenmeyer flask, and then add 5 mL of Sat. aqueous NH_4Cl solution while swirling until the gelatinous precipitate forms.

Purification:

- Set up an extraction apparatus by supporting the separatory funnel on an iron ring with a rubber flask support. Be sure the stopcock is closed.
- Slowly pour the liquid into the separatory funnel. Try to avoid pouring the gelatinous precipitate into the funnel.
- Add 1 mL of 5% HCl to the separatory funnel. Place the stopper in the top of the separatory funnel and twist to ensure a secure fit. Invert the funnel once and ***vent IMMEDIATELY*** before continuing to agitate! Continue to agitate and vent until pressure is no longer released. Allow the layers to separate and draw off the aqueous layer into an Erlenmeyer flask labeled "AQUEOUS WASTE".
- Add 5 mL of 10% $NaHCO_3$ to the separatory funnel. Agitate gently and vent. Draw off the aqueous layer into the previous Erlenmeyer flask labeled "AQUEOUS WASTE".
- Add 5 mL of Sat. NaCl to the separatory funnel. Agitate gently and vent. Draw off the aqueous layer into the previous Erlenmeyer flask labeled "AQUEOUS WASTE".
- Transfer the organic layer from the separatory funnel into a clean 50 mL Erlenmeyer flask. Add a small amount of $MgSO_4$ to dry the organic layer. Using a plastic pipet, transfer the dried solution into a clean 50 mL beaker.
- ***PROCEED TO PRODUCT ANALYSIS.***

Product Analysis:

GC Analysis

- Submit a small sample (5 drops sample + 1 mL GC solvent) for GC analysis. Using chromatographic results identify and determine the degree of purity of your product. Complete Table 15.1 on the POST LAB Assignment (provided online).

Spectral Analysis

- Examine the NMR and IR spectra of the starting alkyl halide and alcohol product (Figures 15.3 and 15.4). Identify and tabulate all characteristic IR absorptions in Table 15.2 on the POST LAB Assignment (provided online). Identify and tabulate all NMR resonances in Tables 15.3 and 15.4 on the POST LAB Assignment (provided online).

SAFETY

****All experiments should be performed in a fume hood with appropriate safety glasses. Gloves will be provided by request.****

All organic solvents used in this experiment are flammable, irritants, and can be toxic if ingested or absorbed through skin. Organic solvents, reagents and products should be kept in sealed containers at all times. Hydrochloric acid is corrosive.

WASTE MANAGEMENT

Unreacted magnesium should be rinsed with water and placed in the container labeled "RECOVERED MAGNESIUM WASTE". The aqueous extracts and washes may be combined and placed in the bottle labeled "LIQUID WASTE—GRIGNARD". The alcohol product should be placed in the container labeled "LIQUID WASTE—GRIGNARD". Used $MgSO_4$ should be placed in the container labeled "SOLID WASTE".

Reference

Klein, David. (2015). *Organic Chemistry*, 2nd ed. Hoboken: John Wiley and Sons.

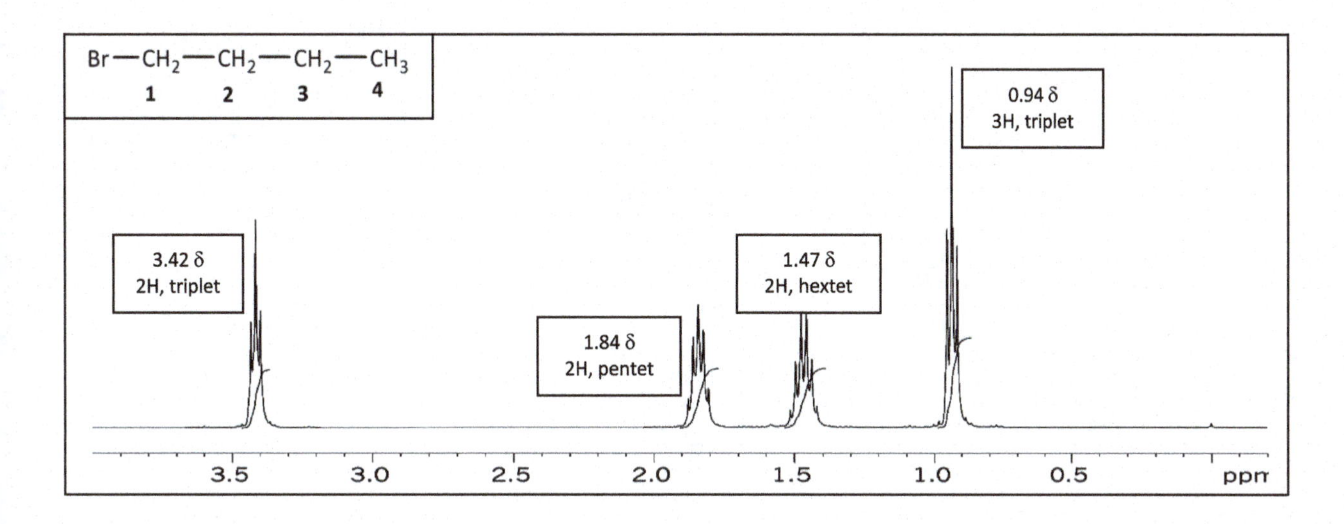

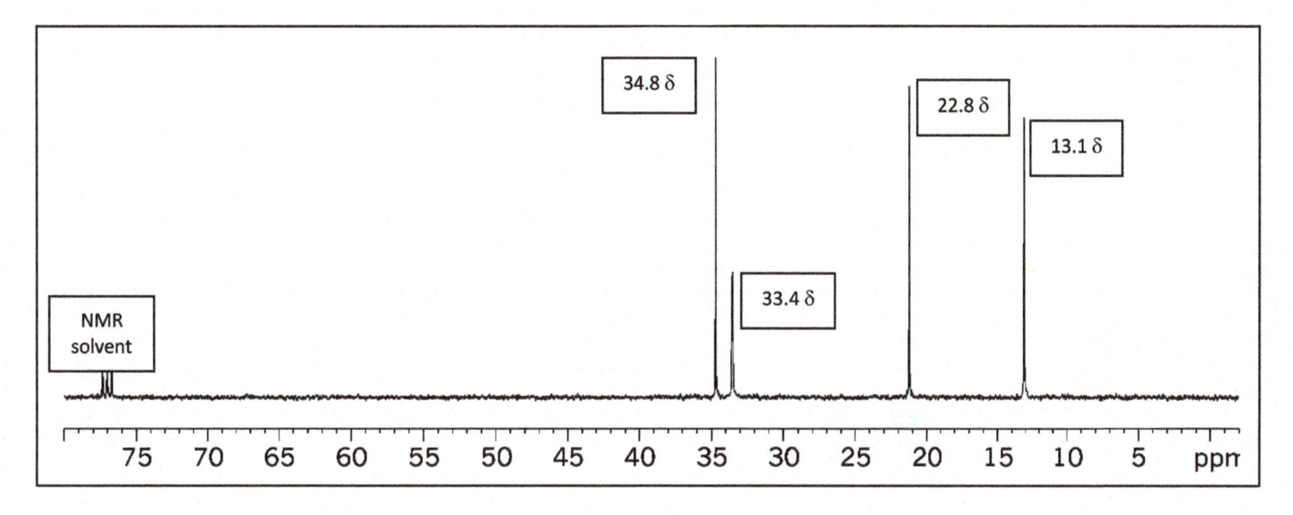

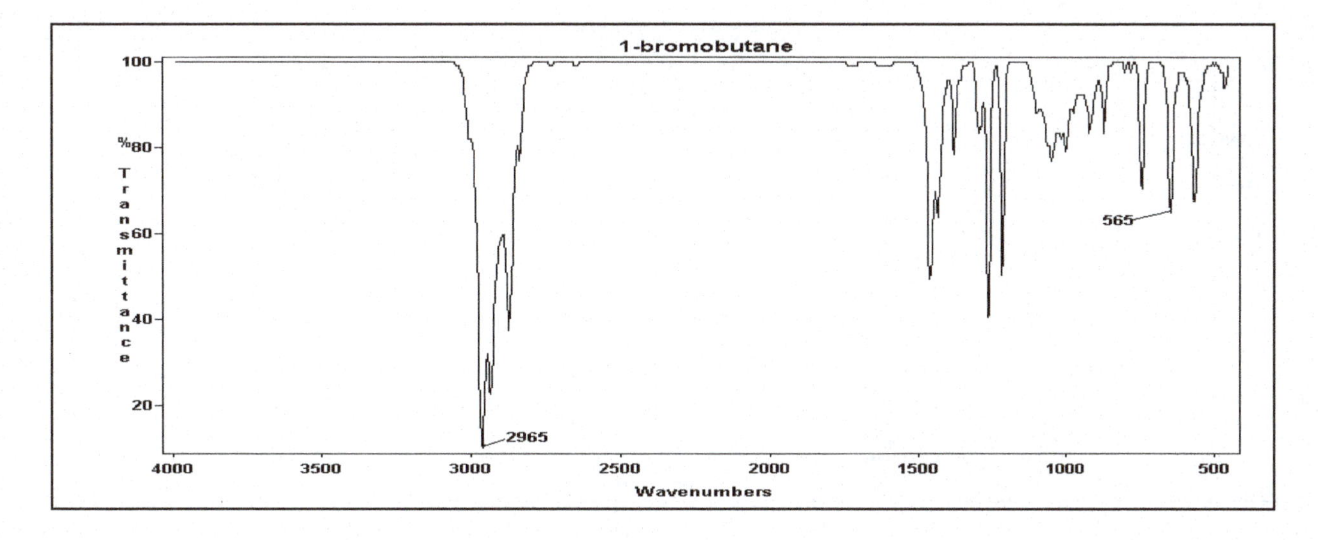

Figure 15.3 ^{1}H, ^{13}C, and IR spectra of 1-bromobutane.

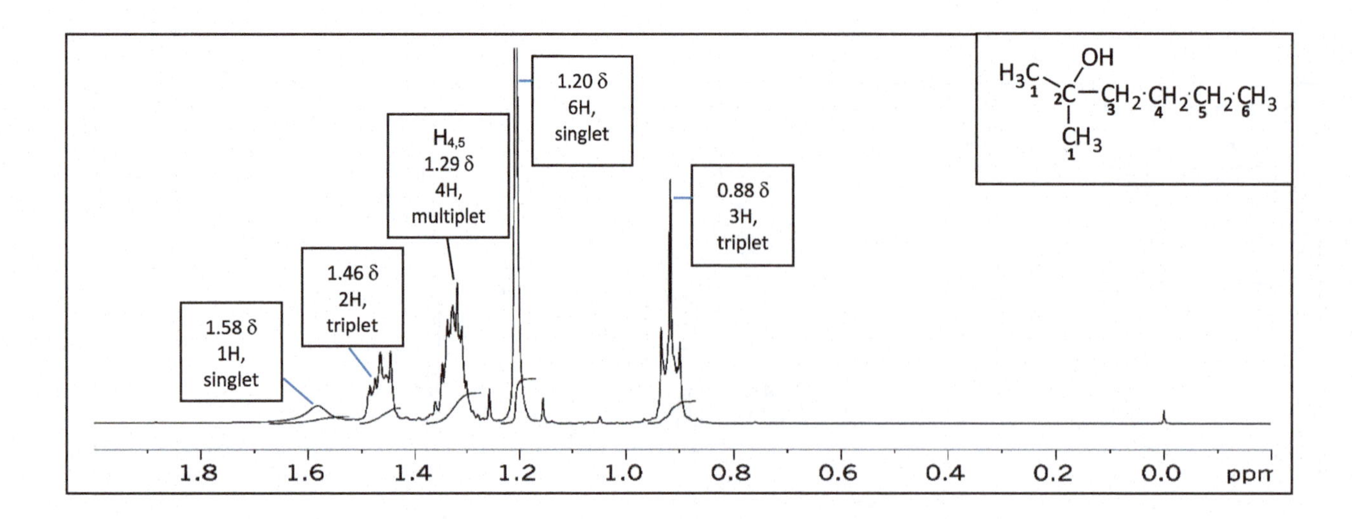

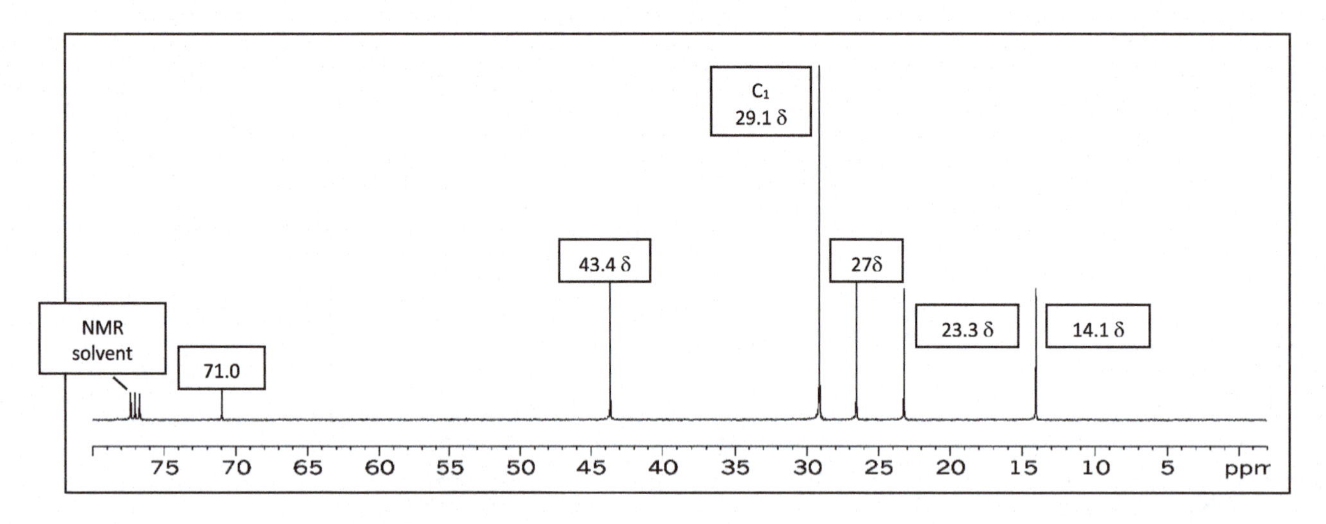

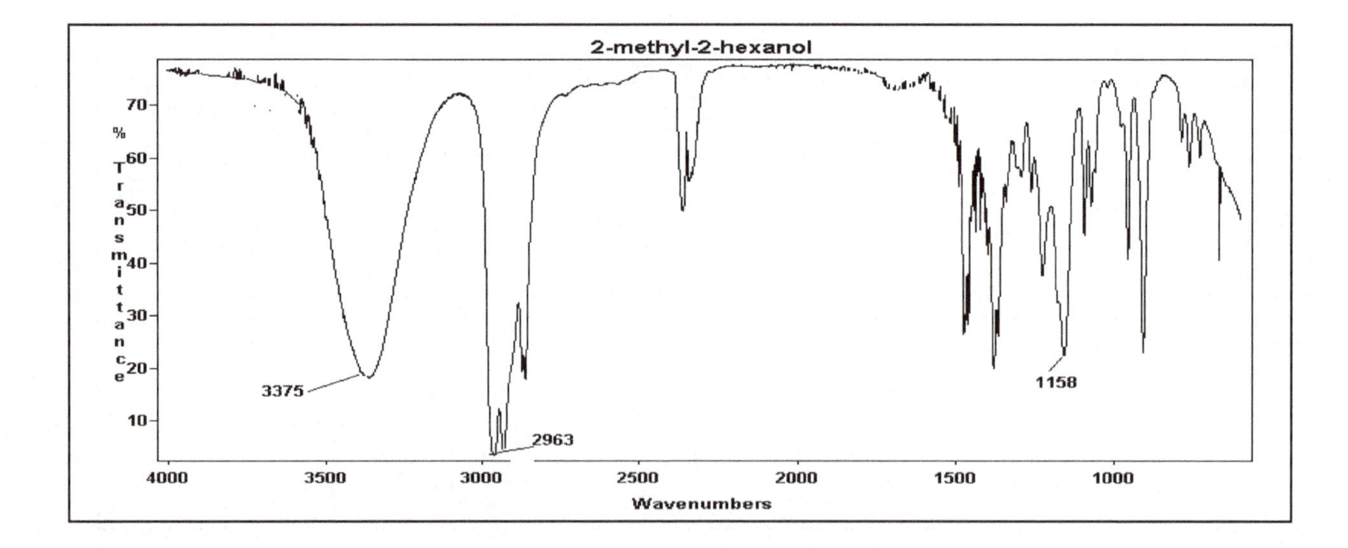

Figure 15.4 ^{1}H, ^{13}C, and IR spectra of 2-methyl-2-hexanol.

Name:			
	Max	Score	Total Grade
Tech	10		
Pre	20		
In	20		
Post	50		

Exp. 15 The Grignard Reaction

PRE-LAB ASSIGNMENT: *(**EACH STUDENT** will complete and submit an original copy at the beginning of the lab period. **Without a complete pre-lab assignment, you will not be allowed to perform the experiment, and will receive a zero for the lab.**) .....max score = 20 pts.*

1. **Objective** *(Write a brief purpose of the experiment in **complete** sentences, addressing all of the following points.)*
   - Name the *specific* compounds combined during the synthesis and the name the product formed.
   - What is the purification technique used in this experiment?
   - What analytical technique will be used during the experiment?
   - What type of spectral analysis will be used to characterize the reactant and products?

2. **Chemical Equation** *(Draw the chemical equation for the synthesis of 2-methyl-2-hexanol using actual chemical structures.)*

3. **Physical Data** *(Complete the following table before coming to lab.)*

Compound	MW (g/mol)	bp (°C)	d (g/mL)
1-bromobutane			
acetone			
magnesium turnings		X	X
diethyl ether	X		
hydrochloric acid	X	X	
methanol	X	X	X
2-methyl-2-hexanol			X

4. Experimental Outline *(Give a brief description of the procedure that will be followed in this experiment in 5 lines or less.)*

1	
2	
3	
4	
5	

5. Pre-Lab Questions *(Answer the following questions prior to lab.)*

A. In this lab, Grignard reactions can be slow to initiate because of the magnesium metal turnings. This is because:

a. the magnesium is flammable.
b. the magnesium is coiled too tightly.
c. the magnesium reacts with air to form a magnesium oxide coating.
d. the magnesium metal reacts with water.

B. List the **3** methods that can be useful to initiate a Grignard reaction:

a.
b.
c.

I have read and understood the experimental procedure for this experiment. I am familiar with the hazards and the required disposal procedures for this experiment.

Sign here: ____

Experiment 16

Diels Alder Cycloaddition

Introduction:

The Diels-Alder cycloaddition reaction is very useful in organic synthesis because it forms two new carbon-carbon bonds in a single step. It is an example of a [4+2] cycloaddition reaction (4 π electrons + 2 π electrons) between a conjugated 1,3-diene and an alkene (the dienophile) and leads to the formation of cyclohexanes.

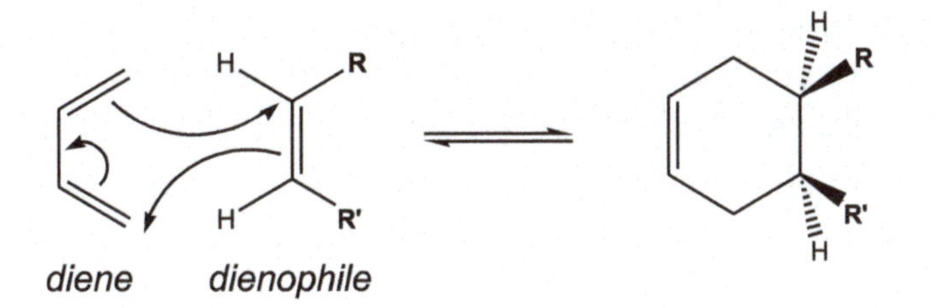

The Diels-Alder cycloaddition reaction takes place in one step and is thus, a concerted reaction. As the three initial π bonds of the diene and the monoene break, both the new C-C single bonds and the C=C π bond are formed simultaneously. The diene component must be in the *cis* conformation to react. The reaction is highly stereospecific and the orientation of the groups on the dienophile are retained in the product. Thus, the two groups that are *cis* on the dienophile will be *cis* in the product.

In this experiment, you will synthesize the Diels-Alder adduct formed from reaction of butadiene with maleic anhydride. Since butadiene is a gas at room temperature, it will be generated from a solid reagent that is easily handled, 3-sulfolene. The product will be purified by crystallization and its melting point will be obtained as a measure of its purity. You will also analyze IR (Figure 16.2).

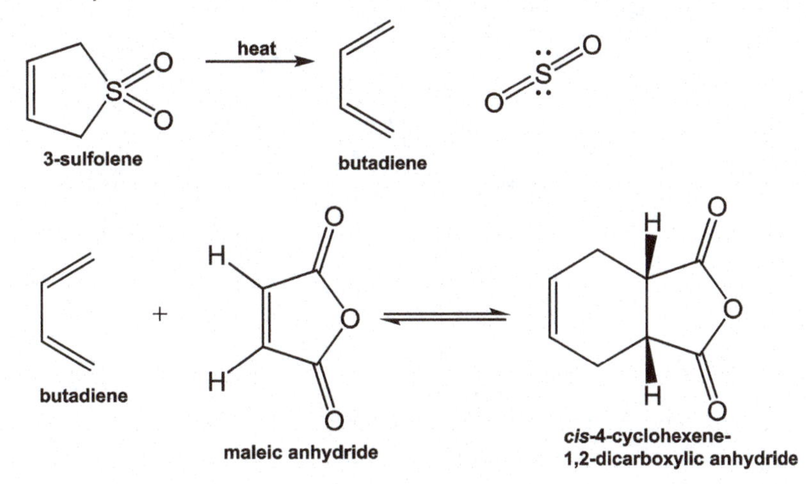

Figure 16.1 Diels-Alder cycloaddition reaction between butadiene and maleic anhydride.

IR Spectroscopy

One characteristic feature in the IR spectrum of anhydrides is the appearance of two strong bands, not necessarily of equal intensities. These two bands represent the C=O stretch, resulting from both asymmetric and symmetric stretches. Note that in the IR spectrum of the maleic anhydride and the product, these two absorptions occur at higher frequencies than expected due to the ring strain present in the five membered ring of the anhydride. Ring strain typically results in a shift in the C=O frequency to higher frequencies, as compared to acyclic anhydrides. Another characteristic feature in the IR spectra of both starting material and product are multiple C-O stretches.

Valuable information can also be determined in the range 650-1000 cm^{-1}, known as the out-of-plane bending region for alkenes. Both the maleic anhydride and the product show two strong bands in this region, and this absorption frequency is also sensitive to ring size. The stereochemistry of the double bond can also be determined using IR spectroscopy. A *cis* arrangement around a double bond shows a strong C-H bend around 700 cm^{-1}. Spectral correlation tables can be found in Appendix I.

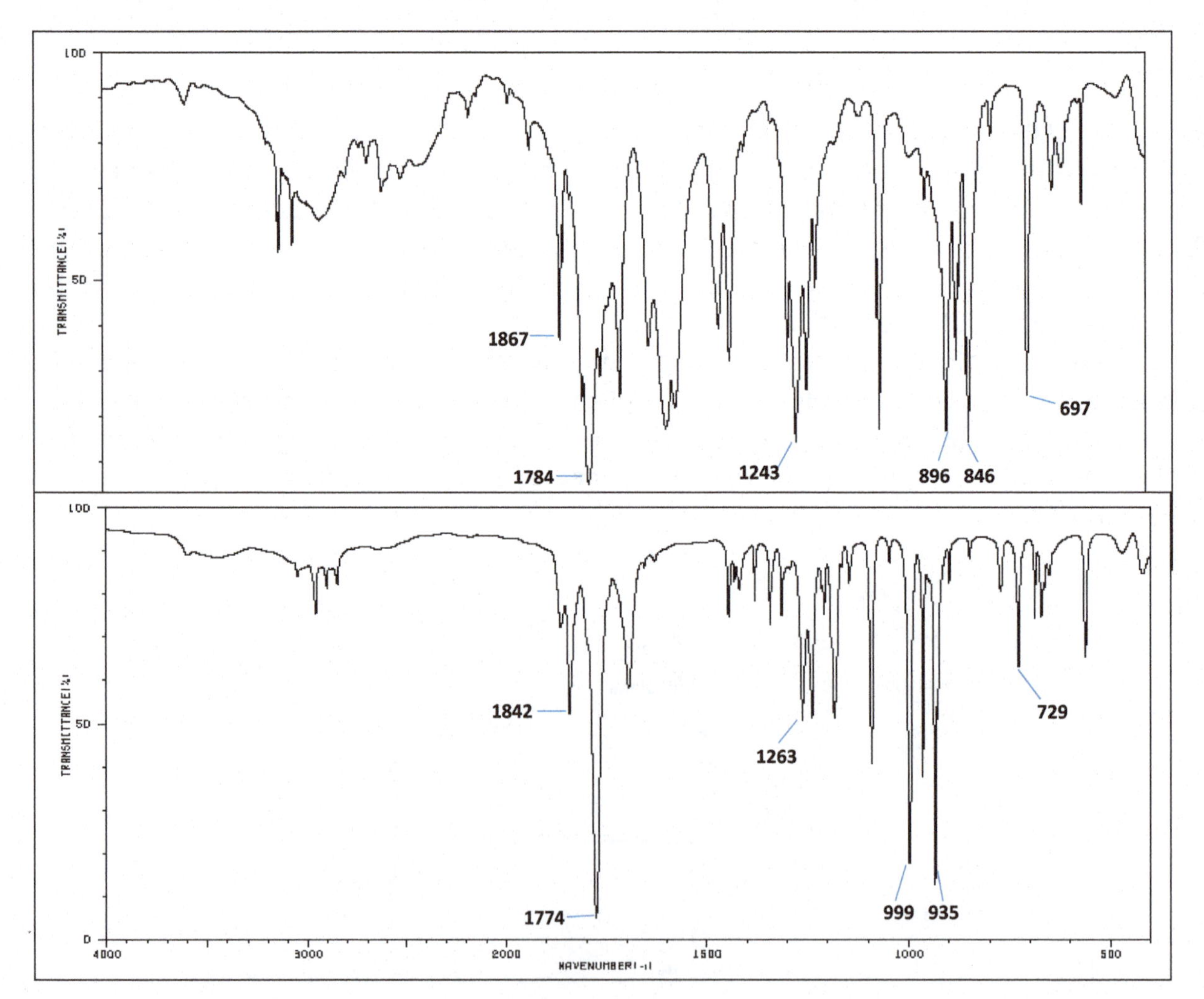

Figure 16.2 IR spectra of maleic anhydride and *cis*-4-cyclohexene-1,2-dicarboxylic anhydride.

Mass Spectrometry

Mass spectrometry can be useful to determine the size and formula of a compound. Recall that when a compound is analyzed using a mass spectrometer, the molecules are bombarded by high energy electrons. This electron bombardment converts some of the molecules to ions by knocking off a single electron. Once the molecules are converted to ions, they are accelerated in an electric field, where they are separated according to their mass-to-charge ratio (*m/z*). These ions are then detected and recorded to product a mass spectrum. The mass spectrum is simply a plot of ion abundance vs. m/z ratio.

The simple removal of a single electron from a molecule yields an ion whose weight is equivalent to the molecular weight of the original molecule. This is called the molecular ion peak ($M^{+\bullet}$). If the $M^{+\bullet}$ peak can be identified on the mass spectrum, it is possible to use the spectrum to determine the molecular weight of an unknown compound.

Once ionized, the molecule becomes unstable and will continue to break into smaller fragments. The amount of each fragment is dependent on the stability of the fragment. The most abundant ion formed gives rise to the tallest peak in the mass spectrum, called the base peak (Figure 16.3). The relative abundances of all other peaks in the spectrum are reported as percentages of the base peak.

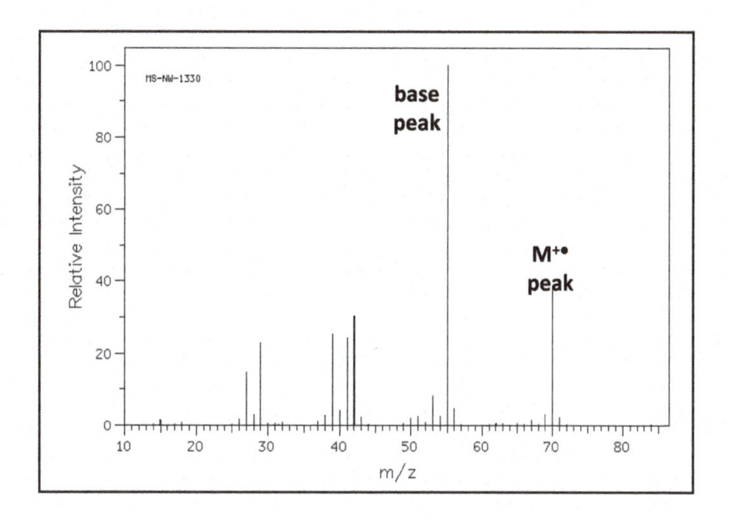

Figure 16.3 Mass spectrum showing base peak and molecular ion peak.

Mass spectrometry can be useful during organic synthesis. If mass spectra of the reactant(s) and product are available, and the $M^{+\bullet}$ is identified for each, it is possible to determine whether or not a reaction was successful. Simply by comparing the $M^{+\bullet}$ peak of the product to that of the reactant(s), along with the use of other characteristic fragmentation patterns, the success of the reaction can be determined. The mass spectra of maleic anhydride and the desired product (*cis*-4-cyclohexene-1,2-dicarboxylic anhydride) are shown in Figure 16.4.

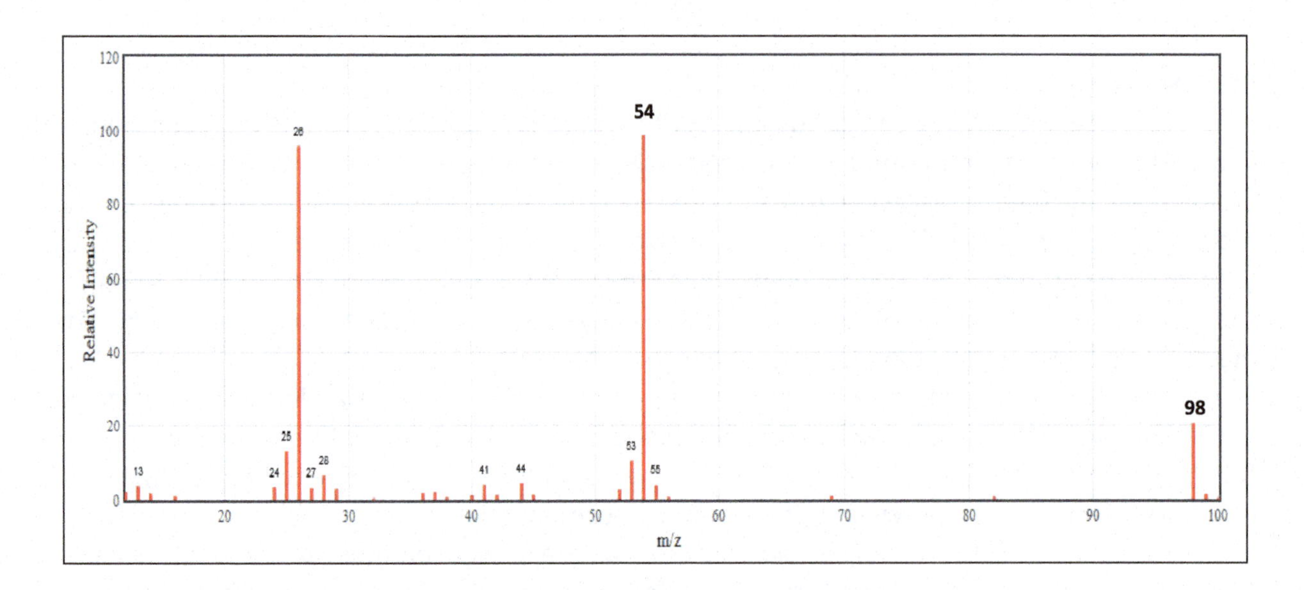

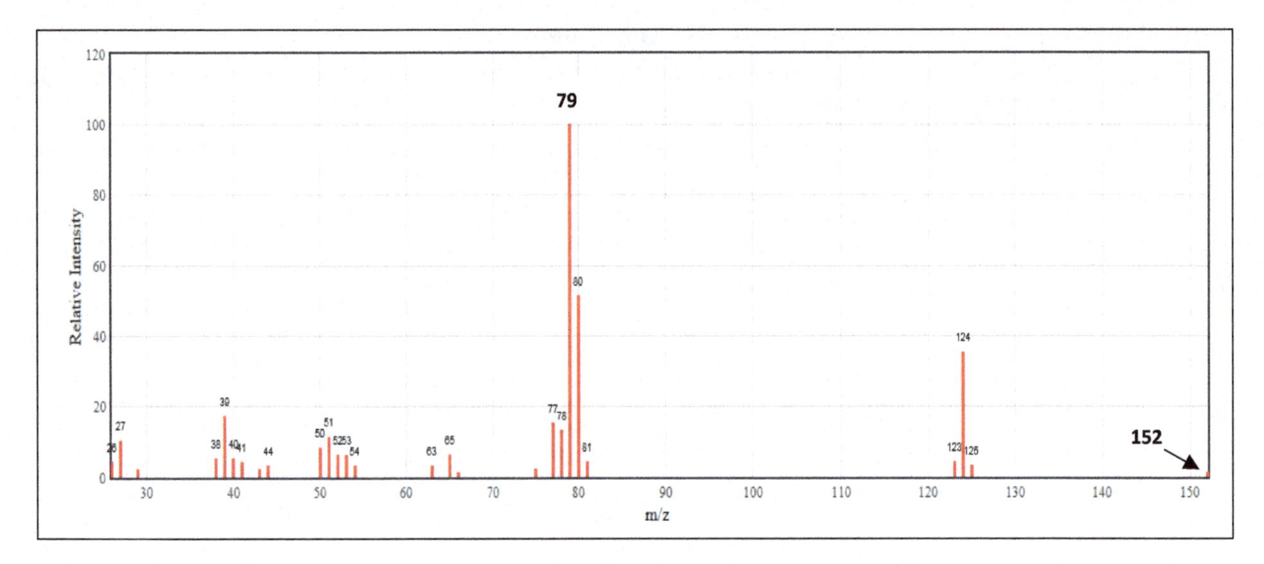

Figure 16.4 Mass spectra of maleic anhydride and *cis*-4-cyclohexene-1, 2-dicarboxylic anhydride.

Objectives

In this experiment, you will perform a Diels-Alder cycloaddition reaction. The solid product will be purified by recrystallization, followed by determination of purity through melting point analysis. The reactant and product will be characterized using mass spectrometry and IR analysis.

Experimental Procedure

Synthesis:

- Weigh ~2.00 g of 3-sulfolene directly into a 25 mL round bottom flask. Clamp the flask to a ring stand several inches from the base of the fume hood, and add 3 boiling chips.
- Weigh ~1.20 g of maleic anhydride on a piece of weigh paper. Transfer this solid to the reaction flask using a short stem funnel. Rinse the funnel with 2.0 mL xylene into the round bottom flask.
- Place a water cooled condenser on the top of the reaction flask, with a $CaSO_4$ drying tube in the top. Begin a gentle water flow.
- Fill the heating mantle ¾ full with sand, and lower the reaction flask into the sand bath.
- Begin heating the reaction flask with a voltage regulator at a setting of 45. Once the solution begins to boil and reflux, reduce the setting to 25, and continue reflux for 30 minutes.
- After 30 min, lower the heating mantle away from the flask and allow the reaction mixture to cool for 5 minutes.

Purification:

- Raise the condenser. Add 10 mL toluene to the reaction flask, along with 0.20 g decolorizing charcoal.
- Raise the heating mantle to the reaction flask and return to a gentle boil (VR setting = 25).
- Once the solution boils, lower the heating mantle away from the flask and carefully remove the condenser and $CaSO_4$ tube from the reaction flask.
- Filter the solution, while hot, through a fluted filter paper into a 50 mL flask.
- Add 2-3 boiling chips and place in a hot water bath prepared in a 250 mL beaker. (The water bath can be placed on a hotplate using a heat setting of 5).
- Heat the reaction filtrate in the hot water bath for 5 minutes, then remove.
- Add petroleum ether in 1 mL increments until it appears cloudy. **DO NOT EXCEED 8 mL of petroleum ether!**
- Place the reaction solution back into the hot water bath and heat for 1-2 minutes to clear the solution. Once clear, remove the flask and cool slowly to room temperature, then in ice bath for 10 minutes. If no crystals appear, scratch the sides of the flask with a glass rod to induce crystallization.
- Set up a suction filtration apparatus to isolate the solid product. Seat the filter paper and rinse the crystals from the flask with ~ 5 mL ice cold petroleum ether.
- Allow the solid to dry under vacuum for 10 minutes.
- Reweigh solid to determine the final product mass and calculate the percent yield.
- ***PROCEED TO PRODUCT ANALYSIS.***

Product Analysis:

Melting Point Analysis

- Prepare a melting point sample of your solid. Insert the melting point capillary into the heating block closed end down.
- Turn on the light switch and adjust the voltage control on the MelTemp apparatus to a setting of 3. As the temperature reaches ~75°C, turn the dial down slightly to slow down the rate of temperature increase.
- Observe the melting of the solid. Record the temperature as the solid just begins to melt. Once the solid has completely melted, record the temperature again. Record these two temperatures, separated by a dash, as the melting range of the solid.
- Record this data in the laboratory notebook and complete Table 16.1 on the POST LAB Assignment (provided online).

IR Analysis

- Using the provided spectra, assign all characteristic absorptions of the maleic anhydride and Diels Alder product. Complete Table 16.2 on the POST LAB Assignment (provided online).

Mass Spectral Analysis

- Using the provided spectra, identify the molecular ion peak of the maleic anhydride and Diels Alder product. Complete Table 16.3 on the POST LAB Assignment (provided online).

SAFETY

****All experiments should be performed in a fume hood with appropriate safety glasses. Gloves will be provided by request.****

All chemicals used in this experiment can be toxic if ingested or absorbed through skin. Maleic anhydride and 3-sulfolene can cause severe skin and eye irritation. Xylene, toluene, and petroleum ether are flammable solvents.

WASTE MANAGEMENT

All liquid waste from crystallization should be placed in the container labeled "LIQUID WASTE—DIELS ALDER". After determining the final product mass and melting point, place your crystals in the container labeled "SOLID WASTE—DIELS ALDER".

References

Klein, David. (2015). *Organic Chemistry*, 2nd ed. Hoboken: John Wiley and Sons.
Mass and IR Spectra are from:
http://riodb01.ibase.aist.go.jp/sdbs/cgibin/cre_index.cgi?lang=eng.

Name:			
	Max	Score	Total Grade
Tech	10		
Pre	20		
In	20		
Post	50		

Exp. 16 Diels Alder Cycloaddition

PRE-LAB ASSIGNMENT: *(**EACH STUDENT** will complete and submit an original copy at the beginning of the lab period. **Without a complete pre-lab assignment, you will not be allowed to perform the experiment, and will receive a zero for the lab.**)* .....*max score = 20 pts.*

1. **Objective** *(Write a brief purpose of the experiment in **complete** sentences, addressing all of the following points)*
   - Name the *specific* compounds combined during the synthesis and the name the product formed.
   - What is the purification technique used in this experiment?
   - What analytical technique will be used during the experiment?
   - What type(s) of analysis will be used to characterize the reactant and products?

2. **Chemical Equation** *(Draw the chemical equation for the synthesis of cis-4-cyclohexene-1,2-dicarboxylic anhydride using actual chemical structures.)*

3. **Physical Data** *(Complete the following table before coming to lab.)*

Compound	MW (g/mol)	mp (°C)	bp (°C)	d (g/mL)
3-sulfolene			X	X
maleic anhydride			X	X
xylene	X	X		
toluene	X	X		
petroleum ether	X	X		
cis-4-cyclohexene-1,2-dicarboxylic anhydride			X	X

4. Experimental Outline *(Give a brief description of the procedure that will be followed in this experiment in 5 lines or less.)*

1	
2	
3	
4	
5	

5. Pre-Lab Questions *(Answer the following questions prior to lab.)*

A. In this experiment, why is 3-sulfolene used as the reactant diene instead of simply using butadiene?

B. The *diene* component in a Diels-Alder cycloaddition must be in the *cis* conformation to react.

a. True

b. False

I have read and understood the experimental procedure for this experiment. I am familiar with the hazards and the required disposal procedures for this experiment.

Sign here: ____

Experiment 17

Oxidation of Fluorene to Fluorenone

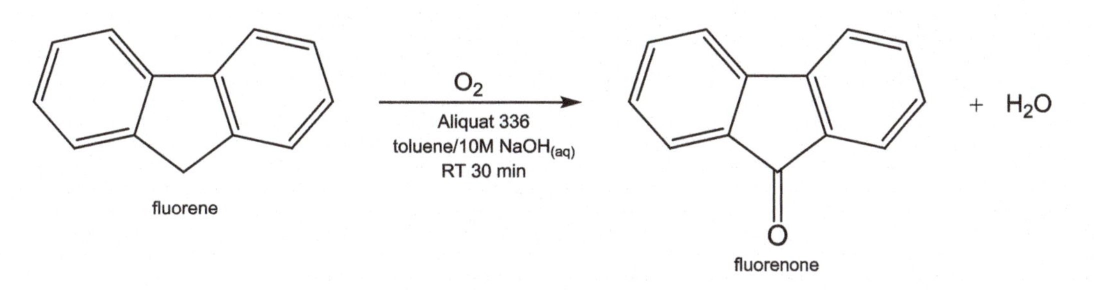

Figure 17.1 Air oxidation fluorene to fluorenone.

Introduction

This experiment is part of a two week experiment. In the first week, you will convert a hydrocarbon to a ketone using air oxidation, in the presence of a phase transfer catalyst. The reaction progress will be monitored by TLC. During the second week of the experiment, the unreacted starting material and the desired product will be separated using a purification technique known as column chromatography. The efficiency of the separation will be determined using HPLC and TLC analysis.

Air Oxidation of Fluorene to 9-Fluorenone

Typical oxidizing reagents such as chromium (VI) compounds and potassium permanganate are often corrosive, toxic, and environmentally damaging; therefore, the development for alternative oxidizing procedures remains an important research goal. One alternate approach is to use atmospheric oxygen as the oxidizing agent in the presence of a phase transfer catalyst such as Aliquat 336 (Stark's catalyst).

In order to oxidize the side chain of an arene, a strong oxidizing agent such as chromium trioxide (CrO_3) or potassium permanganate ($KMnO_4$) must be used. The side chain of fluorene is more reactive than a typical hydrocarbon. The central carbon has an aromatic ring on either side, making the protons on this carbon doubly benzylic. The stability of the benzylic carbocation that forms during the reaction makes these protons particularly reactive. Because of this, the benzylic carbon can be easily oxidized using O_2 and Aliquat 336, along with a strong inorganic base such as NaOH or KOH (Figure 17.1).

Mechanism of Air Oxidation of Fluorene

The mechanism for the air oxidation of fluorene has not been fully established, but the mechanism provided has been proposed (Figure 17.2). Oxygen can exist as an excited state singlet species, or a ground state triplet species. The triplet species of oxygen contains an oxygen-oxygen bond, with two lone pairs of electrons and a radical on each oxygen. Oxygen in the triplet state will readily react with radicals to form new radicals.

O_2 will abstract a benzylic proton radical from fluorene to form a peroxy radical. This is the initiation of a radical reaction process. The peroxy radical can then attach to the fluorene radical to create a peroxy fluorene molecule. Since the oxidation reaction is carried out in aqueous sodium hydroxide, the hydroxide anion abstracts the remaining benzylic proton. As the C-H bond breaks, the bonding electrons shift to form a C-O π bond, releasing a hydroxide anion.

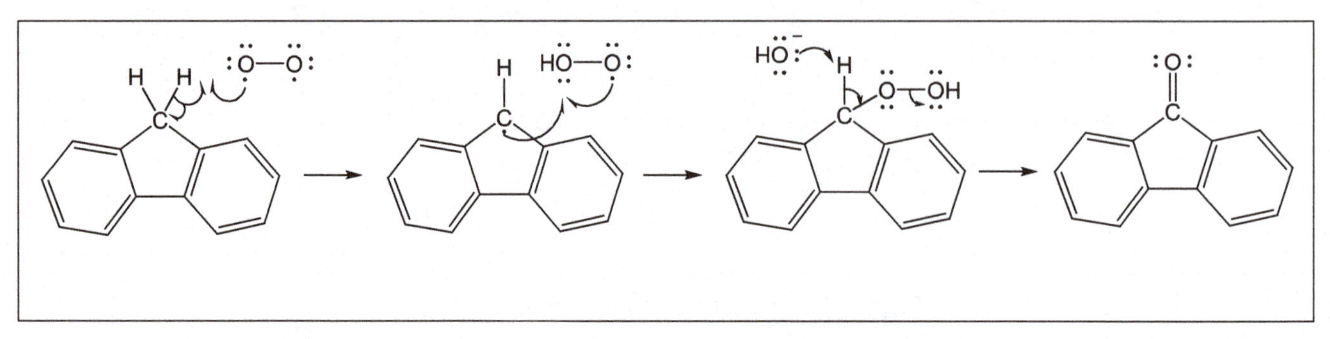

Figure 17.2 Proposed mechanism of oxidation of fluorene.

Phase Transfer Catalysts

Many of the reagents used for oxidation reactions are inorganic salts which are soluble in water, but insoluble in organic solvents. Most organic substrates are insoluble in water, but readily soluble in organic solvents. Aliquat 336 is soluble in both water and organic solvents (Figure 17.3). It is soluble in water because it is ionic, yet still soluble in an organic solvent because it has four nonpolar alkyl chains attached to the central nitrogen atom. The main reason that a phase transfer catalyst is used is to help a reactant migrate from one phase to another where the reaction occurs. This methodology uses a two phase system. An aqueous phase that is a reservoir of reacting anions or base to generate organic anions, and an organic phase containing the organic reactants and catalyst.

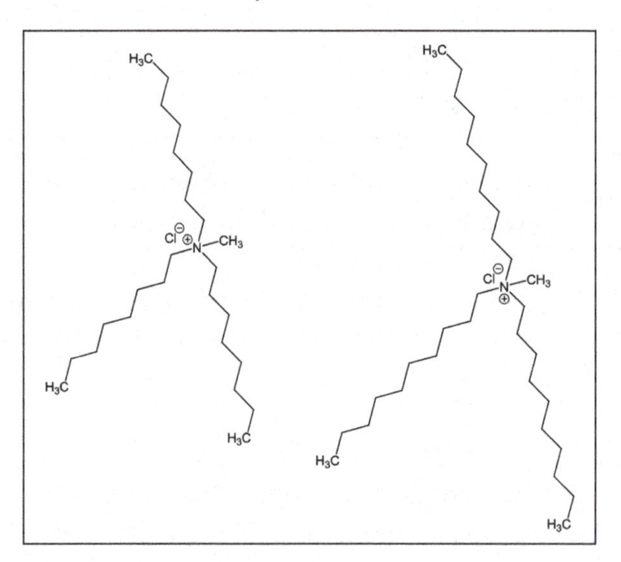

Figure 17.3 Aliquat 336 is a mixture of C_8 and C_{10} quaternary ammonium salts.

There are many benefits to using phase transfer catalysts. Reactions typically proceed faster, generate fewer undesired side products, and achieve higher yields. They are especially important in green chemistry, as a phase transfer catalyst allows the chemist to use water as a reaction solvent in reactions in which an organic solvent would typically be required. The catalyst can be recovered and reused, reducing waste generation.

Objectives

In this experiment you will synthesize 9-fluorenone by air oxidation of fluorene using a phase transfer catalyst. The reaction will be monitored by TLC analysis. You will isolate the product using an extraction technique. Next the reactant, expected product, and product sample will be analyzed using TLC analysis in three different developing solvents in preparation for the column chromatographic separation. The product sample will be saved for purification during the next lab period.

Experimental Procedure

Synthesis:

- Weigh ~0.10 g of fluorene directly into a 50 mL Erlenmeyer flask using a powder funnel. Add a stir bar to the flask.
- Place the reaction flask on a stirrer hotplate. Add 5 mL of 10M NaOH and 5 mL hexane to the flask. Stir vigorously until the solid has dissolved.
- Add 1 drop of Aliquat 336 to the flask while stirring. Stir for ~5 min, then stop the stirring. Perform a TLC analysis on the organic layer ***ONLY*** by dipping the capillary tube into the top hexane layer. Apply this sample solution to the TLC plate alongside the provided standards. Develop the TLC plate in 9:1 hexane/acetone.
- If it appears that your reaction solution contains roughly a 50:50 mixture of reactant to product, remove the flask from the stirrer hotplate. If not, stir another 5 minutes and perform another TLC analysis.

Product Isolation:

- When the reaction appears to be a 50:50 mixture of the reactant: product, transfer the reaction mixture to a separatory funnel. Rinse the reaction flask with 5 mL of hexane, and transfer this rinse to the separatory funnel.
- Drain off the aqueous layer into a 125 mL flask. Wash the organic layer with 10 mL 1.5M HCl. Agitate and vent the separatory funnel. Drain off the aqueous layer. ***Repeat this process two times.***
- Wash the organic layer with 10 mL saturated NaCl. Agitate and vent the separatory funnel. Drain off the aqueous layer. ***Repeat this process two times***.
- Transfer the organic layer to a clean 50 mL flask, and dry over $MgSO_4$. Allow to sit for 5 minutes.
- Weigh a large sample vial containing 2-3 boiling chips. Filter the product solution through a cotton-plugged glass pipet into the large sample vial.
- Transfer 2-3 drops of this solution to a small test tube, and add 1 mL reagent acetone. Set aside this test tube for *PRODUCT ANALYSIS*.

- Clamp this vial to a ring stand, and evaporate the solvent using a hot water bath. When the solvent has evaporated, yielding a yellow solid, reweigh the sample vial to determine the final product mass.
- Complete Tables 17.1 and 17.2 on the POST LAB Assignment (provided online).
- Cap and label the sample vial, then submit the sample to the lab instructor for use during the next lab period. ***PROCEED TO PRODUCT ANALYSIS.***

Product Analysis:

TLC Analysis

- Sketch diagrams of any TLC plates run during the synthesis in the laboratory notebook, including cm measurements of all spots and solvent front. Complete Table 17.3 on the POST LAB Assignment (provided online).

Column Chromatography Preparation

- Prepare 3 TLC plates, with 3 lanes per plate. Apply the fluorene standard to the left lane on each plate, the fluorenone standard on the right lane of each plate, and the previously prepared sample solution in the middle lane of each plate.
- Prepare 3 TLC chambers. In the first chamber, add ~5 mL hexane as the TLC developing solvent. In the second chamber, add ~5 mL of 70:30 hexane/acetone as the developing solvent, and in the third, add ~5 mL of reagent acetone.
- Develop the 3 TLC plates in these solvents. Observe each under a UV lamp after development. Circle and calculate R_f values for any spots which appear on the plate.
- Sketch diagrams of TLC plates in the laboratory notebook, including cm measurements of all spots and solvent front. Complete Table 17.4 on the POST LAB Assignment (provided online).
- Observe the effect that each developing solvent system had on the R_f values of the compounds. This should help understand how a gradient solvent system works during the column chromatography experiment.

SAFETY

****All experiments should be performed in a fume hood with appropriate safety glasses. Gloves will be provided by request.****

Sodium hydroxide and hydrochloric acid are corrosive. Wear gloves when working with them. Hexane and acetone are flammable. Keep away from hot surfaces.

WASTE MANAGEMENT

Place all liquid waste in the container labeled "LIQUID WASTE—OXIDATION". Place used TLC plates in the yellow solid waste trashcan. Place used TLC capillaries and glass pipets in the broken glass container.

References

Klein, David. (2015). *Organic Chemistry*, 2nd ed. Hoboken: John Wiley and Sons.

Lehman, John W. (2009), Multiscale Operational Organic Chemistry, 2nd ed. Upper Saddle River, New Jersey: Pearson/Prentice Hall.

Name:			
	Max	Score	Total Grade
Tech	10		
Pre	20		
In	20		
Post	50		

Exp. 17 Oxidation of Fluorene to Fluorenone

PRE-LAB ASSIGNMENT: *(**EACH STUDENT** will complete and submit an original copy at the beginning of the lab period. **Without a complete pre-lab assignment, you will not be allowed to perform the experiment, and will receive a zero for the lab.**)* *.....max score = 20 pts.*

1. Objective *(Write a brief purpose of the experiment in **complete** sentences, addressing all of the following points.)*

- Name the *specific* compounds combined during the synthesis and name the product formed.
- How is the product isolated in this experiment?
- What analytical technique(s) will be used during the experiment?

2. Chemical Equation *(Draw the chemical equation for the synthesis of 9-fluorenone using actual chemical structures.)*

3. Physical Data *(Complete the following table before coming to lab.)*

Compound	MW (g/mol)	mp (°C)	bp (°C)	d (g/mL)
fluorene			X	X
9-fluorenone			X	X
Aliquat 336	X	X	X	
hexane	X	X		
acetone	X	X		

4. Experimental Outline *(Give a brief description of the procedure that will be followed in this experiment in 5 lines or less.)*

1	
2	
3	
4	
5	

5. Pre-Lab Questions *(Answer the following questions prior to lab.)*

A. Typical oxidizing agents such as $Na_2Cr_2O_7$ and $KMnO_4$ are:

a. toxic
b. environmentally damaging
c. corrosive
d. all of the above

B. Briefly state the main reason that a phase transfer catalyst must be used in this experiment, and give two benefits of using a phase transfer catalyst.

I have read and understood the experimental procedure for this experiment. I am familiar with the hazards and the required disposal procedures for this experiment.

Sign here: ____

Experiment 18

Column Chromatographic Purification of Fluorenone

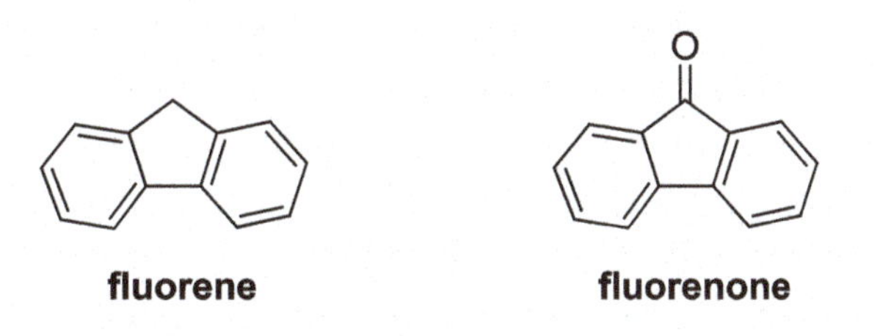

Figure 18.1 Structure of fluorene and fluorenone.

Introduction

In this experiment another important purification technique will be introduced. Column chromatography will be used to isolate the desired product from unreacted starting material remaining after the oxidation of fluorene (Figure 18.1). Although similar in structure, the compounds interact with a polar stationary phase very differently, allowing them to be separated from one another. After the compounds have been separated by column chromatography, the effectiveness of the purification method will be evaluated by TLC and HPLC analysis.

Column Chromatography

Column chromatography is a very common, and extremely valuable, purification technique for synthetic and natural products in organic chemistry. This technique involves the same principle as TLC, but can be used to separate larger quantities than TLC. To understand this technique, we can use the TLC experiment used in the first semester organic chemistry laboratory as an example. In Experiment 5, we separated and analyzed different active ingredients found in over-the-counter analgesics. The TLC experiment allowed us to identify the number and type of compounds in the analgesic, but did not allow us to isolate the compounds and collect them. In order to collect the separated compounds, column chromatography (CC) can be used. In this experiment, column chromatography will be used to separate the desired product from any unreacted starting material in the oxidation of fluorene to fluorenone. Following the column separation, TLC will be used to monitor the success of this separation technique.

There are many ways to do column chromatography, but a general procedure includes preparing the column, loading the column, developing the column, collecting fractions, and analyzing those fractions. The efficiency of separation of the components depends on how the column is prepared.

Choosing a Stationary Phase

Compounds are separated on the column by the same mechanism through which they are separated on a thin layer chromatography plate, through differential intermolecular forces between the components of the mixture with the mobile phase, and between the components with the stationary phase. This is called partitioning. Like TLC, alumina and silica gel are the two most common stationary phases in column chromatography. For these two common stationary phases, partitioning works the same way as TLC. A more polar compound will be retained on the polar stationary phase for longer, whereas a less polar compound will have a lower affinity for the polar stationary phase and travel further with the relatively nonpolar mobile phase (Figure 18.2).

In TLC, the stationary phase is adhered to the surface of a small plate, however in column chromatography, the stationary phase is contained in a glass tube called a column. The mixture of compounds is applied to the stationary phase at the top of the column, solvent is added (mobile phase), and the compounds begin to partition between the stationary and mobile phase, based on their polarity. The mixture of compounds will separate into bands, each compound forming its own band that moves through the column at its own rate. As the bands elute from the column they are collected in separate flasks. Once the compounds have been separated, the solvent is removed. The compounds can be further purified or analyzed spectroscopically.

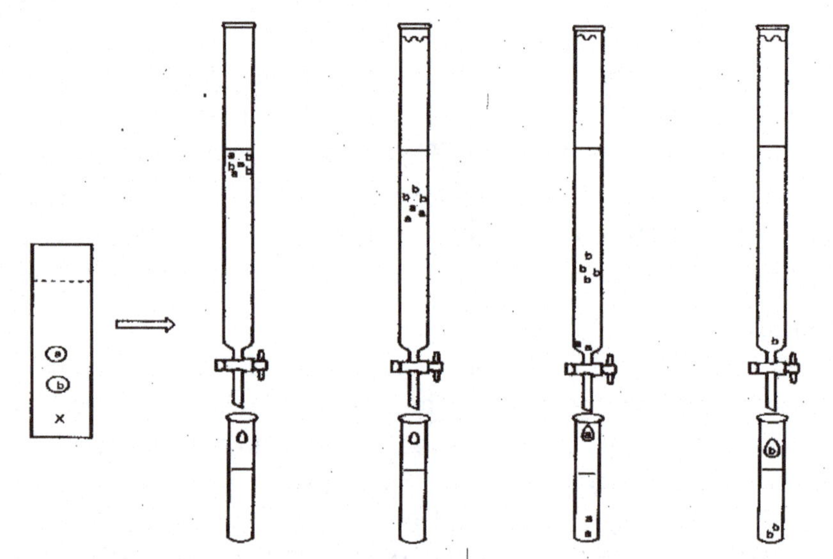

Figure 18.2 Column chromatographic separation of components **a** and **b**.

It is important to note the direction of the solvent flow, and how this affects the order of elution. In TLC, the solvent flows upward, while in CC the solvent flows downward. In TLC, the polar compounds will have lower R_f values, but in CC the polar compounds will be retained in the column for longer. More polar compounds will have lower TLC R_f values, and slower CC elution rates. Less polar compounds will have higher TLC R_f values, and faster CC elution rates.

The type of stationary phase used in column chromatography is typically based on how the compounds separate using that stationary phase in a TLC analysis. This factor, along with the size of the glass column, the polarity of the mobile phase, and the rate of elution can all affect the separation of the compounds. A TLC experiment should always be run prior to using column chromatographic separation to ensure that the proper conditions are used, as it is much less expensive to perform a TLC analysis. If the conditions separate the compounds of interest sufficiently during TLC analysis, the same conditions are generally suitable for column chromatography as well.

Choosing a Solvent System

As mentioned previously, separation of compounds on a silica column parallels their separation on silica TLC sheets. So far, we have used TLC to identify compounds in a mixture and to evaluate the purity of a compound. TLC is also a very valuable technique for determination of a good mobile phase for optimum separation of components using column chromatography, and for analyzing the resulting column fractions.

There are several characteristics of a good solvent system for column chromatographic purification of a component present in a mixture. The optimum solvent will give maximum separation, or resolution, between the compound of interest and closely migrating impurities. Usually maximum separation requires that the components elute neither too quickly nor too slowly from the column. A good solvent for column chromatography is one that results in TLC R_f value between 0.20 to 0.50 for the compound to be purified. Another important characteristic of a good solvent system for column chromatography is that it is easily removed. Low-boiling solvents like acetone, diethyl ether, ethyl acetate and hexane will evaporate quickly, leaving the pure compound for further analysis.

Preparing the Column

A glass column must be uniformly packed with the stationary phase and solvent so that it contains no holes or air bubbles. Two common methods used to pack the column are dry packing and the slurry method. Although the slurry method typically achieves better results, dry packing is the preferred method for a microscale column, and will be used in this experiment.

In order to dry pack the column, the nonpolar solvent is initially added to the column. The powdered stationary phase is added to the column while gently tapping the sides of the column with a pencil or rubber hose. The stationary phase will slowly settle to the bottom of the column. The column must be packed as uniformly as possible, as cracks or air bubbles will greatly affect the quality of the separation. Another way to dry pack a column is to add the stationary phase to the column first, then slowly introducing the nonpolar solvent. This method works best when alumina is used as the stationary phase, since silica gel expands and doesn't pack as well this way.

The slurry method is the preferred method for macroscale columns. In this method, the stationary phase is combined with a small amount of the nonpolar solvent. The solution is mixed until the stationary phase is "suspended" in the solution. Before the stationary phase is able to settle out of solution, it is poured into the column. Regardless of the packing method selected, the most important thing is to create a uniform stationary phase, free of air bubbles and cracks.

Once the column is loaded with the stationary phase the solvent is allowed to run out of the bottom of the column. The solvent should be allowed to flow just to the top of the column packing, but ***NEVER*** below the surface of the packing. Allowing the stationary phase to "run dry" risks the introduction of air bubbles and cracks, resulting in poor separation of compounds.

Loading the Sample

Once the column is packed with stationary phase, the sample can be loaded onto the column. The sample is dissolved in the minimum amount of solvent, so that it is introduced onto the column as a concentrated band. The concentrated sample solution is then added to the surface of the column via pipet, taking great care not to disturb the surface of the column. Once the sample is loaded onto the column, it is allowed to flow into the stationary phase before more solvent is added. The sample flows into the column until the level just reaches the surface of the column packing. A small amount of solvent is then used to rinse any residue from the sides of the column, and this rinse is allowed to add onto the column. Once this rinse has been loaded, the column can be fully developed.

Developing the Column and Sample Collection

To develop the column, the selected solvent (or eluent) is added to the column. This solvent composition can remain constant throughout the entire development stage (isocratic elution) or it can be changed gradually over the course of time (gradient elution).

Normally a separation will begin by preparing the stationary phase, and loading the sample, with a nonpolar or low polarity solvent. This will allow the compounds to adsorb to the stationary phase, and the least polar compound of the mixture will begin to travel slowly down the column. The polarity of the solvent is gradually increased, which will continue to move the less polar compounds, but then begin to move the more polar compounds. The gradual change in solvent polarity allows the separation of components to take place. As the compounds of interest elute through the column, they are collected in separate containers for further analysis. When a gradient solvent system is used, once the less polar compound(s) has been collected, the solvent polarity can be increased to elute the remaining polar components.

Monitoring the Column

If the sample mixture contains colored compounds, the column can be easily monitored. As the colored bands elute down the column, the different colors can be collected in separate containers. However, most organic compounds are colorless, so other detection methods must be employed to monitor the separation.

The objective of a separation technique such as column chromatography is to isolate pure compounds. Ideally, the column chromatographic separation will accomplish this, but in order to determine the quality of this separation other analytical tools, such as TLC and HPLC analysis, must be employed. By performing TLC analyses on the individual fractions collected during the column, the fractions containing pure compounds can be identified and combined.

At times trace amounts of the compounds are not visible during TLC analysis, so further analysis using the HPLC will be used to ensure that the correct fractions were combined in an effort to isolate the pure compounds. The HPLC detector is more sensitive than the human eyes; therefore trace amounts of the compounds may be detected. HPLC can also give the relative purity of a sample based on peak area percent. HPLC analysis of a pure compound will produce a chromatogram containing a single peak, with 100% peak area, whose identity can be determined by comparison to standards run at the same time as the sample (see Appendix E).

Isolation of Compounds

Based on Figure 18.3, it appears that fraction #1 and fraction #2 contain only the higher R_f value compound. These two fractions would be combined in order to isolate this compound. Since fraction #3 appears to contain two compounds, this fraction would be referred to as a "mixed" fraction, and would be discarded, or set aside for further purification. Fraction #4, fraction #5 and fraction #6 could be combined in a separate container to isolate the lower R_f value compound.

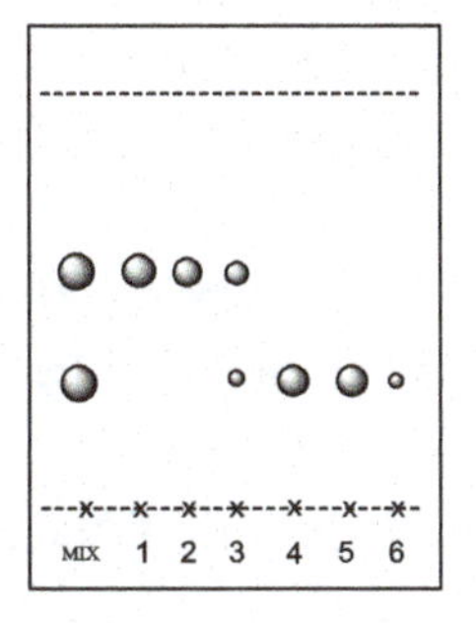

Figure 18.3 TLC diagram of fractions collected during column chromatographic separation.

Objectives

In this experiment the important technique of column chromatography will be introduced. Column chromatography will be used to isolate the desired product from unreacted starting material used in the oxidation of fluorene (Figure 18.1). Although similar in structure, the compounds interact with a polar stationary phase very differently, allowing them to be separated from one another. After the compounds have been separated by column chromatography, the effectiveness of the purification method will be evaluated by TLC and HPLC analysis.

Experimental Procedure

Purification: Column Chromatographic Separation:

Before starting...

- Obtain your sample from the previous lab from your instructor. Dissolve the solid in 1-2 mL hexane.
- Label 4 small test tubes 1-4. Clamp a 5 mL disposable glass pipet (the column) to a ring stand using a 3 prong clamp. Be sure to clamp the column high enough to allow a small test tube underneath.
- Obtain 10 mL hexane in a 25 mL flask. Label and cork the flask.
- Obtain 3 mL 70:30 hexane/acetone in one small test tube, and 3 mL reagent acetone in another. Label and cork both tubes.
- Measure out ~2.5 mL of silica gel into a clean, dry 10 mL graduated cylinder. Take all solvents and the silica gel back to the fume hood.

Preparing the column

- Place the stem of the plastic pipet "funnel" loosely in the top of the column.
- Place the 25 mL hexane flask under the column. Quickly add the solid silica gel to the funnel on top of the column. If the silica gel does not flow freely into the column, the pipet funnel may need to be loosened. As the silica enters the column, tap gently on the sides of the column with a rubber hose to pack the silica gel.
- Fill the column to the "0" mark on the glass pipet column with hexane.

- Using ***gentle*** air pressure, carefully flush the silica gel with hexane. Just as the solvent meniscus approaches the top of the silica bed, refill the column with hexane. Repeat this step two more times. ***DO NOT ALLOW THE SOLVENT LEVEL TO FALL BELOW THE SURFACE OF THE SILICA GEL!***
- After successfully flushing the silica gel with hexane three times, the silica plug should be packed. Add a small layer of sand to the top of the silica gel.
- Just as the solvent meniscus approaches the top of the silica bed, prepare to load the sample solution. ***DO NOT ALLOW THE SOLVENT LEVEL TO FALL BELOW THE SURFACE OF THE SILICA GEL!***

Loading the column

- Using a plastic pipette, add 15-20 drops of the fluorene/fluorenone sample solution. Allow this solution to add into the column as a concentrated band.
- As the sample is loading onto the silica bed, prepare to rinse the insides of the column.
- Just as the sample meniscus just reaches the top of the silica bed, slowly rinse the inside walls of the column with 1 mL of hexane using a swirling motion to remove any sample residue. Allow this column rinse to load into the silica gel. ***DO NOT ALLOW THE SOLVENT LEVEL TO FALL BELOW THE SURFACE OF THE SILICA GEL!***
- As the column rinse is loading onto the silica gel, prepare to add more hexane.

Developing the column

- Just as the column rinse meniscus just reaches the top of the silica bed, slowly fill the column with 5 mL of hexane, and switch the collection flask to **TEST TUBE #1**. Collect all of this eluent in this test tube.
    - Add the first 1-2 mL ***SLOWLY*** so that the top of the silica bed is not disturbed.
    - All of the hexane may not fit into the column at once. If not, continue to add in small amounts until the 5 mL total has been added.
- Just as the solvent meniscus approaches the top of the silica bed, add 3 mL of 70:30 hexane/acetone to the column. Switch to **TEST TUBE #2**. Continue to collect the eluent in this test tube until the yellow band approaches the bottom of the column.
- As the yellow band reaches the cotton at the bottom of the column, switch to **TEST TUBE #3**.
- As the solvent meniscus reaches the top of the silica bed, add 3 mL reagent acetone and switch to **TEST TUBE #4**. Air can be used to evacuate the remaining solvent to complete the column.

Analyzing fractions by TLC

- Prepare a TLC plate. Measure ~1 cm from the bottom of the plate and make a fine line using a pencil. This is the origin line on which you will apply the sample solutions (fractions).
- Lightly draw 5 small marks on the origin line equidistant apart. Label the lanes M (original sample), 1, 2, 3 and 4.
- Apply the original sample solution, along with the fractions collected, on the TLC plate. Be sure to use a clean capillary tube for each solution applied.

- Using a UV light, check the concentration of the sample solutions. As long as a dark spot appears on the origin line, the solution should be concentrated enough to continue. If any lane shows only a faint spot, apply a small amount more of the required sample to the lane and check again.
- To prepare the TLC chamber, place a small truncated filter paper into the TLC chamber. Add 5 mL of the TLC developing solvent (90:10 hexane/acetone). Cap the TLC chamber securely and swirl the chamber to saturate the filter paper with solvent.
- Using tweezers, carefully place the TLC plate into the chamber. Place the cap back onto the chamber, and do not disturb the TLC chamber during development.
- Allow the solvent to wick up the plate until the solvent reaches ~1 cm from the top of the plate. Remove the plate and quickly mark the level that the solvent traveled (solvent front) with a pencil.
- Once the TLC plate has dried of solvent, visualize the plate using the UV lamp. Circle all spots observed in each lane. Mark the center of each spot with a small pencil mark.
- Measure the solvent front distance, as well as the distance traveled by each spot, in cm. Use these values to calculate the R_f value of any spot present in each fraction lane.
- Sketch the TLC plate in your laboratory notebook. Be sure to indicate the cm measurements for all spots and the solvent front in your sketch. Complete Table 18.1 on POST LAB Assignment (provided online) using this data.

<u>Combining fractions and isolating compounds</u>

- Once the TLC plate has been developed, identify which fractions contain pure compounds. Combine the fractions containing pure fluorene in a 50 mL flask containing 2-3 boiling chips.
- Place the flask in a 70°C water bath to evaporate the solvent. **<u>DO NOT PUT THE FLASK DIRECTLY ON THE HOTPLATE!</u>**

Combine the fractions that contain pure fluorenone in a clean 50 mL flask containing 2-3 boiling chips. Evaporate the solvent using the hot water bath. ***<u>PROCEED TO PRODUCT ANALYSIS.</u>***

<u>Product Analysis:</u>

<u>HPLC Analysis</u>

- Prepare separate HPLC samples of the purified fluorene and fluorenone. Add 1 mL of HPLC solvent (90:10 hexane/acetone) to the flask to re-dissolve the solid sample.
- Transfer 5 drops of this sample to a small auto analyzer vial. Add another 1 mL of HPLC solvent directly to the sample vial. Cap the vial and shake to ensure that all solid is completely dissolved.
- Once the HPLC chromatograms have been returned, identify the compounds by comparing sample retention times to standard retention times. Complete Table 18.2 on POST LAB Assignment (provided online) using this data.
- Your separation was successful if there is only **<u>ONE</u>** compound present per sample chromatogram.

SAFETY

All experiments should be performed in a fume hood with appropriate safety glasses. Gloves will be provided by request.

Hexane and acetone are flammable. Keep away from hot surfaces. Silica gel can irritate lungs if inhaled. Be sure to keep silica gel in the fume hood at all times.

WASTE MANAGEMENT

Place all liquid organic waste into the container marked "LIQUID WASTE—COLUMN." Place used TLC capillaries in the broken glass container. Place used TLC plates in the yellow solid waste trashcan under the supply hood. Place used columns in the box in the waste hood for reuse.

References

Klein, David. (2015). *Organic Chemistry*, 2nd ed. Hoboken: John Wiley and Sons.
Lehman, John W. (2009), Multiscale Operational Organic Chemistry, 2nd ed. Upper Saddle River, New Jersey: Pearson/Prentice Hall.

Name:			
	Max	Score	Total Grade
Tech	10		
Pre	20		
In	20		
Post	50		

Exp. 18 Column Chromatographic Purification of Fluorenone

PRE-LAB ASSIGNMENT: *(**EACH STUDENT** will complete and submit an original copy at the beginning of the lab period. **Without a complete pre-lab assignment, you will not be allowed to perform the experiment, and will receive a zero for the lab.**) .....max score = 20 pts.*

1. **Objective** *(Write a brief purpose of the experiment in **complete** sentences, addressing all of the following points)*
   - What purification technique will be introduced in this experiment?
   - What goal will be accomplished using this technique?
   - How will success of the purification be determined?

2. **Chemical Structures** *(Draw the chemical structures of all compounds below)*

fluorene	fluorenone	*n*-hexane	acetone

3. **Physical Data** *(Complete the following table before coming to lab.)*

Compound	MW (g/mol)	bp (C°)	d (g/mL)
acetone			
n-hexane			
fluorene		XXX	XXX
fluorenone		XXX	XXX

4. **Experimental Outline** *(Give a brief description of the procedure that will be followed in this experiment in 5 lines or less.)*

1	
2	
3	
4	
5	

5. **Pre-Lab Questions** *(Answer the following questions prior to lab.)*

A. List the specific identities of the ***three*** solvent systems used to develop the column, in order from least to most polar:

a. **LEAST**:
b. **MID**:
c. **MOST**:

B. How would the column separation be affected if you prepared and loaded the column with the ***MOST*** polar solvent first, instead of the least polar solvent?

I have read and understood the experimental procedure for this experiment. I am familiar with the hazards and the required disposal procedures for this experiment.

Sign here: ____

Experiment 19

Substituent Effects on the Rate of Electrophilic Aromatic Substitution

Introduction

This experiment is designed to show in a qualitative way which different substituent groups (-G) on an aromatic ring affect the rate of electrophilic aromatic substitution. The substitution reaction that will be investigated is a bromination using a solution of bromine in acetic acid, selected because the progress of the reaction can be followed easily by a color change (Figure 19.1). The more reactive the aromatic ring is, the faster the color will disappear.

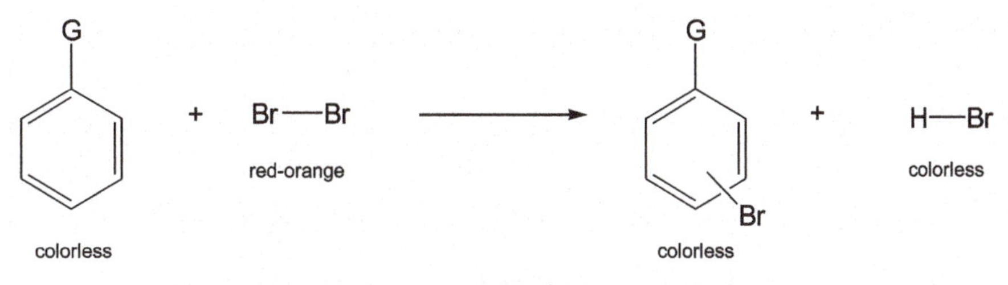

Figure 19.1 Bromination of monosubstituted aromatic ring.

Substituents affect the reactivity of the aromatic ring. Substituents can make the ring more reactive than benzene or less reactive than benzene in reaction with electrophiles, depending on how they interact with the carbocation intermediate (stabilizing it or destabilizing it relative to H).

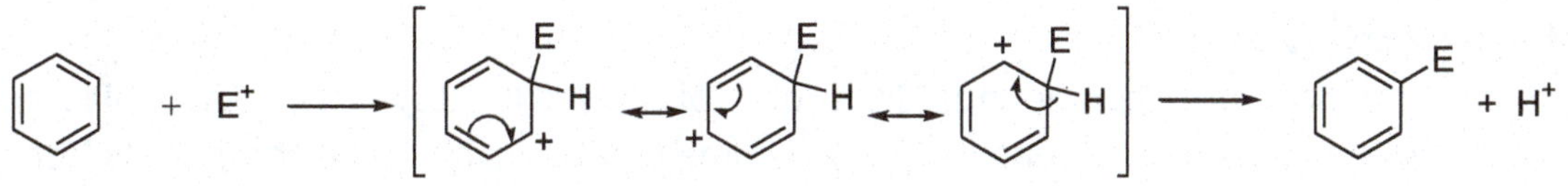

Figure 19.2 Formation of carbocation intermediate in aromatic ring.

Substituents can also affect the orientation of the reaction (*o*-, *m*-, or *p*-). The substituent *already on* the aromatic ring determines the position and rate of substitution of the second (*incoming*) electrophile. If we use the reactivity of benzene (substituent = H) as a reference point, activating substituents are all electron-donating groups, and their relative activation strengths are:

H < Phenyl < CH_3 < $NHCOCH_3$ < OCH_3 < OH < NH_2

Less Reactive ⟹ More Reactive

Deactivating groups are electron-withdrawing groups and their activities relative to hydrogen are:

NO_2 < COR < CHO < I < Br < Cl < F < **H**

Less Reactive ⟹ More Reactive

Objectives

In this experiment you will explore how different substituent groups on an aromatic ring affect the rate and orientation of electrophilic aromatic substitution using a qualitative bromine test. You will then determine the directing ability of the acetyl amino group using TLC analysis.

Experimental Procedure

Synthesis:

- Prepare a water bath in a 250 mL beaker. Place this beaker on a hotplate and begin heating to a temperature of 70-80°C (hotplate on 4).
- Label 6 small test tubes #1-6. Place 0.5 mL of the following aromatic compounds in the appropriate test tube:
  1. ethylbenzene
  2. benzaldehyde
  3. phenol
  4. anisole
  5. nitrobenzene
  6. acetanilide
- Add 0.5 mL of the bromine solution to each tube, noting the time of addition.
- Watch the tubes for a color change, and record in a table the time it takes for the red-orange color to disappear. If no color change is noticeable within 10 minutes, place the test tubes in a 70–80°C water bath to increase the rate of the reaction. Continue to watch for the color change for 60 minutes. ***PROCEED TO PRODUCT ANALYSIS.***

Product Analysis:

Rate of Reactivity

- Rank the aromatic compounds in order of reactivity toward electrophilic aromatic substitution (list the most reactive first). If some test tubes have not completely lost the red-orange color at the end of 60 minutes, estimate the order of reactivity by the relative amount of color loss. Record this data in the laboratory notebook, and complete Table 19.1 of the POST LAB Assignment (provided online).

TLC Analysis

- Investigate the orientation effect of an amide group.
- After the acetanilide tube has turned pale yellow, allow it to cool to room temperature.
- Add 10 drops of 20% NaOH to the acetanilide tube, then stir gently with the glass rod to mix. Test the pH. Continue adding the base in 10 drop increments until pH ~5. Then, reduce to 1-2 drop additions, testing pH after each addition until the solution is just basic.
- Add 3 mL of ethyl acetate to the test tube. Place a small cork on the test tube and agitate. Allow the layers to separate while you prepare for TLC analysis.
- Prepare a single TLC plate with four lanes. Add ~5 mL of 3:1 ethyl acetate: hexane to the TLC chamber, along with a truncated filter paper. Allow the solvent to completely saturate the filter paper to equilibrate the system. To the TLC plate, apply the provided standard solutions of *o-, m-* and *p*-bromoacetanilide. In the fourth lane, apply a small amount of the TOP LAYER of the acetanilide sample solution from the test tube.

- Develop the TLC plate. Visualize under the UV lamp. Calculate R_f values of spots. Record this data and sketch the TLC plate in the laboratory notebook. Complete Table 19.2 on the POST LAB Assignment (provided online).

SAFETY:

All experiments should be performed in a fume hood with appropriate safety glasses. Gloves will be provided by request.

All chemicals used in this experiment can be toxic if ingested or absorbed through skin. Bromine is especially toxic and corrosive, as are sodium hydroxide and acetic acid. Gloves are highly recommended when using these compounds. Ethyl acetate and hexane are flammable.

WASTE MANAGEMENT

Place all liquid waste in container labeled "LIQUID ORGANIC WASTE—EAS" Located in waste hood. Place used TLC capillaries in broken glass container and TLC plates in yellow trashcan.

Reference

Klein, David. (2015). *Organic Chemistry*, 2nd ed. Hoboken: John Wiley and Sons.

Name:			
	Max	Score	Total Grade
Tech	10		
Pre	20		
In	20		
Post	50		

Exp. 19 Substituent Effects on the Rate of Electrophilic Aromatic Substitution

PRE-LAB ASSIGNMENT: *(**EACH STUDENT** will complete and submit an original copy at the beginning of the lab period.* ***Without a complete pre-lab assignment, you will not be allowed to perform the experiment, and will receive a zero for the lab.****)* *.....max score = 20 pts.*

1. **Objective** *(Write a brief purpose of the experiment in **complete** sentences, addressing all of the following points.)*
   - Name the *specific* aromatic compounds used in this experiment.
   - What concept will be explored in this experiment, and how?
   - What is the purification and what analytical technique will be used on the acetanilide product during the experiment?

2. **Chemical Equation** *(Draw the chemical equation for bromination of a generic monosubstituted aromatic ring using actual chemical structures.)*

3. **Physical Data** *(Complete the following table before coming to lab.)*

Compound	ethyl benzene	anisole	acetanilide	phenol	benzaldehyde	nitrobenzene
Structure						

4. Experimental Outline *(Give a brief description of the procedure that will be followed in this experiment in 5 lines or less.)*

1	
2	
3	
4	
5	

5. Pre-Lab Questions *(Answer the following questions prior to lab.)*

A. In a separatory funnel containing ethyl acetate and 10% aqueous sodium bicarbonate ($NaHCO_3$), which layer would contain the brominated aromatic compound?

a. Top layer
b. Bottom layer

B. Using the TLC diagram shown, calculate the R_f value of the ***most*** polar spot in the sample lane:

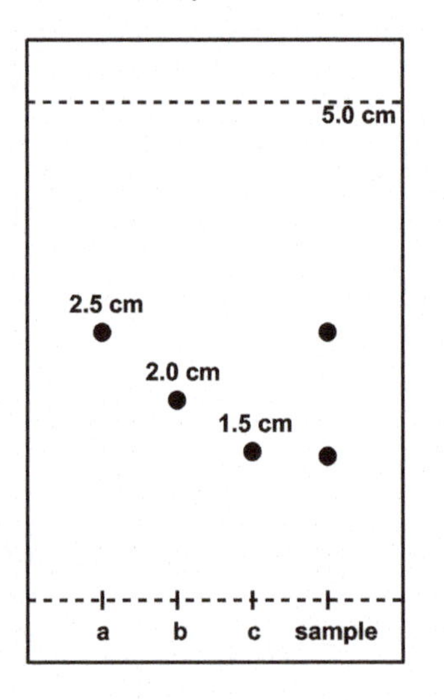

I have read and understood the experimental procedure for this experiment. I am familiar with the hazards and the required disposal procedures for this experiment.

Sign here: ______________________________

Experiment 20

Nitration of Acetanilide

Introduction:

Substituents on aromatic rings can be introduced by electrophilic or nucleophilic substitution reactions. Understanding the reactivity of aromatic compounds is important in the design and synthesis of new chemicals, especially in pharmaceutical, dye, and polymer chemistry. Over the course of two lab periods, you will perform an electrophilic aromatic nitration of acetanilide. You will purify and isolate your product by recrystallization. During the second lab period, identification and purity of the product will be determined based on melting point analysis, TLC analysis, and HPLC analysis compared to standards. Finally, the reactant and major product will be characterized using provided with ^{1}H- and ^{13}C-NMR spectra.

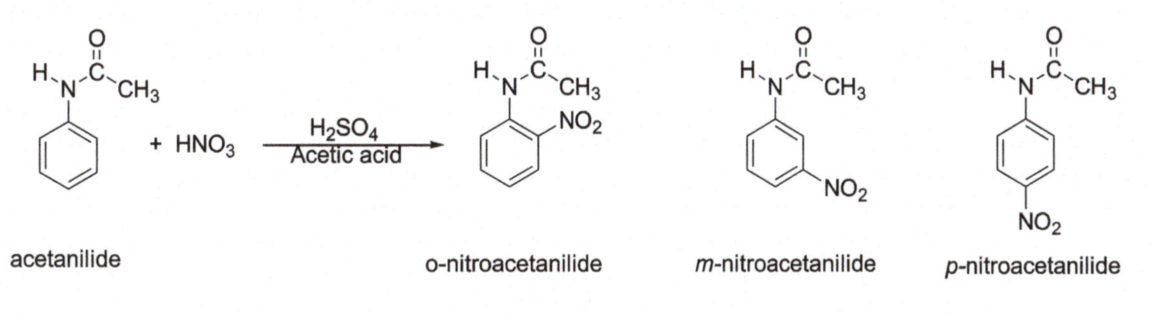

Figure 20.1 Nitration of acetanilide to form three isomeric nitroacetanilide products.

Electrophilic aromatic substitution

Nitration of acetanilide is an example of an electrophilic aromatic substitution reaction. The reaction takes place in two steps, initial reaction of an electrophile (E^{+}) with the aromatic ring, followed by loss of H^+ from the resonance stabilized carbocation intermediate to regenerate the aromatic ring (Figure 20.2). Substituents on the aromatic ring affect both the reactivity of the ring toward substitution and the orientation of that substitution. Is the acetyl amino functionality of the acetanilide an *activating* or a *deactivating* substituent? You should be able to draw resonance structures for the intermediate cation to help explain your answer.

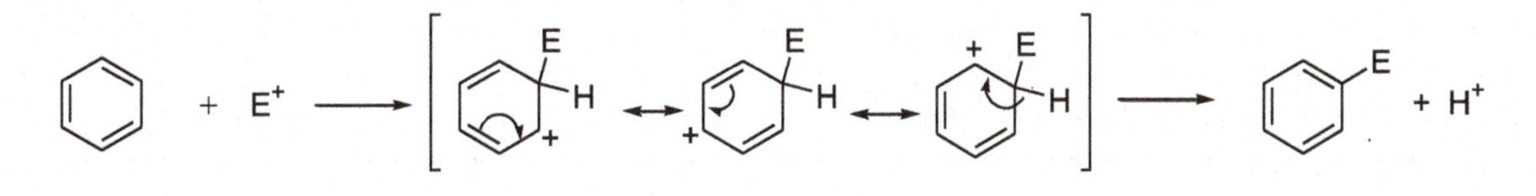

Figure 20.2 Electrophilic aromatic substitution.

In electrophilic nitration of an aromatic ring, the electrophile is the nitronium ion, NO_2^+, which is generated from nitric acid (70% HNO_3 and 30% H_2O) after protonation by sulfuric acid and loss of water. The nitronium ion reacts with the aromatic ring to yield a resonance stabilized carbocation intermediate. Loss of H^+ gives the substitution product.

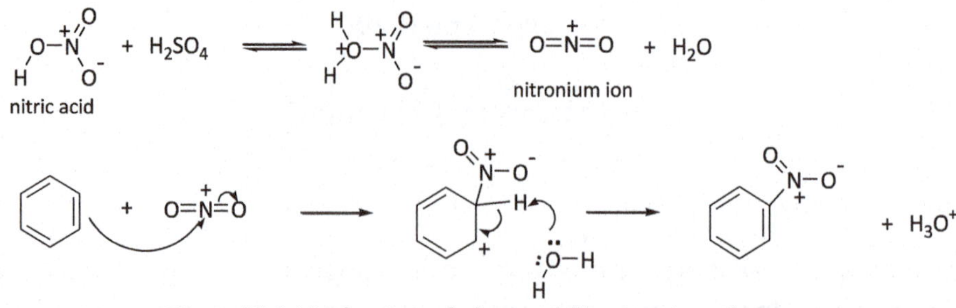

Figure 20.3 Mechanism of electrophilic nitration of benzene.

NMR Spectroscopy

The NMR spectra of acetanilide and the major nitration product from this experiment are provided in Figures 20.4 and 20.5. Note that the major changes occur in the aromatic region. The symmetry observed in a para substituted aromatic compound results in fewer signals than expected in the aromatic region. However, once the nitro group has added onto the ring, this symmetry no longer exists. Each aromatic carbon atom and hydrogen atom generates a distinct signal.

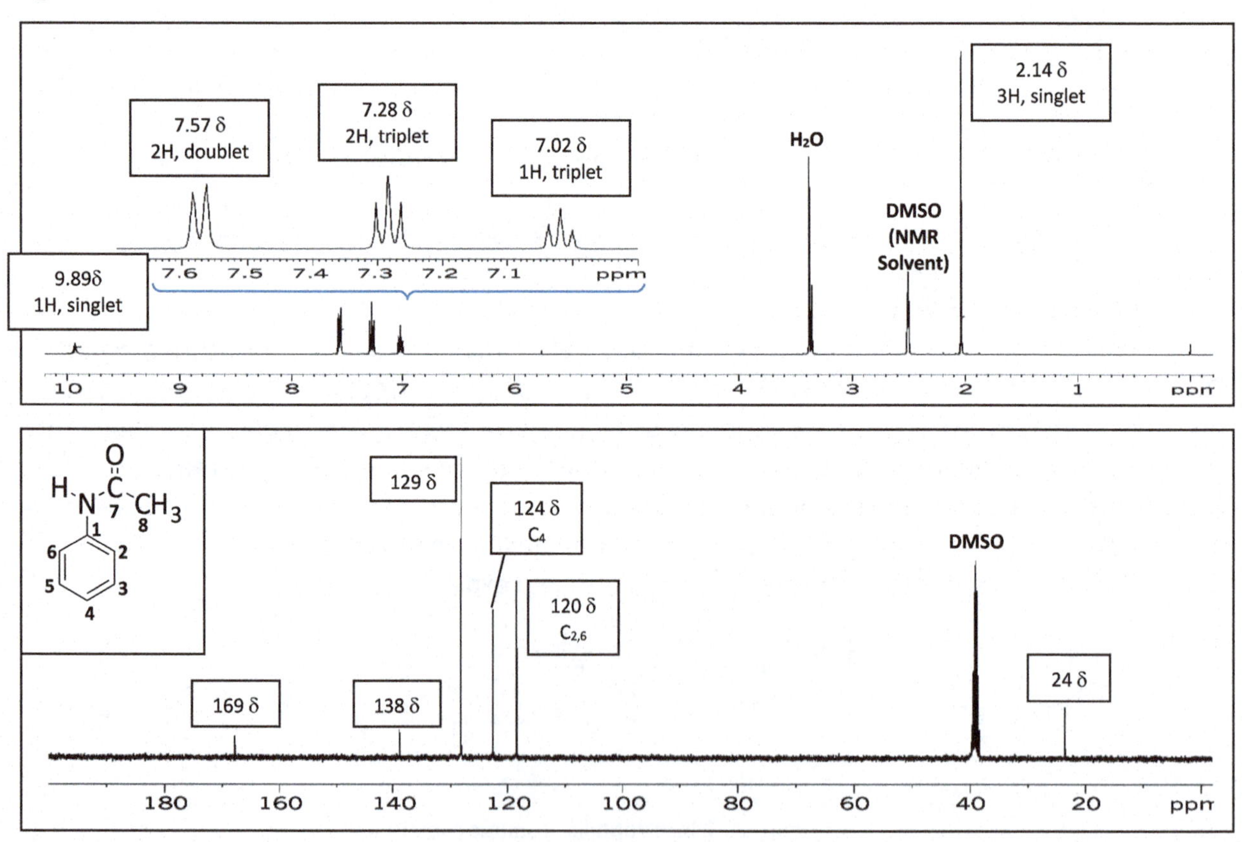

Figure 20.4 ^{1}H and ^{13}C NMR of acetanilide.

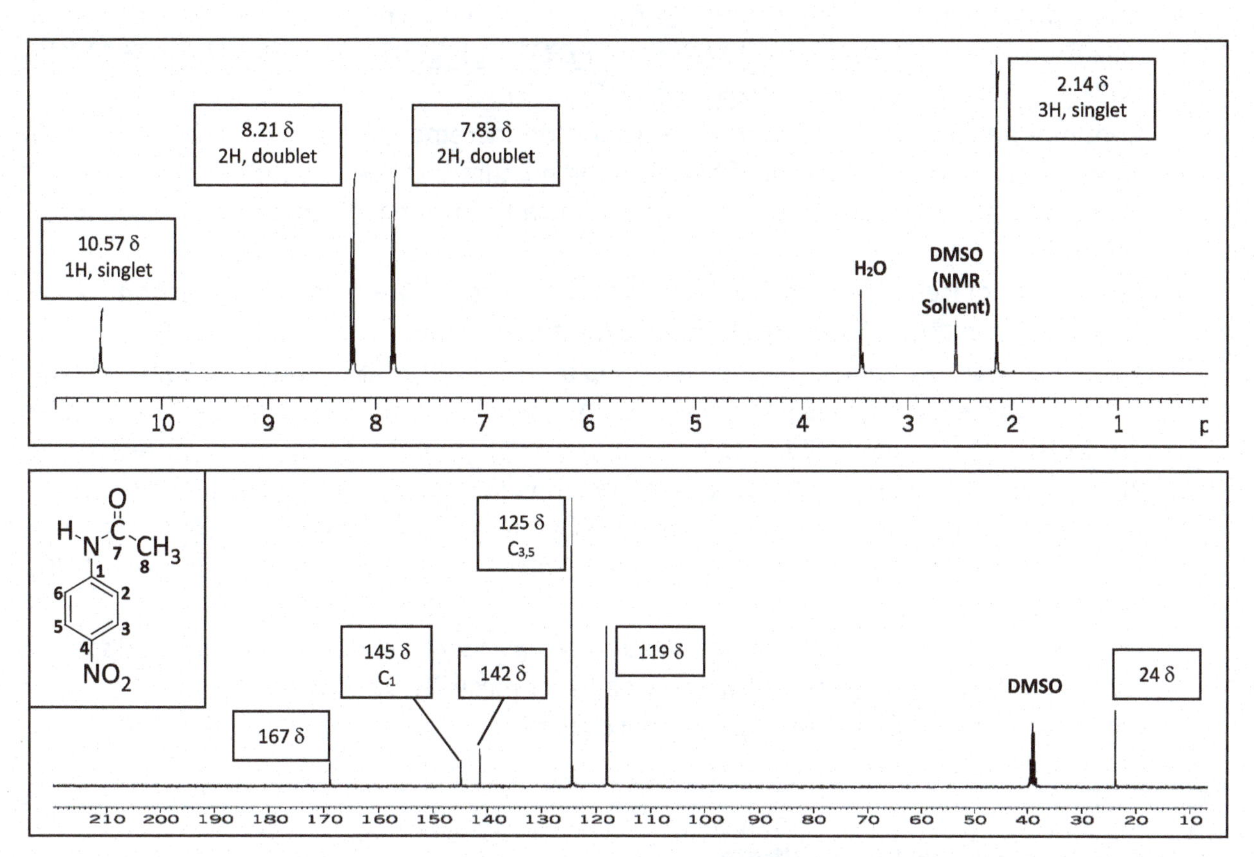

Figure 20.5 ^{1}H and ^{13}C NMR of major product formed from nitration of acetanilide.

Objectives

This is a two-week experiment. In lab the first week, you will perform a nitration of acetanilide. Once synthesized, the crude solid product will be isolated using vacuum filtration. The liquid filtrate will be neutralized and extracted for further analysis. The crude solid will then be purified by recrystallization. TLC and HPLC samples of the crude solid, neutralized filtrate, and recrystallized solid will be submitted for analysis. During the second week, the purity of the crude solid, filtrate, and recrystallized solid will be determined using melting point, TLC, and HPLC analysis. Finally, NMR spectroscopy will be used to distinguish between reactants and products.

Experimental Procedure

Synthesis:

- Add ~1.0 g of ground acetanilide to 2 mL of glacial acetic acid in a 50 mL Erlenmeyer flask. Warm this flask on a warm hotplate until the acetanilide is just dissolved.
- Clamp the reaction flask to a ring stand. Lower the flask into a beaker of tap water.
- Carefully add 2 mL of concentrated H_2SO_4 to the reaction flask. Stir gently with a glass rod.
- Mix 1 mL concentrated HNO_3 and 1 mL concentrated H_2SO_4 together. This constitutes the nitrating mixture. Cool this nitrating solution in the ice water bath for a few seconds before use. ***USE EXTREME CAUTION WHEN USING THIS SOLUTION!***

- Add 5 drops of the nitrating mixture to the reaction flask while in the water bath. Stir to mix the solution. ***(CAUTION!*** ***KEEP PIPET PERFECTLY VERTICAL WHEN TRANSFERRING NITRATING MIXTURE TO PREVENT INJURY!)***
- Continue adding the nitrating mixture in 5 drop increments to the reaction flask, with intermittent stirring. Once all of the nitrating mixture has been added, remove the flask from the water bath, and allow the reaction mixture to stand at room temperature for 10 minutes.
- Pour 30 mL of ice-cold deionized water into the reaction flask. Immerse reaction flask in an ice water bath for 5 minutes to induce crystal formation.
- While the reaction solution is cooling, set up a suction filtration apparatus. Seat the filter paper with ice-cold deionized water, and apply the vacuum. Filter the reaction mixture to isolate the crude solid. Use as much cold deionized water as necessary to quantitatively transfer the solid from the reaction flask and rinse the solid.
- Prepare an HPLC sample of your *crude solid* by placing a few crystals of your solid in an auto sampler vial and adding ~1 mL of HPLC solvent, 2:1 ethyl acetate/hexane. (**CRUDE HPLC SAMPLE**).
- Prepare a TLC sample of your *crude solid* by placing a few crystals in a small labeled sample vial. Add ~1 mL of reagent acetone and shake to mix (**CRUDE TLC SAMPLE**).
- Set crude solid aside to purify by recrystallization.

Purification:

Neutralization and Extraction of Filtrate

- Transfer ~2 mL of the liquid filtrate from the suction flask to a small test tube.
- Add 5 drops of 20% NaOH to the filtrate to neutralize. Check pH of solution with pH-Hydrion paper. Continue adding the base in five drop increments until the solution is pH 5, checking the pH after each addition. At pH 5, add only single drops of the NaOH, checking the pH after every addition.
- Once neutral, add 3 mL of ethyl acetate. Seal the test tube with a cork and shake to mix. Allow layers to separate completely.
- Prepare a TLC sample of the *liquid filtrate* by filling a labeled sample vial ½ full with the top organic layer (**FILTRATE TLC SAMPLE**).
- Prepare an HPLC sample of the *liquid filtrate* by filling an auto sampler vial ½ full with the top organic layer (**FILTRATE HPLC SAMPLE**).

Purification of Crude Solid

- Transfer the crude solid from above to a clean 50 mL flask using a powder funnel. Rinse any remaining solid from the filter paper by holding the filter over the funnel with tweezers, and rinsing with ~5 mL of 95% ethanol.
- Place the flask on a warm hotplate (**NO HIGHER THAN 3!**) and heat until the solvent just begins to boil. Add more 95% ethanol only if necessary to completely dissolve the crystals, but **DO NOT EXCEED 10 mL.**
- Remove the flask from the hotplate and allow the flask to cool to room temperature slowly. If crystals do not appear by the time the solution reaches room temperature, it

may be necessary to scratch the inside of the flask with a glass rod to induce crystallization.

- Place the flask in an ice water bath for a *minimum* of 10 minutes to ensure complete crystallization.
- Set up a suction filtration apparatus. Seat the filter paper with ice-cold ethanol. Filter the reaction mixture to isolate the crude solid. Use as much cold ethanol as necessary to quantitatively transfer the solid from the flask and rinse the solid.
- Prepare an HPLC sample of your *recrystallized solid* by placing a few crystals of your solid in an auto sampler vial and adding ~1 mL of HPLC solvent. (**RECRYSTALLIZED HPLC SAMPLE**). Be sure that all solid is dissolved; otherwise the sample will be discarded.
- Prepare a TLC sample of your *recrystallized solid* placing a few crystals in a small labeled sample vial. Add ~1 mL of reagent acetone and shake to mix (**RECRYSTALLIZED TLC SAMPLE**).
- Transfer the small filter paper with the purified product to a larger preweighed filter paper and submit to instructor to dry until the next lab period.
- Complete Tables 20.1 and 20.2 on the POST LAB Assignment (provided online).
- At the beginning of the next lab period, obtain the dried solid, determine final mass and calculate percent yield. Record this data in the laboratory notebook, and complete Tables 20.4 and 20.5 on the POST LAB Assignment (provided online).
- ***PROCEED TO PRODUCT ANALYSIS.***

Product Analysis:

HPLC Analysis

- Obtain the chromatographic results of the crude solid, filtrate, and recrystallized solid.
- Using the standard chromatogram, identify the compounds present in your sample.
- Once the compounds are identified, determine the degree of purity of your product and identify possible sources of product loss.
- Complete Table 20.3 on the POST LAB Assignment (provided online).

Melting Point Analysis

- After the product is sufficiently dry, prepare a melting point capillary of your recrystallized solid.
- Determine the experimental melting range and compare this value with the literature value of the melting point of the major product in order to determine the degree of purity. Record the melting range in Table 20.4 on the POST LAB Assignment.

TLC Analysis

- Perform a TLC analysis on the previously prepared samples against provided standards. Prepare a single TLC plate with 6 lanes.
- Apply the provided standards of acetanilide, *o*-nitroacetanilide and *p*-nitroacetanilide. Apply the previously prepared crude solid, filtrate, and recrystallized solid samples.

- Check the TLC plate under a UV lamp prior to development to ensure that samples will be visible.
- Develop the TLC plate in 2:1 ethyl acetate/hexane. Visualize spots using UV lamp.
- Circle all spots, and determine the TLC R_f values. Sketch a diagram of the TLC plate in the lab notebook, with all spots identified and cm measurements of all spots and solvent front given. Complete Table 20.6 on the POST LAB Assignment (provided online).

NMR Characterization

- Using the provided ^{1}H-NMR and ^{13}C-NMR spectra of the reactant and major product (Figures 20.4 and 20.5), identify signals of the assigned protons and carbons. Complete Table 20.7 and Table 20.8 on the POST LAB Assignment (provided online).

SAFETY

****All experiments should be performed in a fume hood with appropriate safety glasses. Gloves will be provided by request.****

All organic solvents used in this experiment are flammable, irritants, and can be toxic if ingested or absorbed through skin. Sulfuric acid, nitric acid, and acetic acid are extremely corrosive. The solid reactant and products are irritants, and can be toxic if ingested or absorbed through skin.

WASTE MANAGEMENT

Place all liquid waste in the bottle labeled "LIQUID WASTE—NITRATION". Place solid product in bottle labeled "SOLID ORGANIC WASTE". Place all used TLC spotters and melting point capillary tubes in the broken glass container.

Reference

Klein, David. (2015). *Organic Chemistry*, 2nd ed. Hoboken: John Wiley and Sons.

Name:			
	Max	Score	Total Grade
Tech	10		
Pre	20		
In	20		
Post	50		

Exp. 20 Nitration of Acetanilide

PRE-LAB ASSIGNMENT: *(**EACH STUDENT** will complete and submit an original copy at the beginning of the lab period. **Without a complete pre-lab assignment, you will not be allowed to perform the experiment, and will receive a zero for the lab.**) .....max score = 20 pts.*

1. **Objective** (*Write a brief purpose of the experiment in **complete** sentences, addressing all of the following points.*)
   - Name the *specific* compounds combined during the synthesis and name the product(s) formed.
   - What is the purification technique used in this experiment on the solid?
   - What analytical techniques will be used during the experiment to identify determine purity of product?
   - What type of spectral analysis will be used to characterize the reactant and product?

2. **Chemical Equation** (*Draw the chemical equation for nitration of acetanilide using actual chemical structures.*)

3. **Physical Data** (*Complete the following table before coming to lab.*)

Compound	MW (g/mol)	mp (°C)	bp (°C)	d (g/mL)
acetanilide			X	X
70% nitric acid		X		
sulfuric acid	X	X	X	
acetic acid	X	X		
ethanol	X	X		
ethyl acetate	X	X		
hexane	X	X		
2-nitroacetanilide			X	X
4-nitroacetanilide			X	X

4. Experimental Outline *(Give a brief description of the procedure that will be followed in this experiment in 5 lines or less.)*

1	
2	
3	
4	
5	

5. Pre-Lab Questions *(Answer the following questions prior to lab.)*

A. Three acids are used in the synthesis of the product, all having a different function. ***Circle*** the correct function of the acids below:

a. Sulfuric acid: REACTANT CATALYST SOLVENT

b. Nitric acid: REACTANT CATALYST SOLVENT

c. Acetic acid: REACTANT CATALYST SOLVENT

B. What is the ***directing*** effect of the amide group on the aromatic ring?

a. *ortho/para*

b. *meta*

I have read and understood the experimental procedure for this experiment. I am familiar with the hazards and the required disposal procedures for this experiment.

Sign here: ______________________________

Experiment 21

Esters: Synthesis and Fragrance

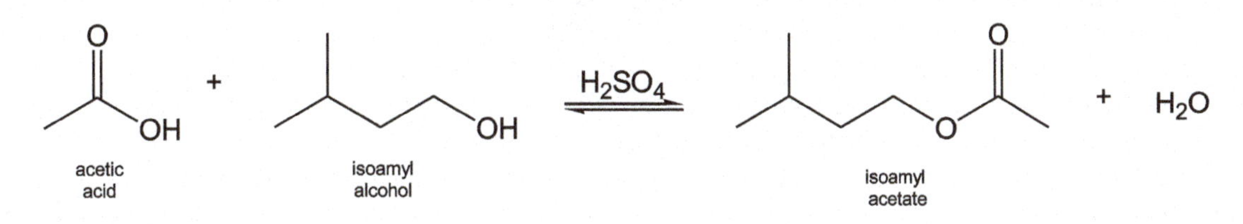

Figure 21.1 Acid-catalyzed esterification of an alcohol.

Introduction

Conversion of carboxylic acids to carboxylic esters is an important reaction in organic chemistry. Several methods for ester preparation include nucleophilic acyl substitution between RCO_2H and ROH (Figure 21.1), an S_N2 reaction between the carboxylate anion (RCO_2^-) and a primary alkyl halide, and a reaction of the acid with diazomethane.

Free carboxylic acids are typically not reactive enough for nucleophilic acyl attack, but are more reactive in the presence of strong mineral acid (H_2SO_4). The net effect is substitution of an -OH group by -OR'. In such acid-catalyzed esterification reactions, *all steps are reversible* and the reaction can be driven in either direction by changing conditions (Figure 21.2). Ester (RCO_2R') formation is favored when a large excess of the alcohol (ROH) is used. Acid (RCO_2H) formation is favored when a large excess of H_2O is present.

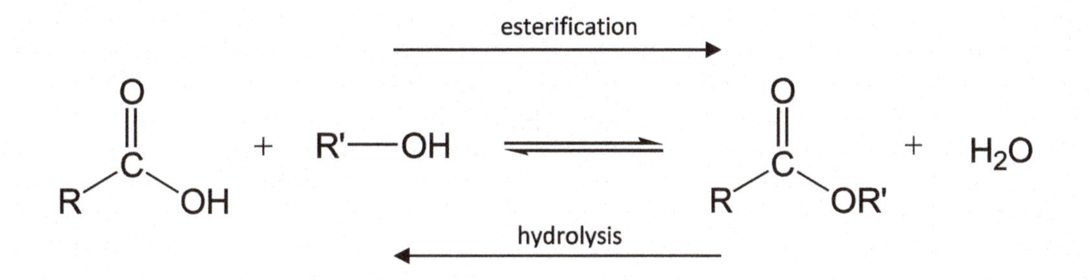

Figure 21.2 Esterification vs. hydrolysis.

You will synthesize an ester through an acid-catalyzed esterification. You have learned that in an acid-catalyzed esterification all reaction steps are reversible and the reaction can be driven in either direction by changing conditions. Ester (RCO_2R') formation is favored when a large excess of either the alcohol (ROH) or acid (RCO_2H) is used. However, separating the ester from excess starting alcohol would require fractional distillation. In contrast, excess acetic acid can be removed through extraction with aqueous sodium bicarbonate. After drying, the odor of the compound will provide your first clue as to the identity of your starting alcohol. The mechanism for acid-catalyzed esterification is shown in Figure 21.3. You should be able to write a complete mechanism for both acid-catalyzed esterification and hydrolysis.

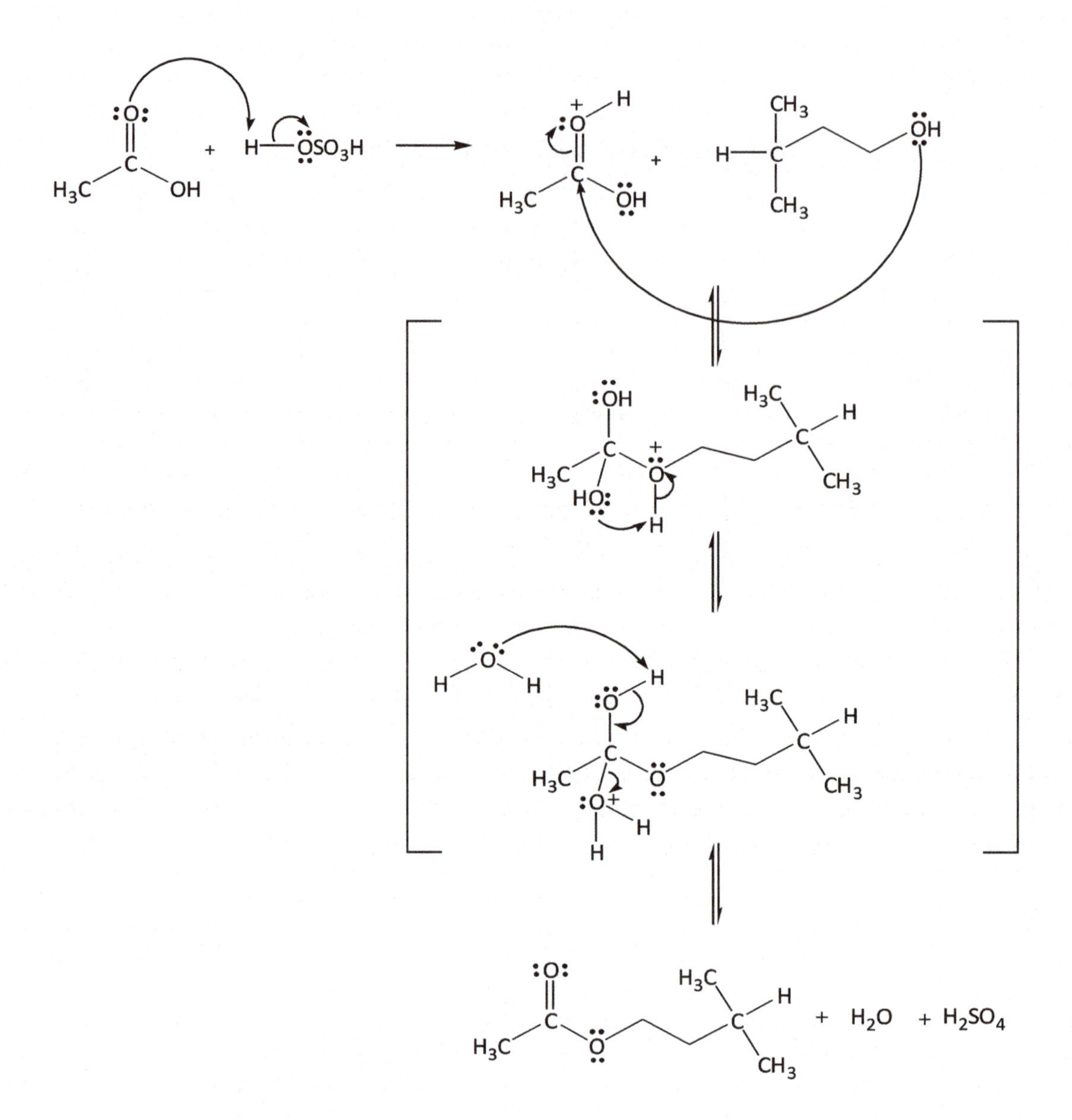

Figure 21.3 Mechanism of acid-catalyzed esterification of acetic acid with isoamyl alcohol.

Flavors and Fragrances

Many esters have characteristic flavors and fragrances (Figure 21.4). Different substances have different odors because there are different receptor sites for each. An odor is detected when a molecule fits into an olfactory receptor site for that odor and activates it. Therefore, the molecule's functional groups, size, and three-dimensional shape are defining criteria for whether the substance has a specific odor.

Methyl salicylate is responsible for the odor of wintergreen. However, salicylic acid itself has very little odor. Not only is the carboxylic acid less volatile than the methyl ester, but the salicylic acid molecule does not activate the odor receptor to the degree that methyl salicylate does. Therefore, the ester is necessary for the odor. In addition, although the *ethyl* ester of salicylic acid does have the wintergreen odor, it is much less intense than the odor of methyl salicylate.

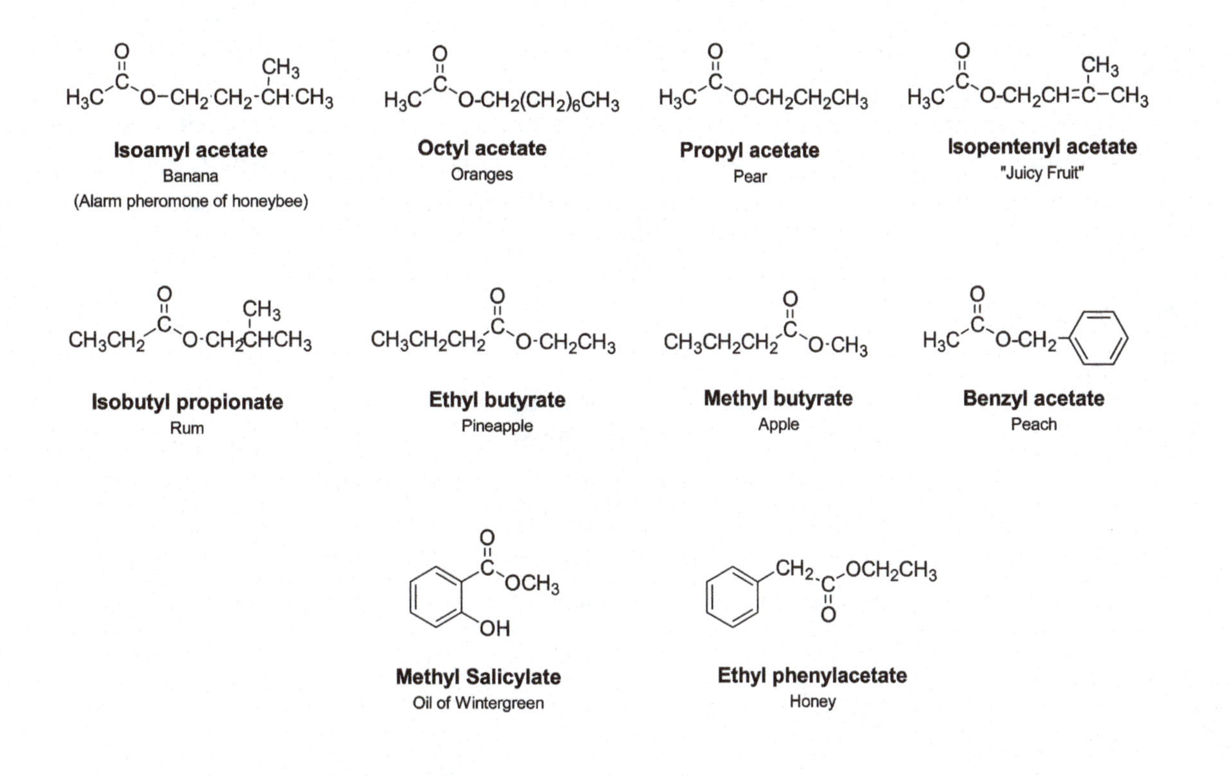

Figure 21.4 Ester flavors and fragrances.

IR Spectroscopy

The product can easily be distinguished from the reactants using IR spectroscopy. Note that both the carboxylic acid and the alcohol contain OH bonds. In carboxylic acids, this absorption is *extremely broad*. At times this broad absorption even overlaps the C-H absorptions. In alcohols, the appearance and frequency of this peak varies, depending on the preparation of the sample. If the compound is dissolved in a solvent, it typically appears as a *sharp* peak. If the compound is not dissolved in a solvent, the O-H stretch appears as a *broad* peak, typically. In the product ester, there is no OH group present; therefore, this absorption is absent in the IR spectrum of an ester. Another characteristic of esters is the presence of **two** C-O bands, one stronger and broader than the other. The IR spectra of the reactants and products are shown in Figure 21.5. Spectral correlation tables can be found in Appendix I.

NMR Spectroscopy

^{1}H-NMR is another tool that can be used to distinguish between the product and reactants. In Figure 21.6, the ^{1}H-NMR spectra of the reactants and product are shown. Notice that both reactants contain a hydroxyl group. The signals for these hydroxyl groups are very different. The alcohol reactant contains an alcoholic hydroxyl, while the acid reactant contains an acid hydroxyl which appears at a much higher chemical shift. The product ester does not contain a hydroxyl group at all.

Objectives

In this experiment you will perform a Fischer esterification of an alcohol with a carboxylic acid. The product will be purified by extraction and simple distillation. Once isolated, the product will be identified and the purity determined using GC analysis. Finally, IR and NMR analysis will be used to distinguish between reactants and products.

Experimental Procedure

Synthesis:

- Combine 3.0 mL of acetic acid and 3.0 mL of isoamyl alcohol in a 25 mL round bottom flask containing two or three boiling chips. Add two drops of concentrated H_2SO_4. Swirl to mix.
- Set up a reflux apparatus (see Appendix A). Place a $CaSO_4$ drying tube in the top of the condenser. Begin water flow, apply heat (VR @ 45), and reflux for 30 minutes.
- Cool the reaction flask slightly in a tap water bath. ***SLOWLY*** add 20 mL of 10% $NaHCO_3$, 1 mL at a time to prevent bubbling over. Transfer this mixture to a separatory funnel.

Purification:

Extraction

- Rinse the reaction flask with 20 mL of diethyl ether and transfer this rinse to the separatory funnel. Agitate the mixture, and draw off the bottom aqueous layer into a small flask.
- Add 10 mL of 10% $NaHCO_3$ to the separatory funnel and re-extract the organic layer. Agitate, vent, and draw off the aqueous layer.
- Transfer the organic layer to a clean 50 mL Erlenmeyer flask. Dry over $MgSO_4$. Using a plastic pipette, transfer the liquid to a preweighed 50 mL round bottom containing three boiling chips.

Distillation

- Set up for simple distillation using the 25 mL round bottom as a receiving flask. Begin water flow, apply heat (VR @ 20), and begin collecting the ether. Collect all liquid that boils under 40°C (mainly ether). Be sure to record the distillation range.
- Lower the heating mantle and allow the reaction flask to cool to room temperature. Reweigh the reaction flask to determine the final product mass. Calculate the percent yield. Record this data in the laboratory notebook, and complete Table 21.1 on the POST LAB Assignment (provided online).
- ***PROCEED TO PRODUCT ANALYSIS.***

Product Analysis:

GC Analysis

- Prepare and submit a sample of your product for chromatographic analysis by placing five drops of your sample in an autoanalyzer vial and 1 mL of GC solvent (methanol).
- Using the chromatographic results, propose an identity and determine the degree of purity of your product. Complete Table 21.2 on the POST LAB Assignment (provided online).

IR Analysis

- Using the spectra provided, identify all characteristic absorptions in the spectra of the reactants and product. Complete Table 21.3 on the POST LAB Assignment (provided online).

NMR Analysis

- Using the provided spectra, identify all characteristic resonances of reactants and products. Complete Table 21.4 on the POST LAB Assignment (provided online).

SAFETY

****All experiments should be performed in a fume hood with appropriate safety glasses. Gloves will be provided by request.****

All organic solvents used in this experiment are flammable, irritants, and can be toxic if ingested or absorbed through skin. Organic solvents, reagents and products should be kept in sealed containers at all times. Sulfuric acid is extremely corrosive and will burn skin upon contact.

WASTE MANAGEMENT

Rinse the aqueous washes from the extraction down the drain with plenty of water. Place the distillate collected and your product in the container labeled "LIQUID WASTE—ESTERS."

Reference

Klein, David. (2015). *Organic Chemistry*, 2nd ed. Hoboken: John Wiley and Sons.

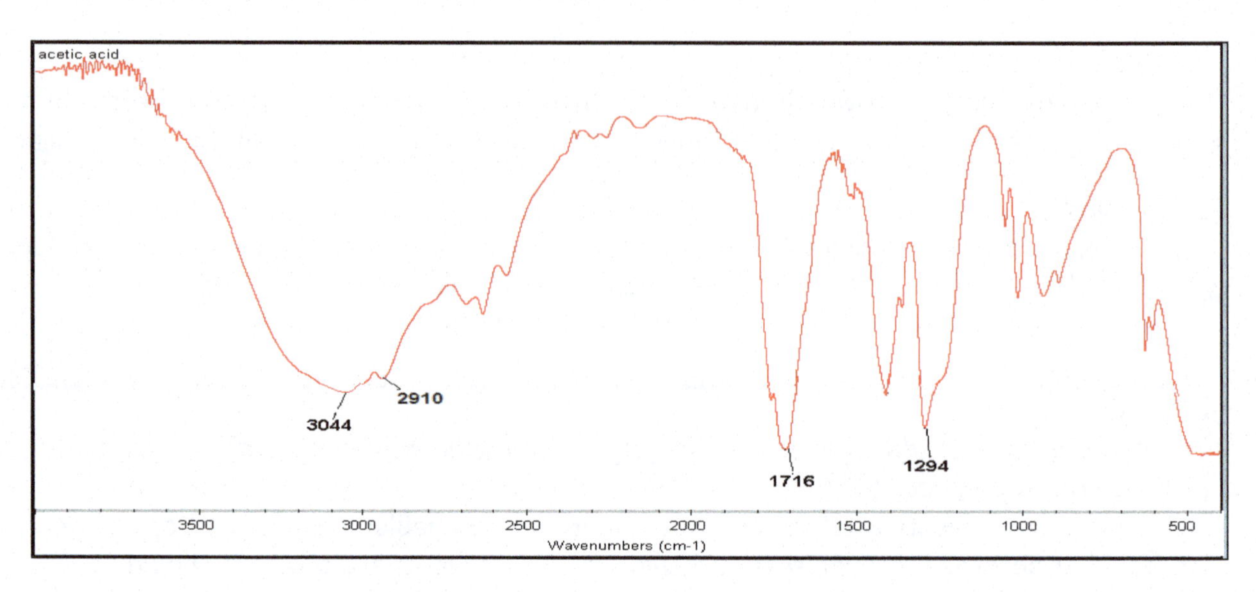

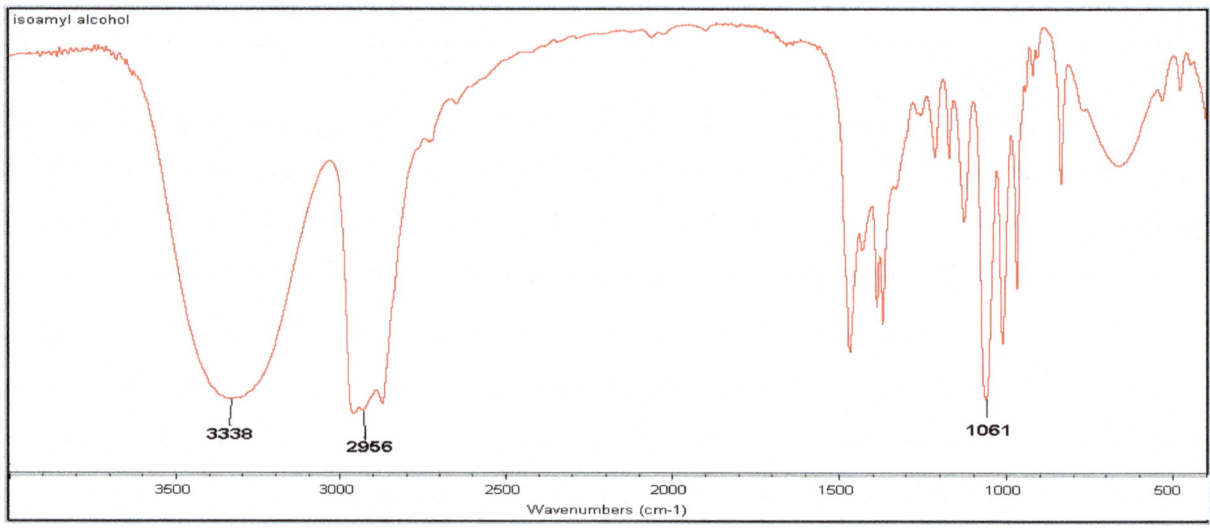

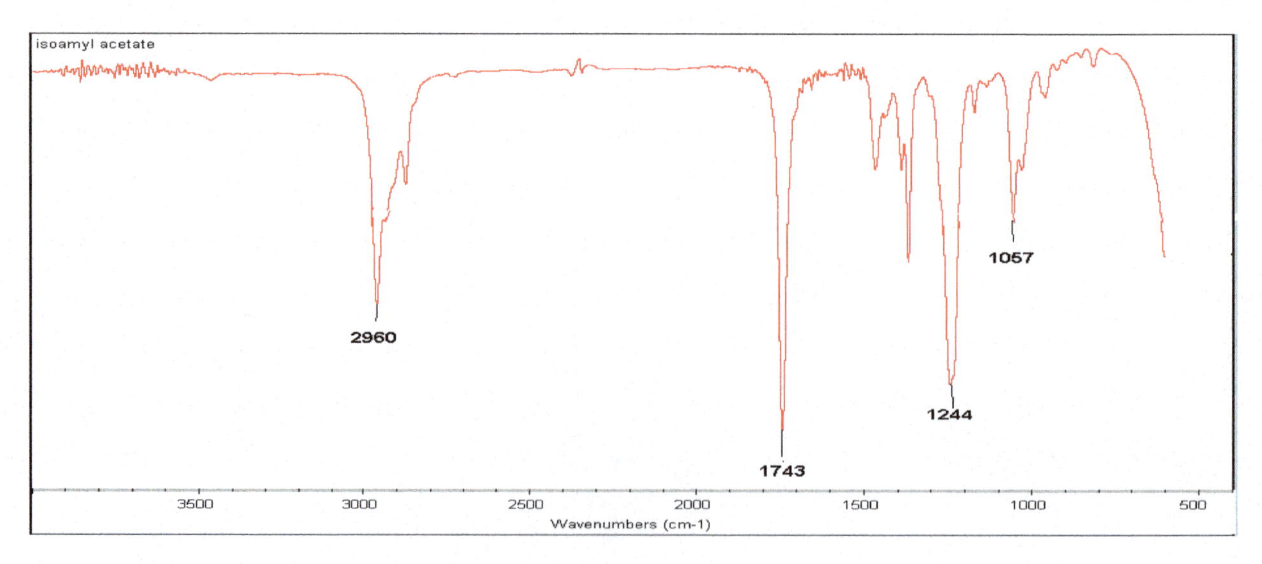

Figure 21.5 IR spectra of reactants and products.

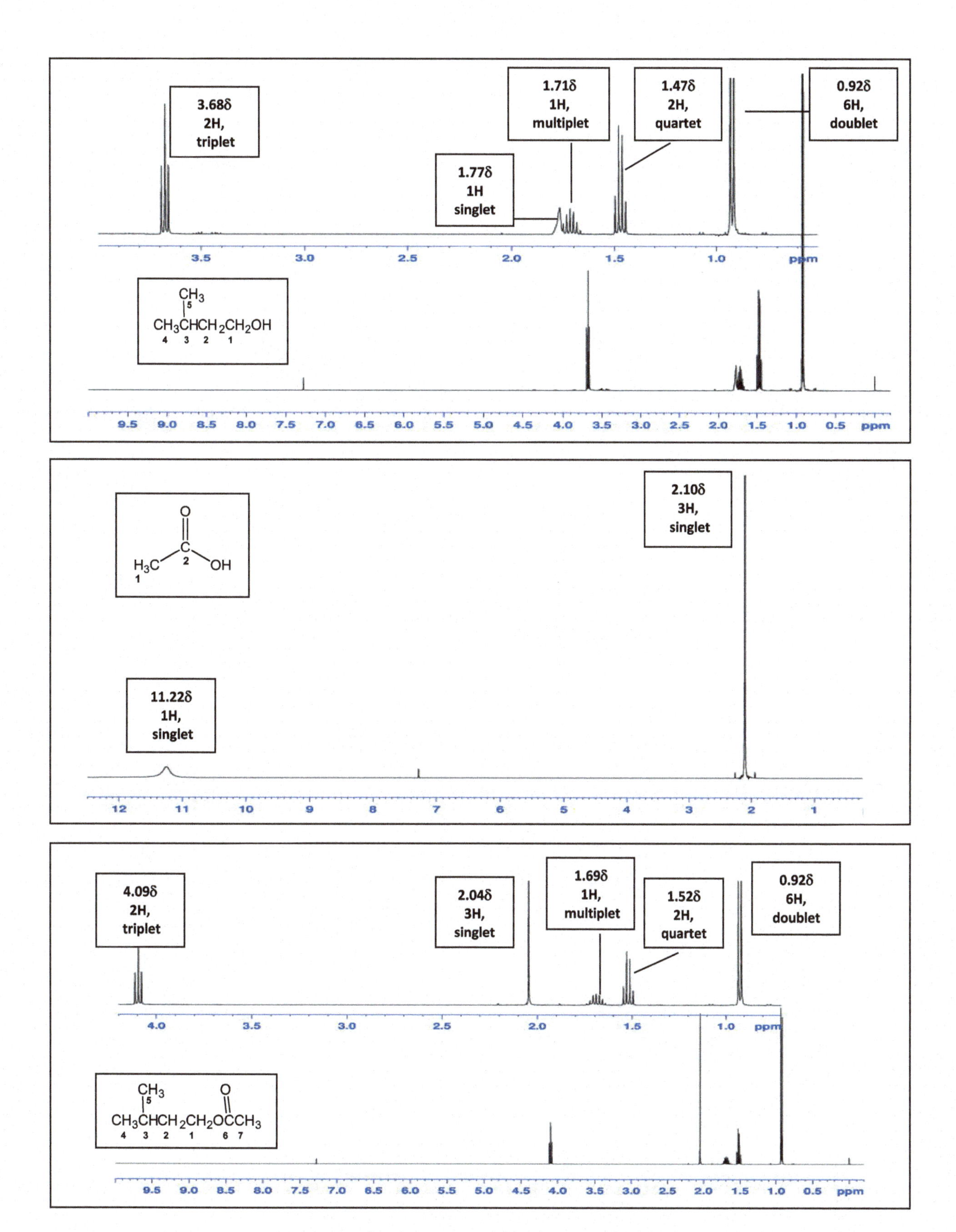

Figure 21.6 ^{1}H-NMR spectra of reactants and product.

Name:			
	Max	Score	Total Grade
Tech	10		
Pre	20		
In	20		
Post	50		

Exp. 21 Esters: Synthesis and Fragrance

PRE-LAB ASSIGNMENT: *(**EACH STUDENT** will complete and submit an original copy at the beginning of the lab period. **Without a complete pre-lab assignment, you will not be allowed to perform the experiment, and will receive a zero for the lab.**) .....max score = 20 pts.*

1. **Objective** *(Write a brief purpose of the experiment in **complete** sentences, addressing all of the following points.)*
   - Name the *specific* compounds combined during the synthesis and the name the product formed.
   - What are the purification techniques used in this experiment?
   - What analytical technique will be used during the experiment?
   - What type of spectral analysis will be used to characterize the reactant and products?

2. **Chemical Equation** *(Draw the chemical equation for the synthesis of isoamyl acetate using actual chemical structures.)*

3. **Physical Data** *(Complete the following table before coming to lab.)*

Compound	MW (g/mol)	bp (°C)	d (g/mL)
isoamyl alcohol			
isoamyl acetate			
acetic acid			
sulfuric acid	X	X	X
diethyl ether	X		
methanol	X		

4. **Experimental Outline** *(Give a brief description of the procedure that will be followed in this experiment in 5 lines or less.)*

1	
2	
3	
4	
5	

5. **Pre Lab Questions:** *(Answer the following questions prior to lab.)*

A. The limiting reagent for this reaction is the sulfuric acid catalyst.
   a. True
   b. False

B. All steps in an acid catalyzed esterification are reversible. When a large excess of water is present, hydrolysis of the ester is favored.
   a. True
   b. False

I have read and understood the experimental procedure for this experiment. I am familiar with the hazards and the required disposal procedures for this experiment.

Sign here: ____

Experiment 22

Solventless Aldol Condensation

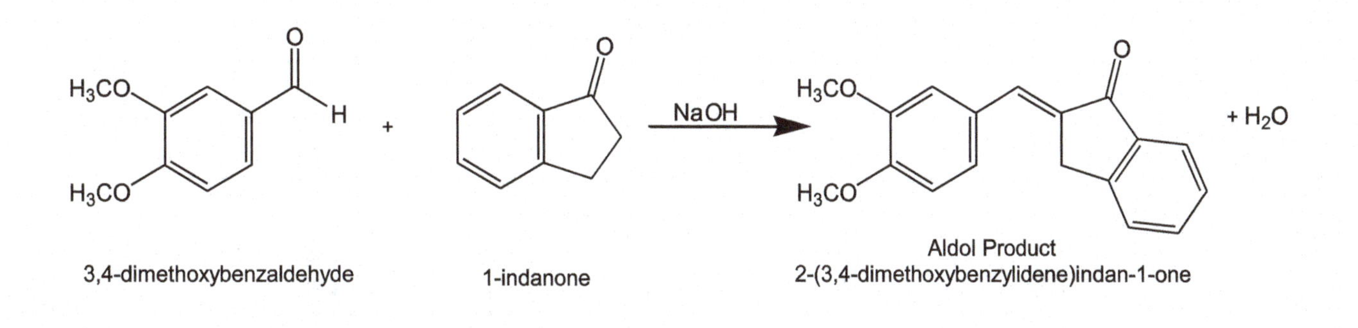

Figure 22.1 Aldol condensation.

Introduction

An aldol condensation is a condensation reaction between two carbonyl compounds. The base-catalyzed reaction occurs by nucleophilic addition of the enolate ion of the donor molecule to the carbonyl group of the acceptor molecule (Figure 22.1). The first step is generation of an enolate from a ketone. The second step is nucleophilic addition of the enolate to the carbonyl carbon of the acceptor molecule (aldehyde). Protonation and dehydration (step 3) gives the α,β-unsaturated ketone. The aldol condensation represents a powerful general method for construction of carbon-carbon bonds.

In contrast to typical experimental procedures for aldol condensation reactions, this reaction will be carried out without solvent, representing the best possible solution to choice of a benign solvent. You will confirm the identity of your product by TLC and melting range analysis, as compared with a standard. TLC and melting range analysis will also allow you to evaluate the purity of your product. IR spectroscopy can be used to differentiate between reactants and products.

The Solvent-free Reaction

As you will observe, when the two solids are mixed, they actually melt, allowing the reaction to occur in the liquid state (solvent-free), therefore, reducing solvent waste. This is a vivid example of how impurities lead to lower melting points. Addition of a small amount of powdered NaOH catalyzes the aldol condensation reaction. The resulting solid will be purified by recrystallization from propanol/water. This workup is much less complicated than the typical workup required in an aldol condensation performed with a solvent. Another benefit of solid-state reactions includes the high efficiency of the reaction, producing almost 100% atom economy. Also important is the fact that unlike the solution phase reaction, the solid-state reaction is irreversible, leading to greater percent yields.

Mechanism

In this base-catalyzed aldol condensation, the carbon alpha (α) to the carbonyl of 1-indanone is deprotonated to form a resonance stabilized enolate ion, which then carries out nucleophilic attack at the carbonyl of 3,4-dimethoxybenzaldehyde (Figure 22.2). After protonation, elimination of water affords the product.

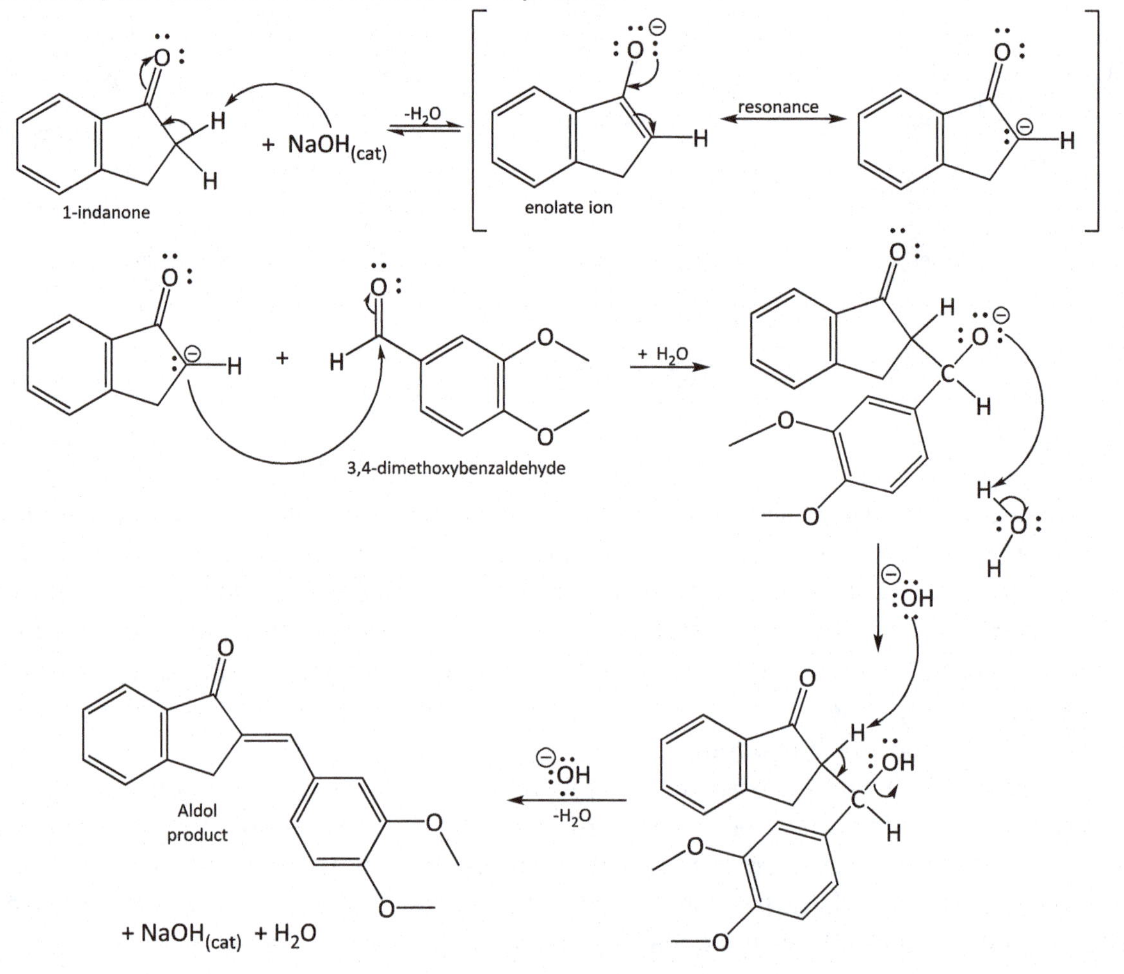

Figure 22.2 Mechanism of the aldol condensation between 3,4-dimethoxybenzaldehyde and 1-indanone.

IR Spectroscopy

Several common absorptions appear in the IR spectra of both reactants and the product (Figure 22.3). The sp^3 C-H stretches and the sp^2 C-H stretch appears in both reactants and the product. Other absorptions common in the IR spectra are two aromatic C=C stretches the C=O stretches. The reactant aldehyde and aldol product spectra contain a C-O stretch due to the presence of the ethoxy substituent. Aldehydes have ***two*** characteristic C-H stretches which do not appear in the spectra of the starting ketone, or aldol product. Spectral correlation tables are provided in Appendix I.

Objectives

In this experiment you will perform an aldol condensation between an aldehyde and a ketone under solvent-free conditions. You will purify your product by recrystallization and isolate it by vacuum filtration. The product will be identified and the purity determined using TLC and melting point analysis. Finally, IR spectroscopy will be used to distinguish between reactants and products.

Experimental Procedure

Synthesis:

- Place ~0.25 g of 3,4-dimethoxybenzaldehyde and ~0.20 g of 1-indanone in 25 mL Erlenmeyer flask.
- Using a glass rod, scrape and crush the two solids together until they melt and become a brown oil.
- Add ~0.05 g of finely ground NaOH to the reaction mixture. Continue stirring with the glass rod until the reaction mixture forms a tacky solid and no more liquid is visible.
- Allow the mixture to stand undisturbed for 10 minutes.

Purification:

- Add 2 mL 10% HCl to the reaction mixture and swirl to mix. Be sure to thoroughly mix all solid with the acid. Check to make sure the solution is acidic.
- Suction filter to isolate the crude solid, then transfer the solid to a clean 50 mL flask.
- Add 2-3 mL of 90% ethanol/water (recrystallizing solvent) directly to the flask.
- Place the flask directly on a warm hotplate (**SETTING = 3, NO HIGHER!**), and heat until the majority of the solid has gone into solution. If necessary, add an additional 1-2 mL of solvent, but **DO NOT EXCEED 5 mL!**
- Remove the flask from the hotplate and allow the contents to cool to room temp slowly, and then place in ice bath for a minimum of 10 minutes.
- Suction filter to isolate the pure solid. Seat the ***PREWEIGHED*** filter paper with the ice-cold ethanol/water. Rinse the crystals in the funnel with an additional 1–2 mL of ice-cold ethanol/water.
- Allow the crystals to dry under vacuum for 5 minutes, and then transfer the product to a labeled and preweighed watch glass. Place the pure product in the oven for 10 minutes to dry.
- Remove product from oven and allow it to cool to room temperature. Determine final product mass and calculate the percent yield. Record this data in the laboratory notebook and complete Table 22.1 and Table 22.2 on the POST LAB Assignment (provided online).
- ***PROCEED TO PRODUCT ANALYSIS.***

Product Analysis:

TLC Analysis

- Prepare a TLC sample of your pure solid. Transfer two or three crystals to a small test tube and dissolve in 1 mL of reagent acetone.

- Prepare a TLC plate by applying the sample solution along with the provided standards of the 3,4-dimethoxybenzaldehyde, the 1-indanone, and the expected product.
- Develop the TLC plate in 1:2 diethyl ether: hexane. Visualize spots using a UV lamp. Circle and measure distances migrated by all spots present on plate.
- Sketch the TLC plate in the laboratory notebook. Calculate all R_f values and determine purity of your sample. Complete Table 22.3 on the POST LAB Assignment (provided online).

Melting Point Analysis

- Prepare a melting point sample of your pure product. Determine the experimental melting point and compare to the literature melting point of the product (180–181°C) to determine purity.
- Record the melting range (T_i-T_f) in the laboratory notebook. Record this value in Table 22.1 on the POST LAB Assignment (provided online).

IR Analysis

- Using the spectra provided, identify all characteristic absorptions in the IR spectra of the reactants and product. Complete Table 22.4 on the POST LAB Assignment (provided online).

SAFETY

****All experiments should be performed in a fume hood with appropriate safety glasses. Gloves will be provided by request.****

All organic solvents used in this experiment are flammable, irritants, and can be toxic if ingested or absorbed through skin. Organic solvents, reagents and products should be kept in sealed containers at all times. Hydrochloric acid and sodium hydroxide are corrosive.

WASTE MANAGEMENT

Place all liquid waste in the container labeled "LIQUID WASTE—ALDOL" located in the waste hood. Place all used TLC spotters and melting point capillary tubes in the broken glass container. Place solid waste in the container labeled "SOLID ORGANIC WASTE."

References

Kirchhoff, M.,Ryan, M. (2002) *Greener Approaches to Undergraduate Chemistry Experiments*. Washington. American Chemical Society.

Doxsee, Kenneth M., & Hutchinson, James E. (2003). *Green Organic Chemistry, Strategies, Tools and Laboratory Experiments*. Belmont: Thomson/Brooks/Cole.

Klein, David. (2015). *Organic Chemistry*, 2nd ed. Hoboken: John Wiley and Sons.

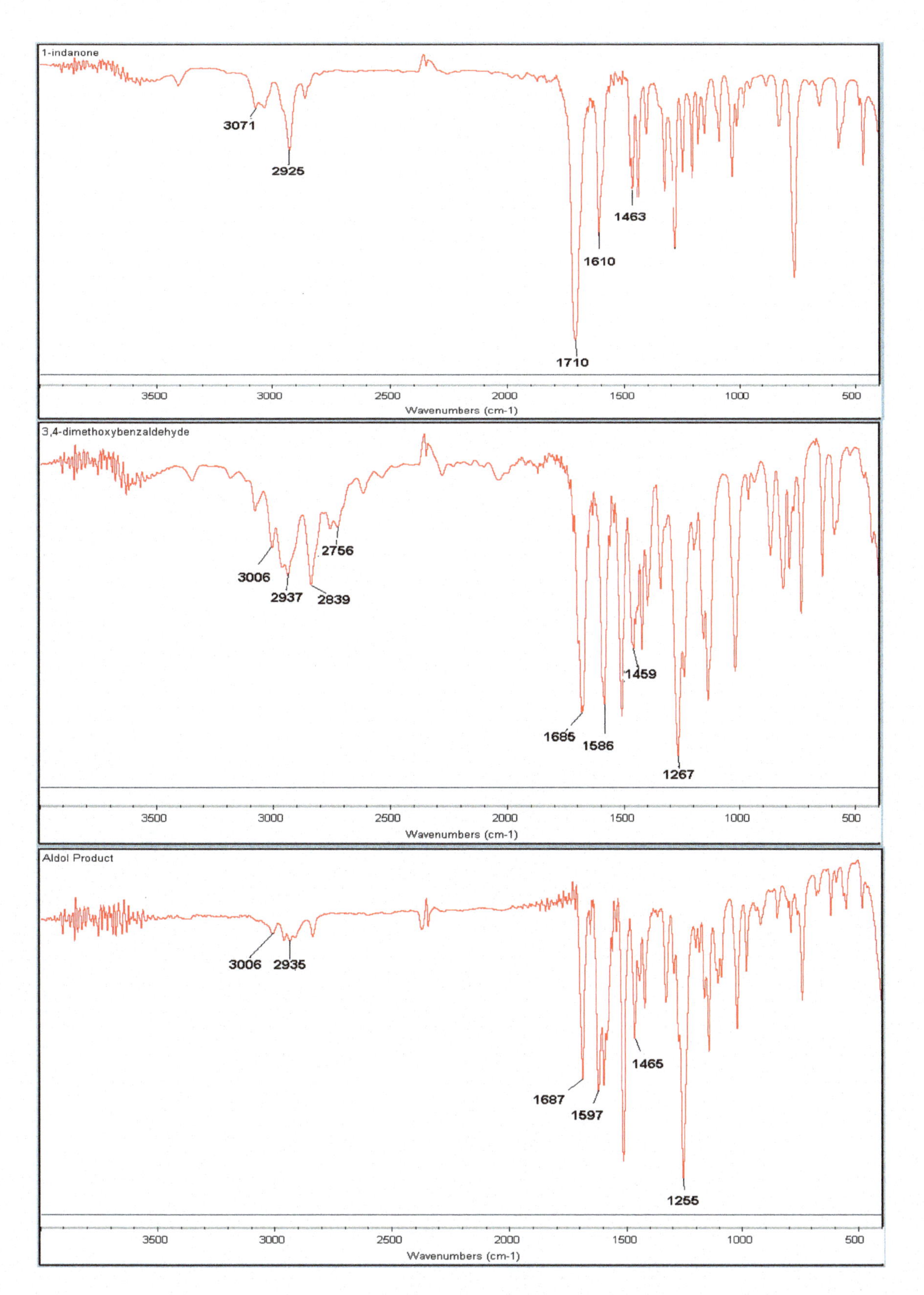

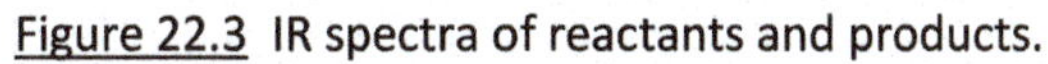

Figure 22.3 IR spectra of reactants and products.

Name:			
	Max	Score	Total Grade
Tech	10		
Pre	20		
In	20		
Post	50		

Exp. 22 Solventless Aldol Condensation

PRE-LAB ASSIGNMENT: *(**EACH STUDENT** will complete and submit an original copy at the beginning of the lab period. **Without a complete pre-lab assignment, you will not be allowed to perform the experiment, and will receive a zero for the lab.**) .....max score = 20 pts.*

1. Objective *(Write a brief purpose of the experiment in **complete** sentences, addressing all of the following points.)*

- Name the *specific* compounds combined during the synthesis and name the product formed.
- What is the purification technique used in this experiment?
- What analytical technique(s) will be used during the experiment?
- What type of spectral analysis will be used to characterize the reactant and products?

2. Chemical Equation *(Draw the chemical equation for the synthesis of the aldol product using actual chemical structures.)*

3. Physical Data *(Complete the following table before coming to lab.)*

Compound	MW (g/mol)	mp (°C)	bp (°C)	d (g/mL)
3,4-dimethoxy-benzaldehyde			X	X
1-indanone			X	X
sodium hydroxide		X	X	X
ethanol	X	X		
diethyl ether	X	X		
hexane	X	X		
aldol product	280.32	180-181	X	X

4. Experimental Outline *(Give a brief description of the procedure that will be followed in this experiment in 5 lines or less.)*

1	
2	
3	
4	
5	

5. Pre-Lab Questions *(Answer the following questions prior to lab.)*

A. The solvent used in today's synthesis is:

a. an aldehyde
b. an aqueous base
c. a ketone
d. none of the above

B. List the **3** benefits of solid-state reactions such as the reaction performed in this experiment.

a.

b.

c.

I have read and understood the experimental procedure for this experiment. I am familiar with the hazards and the required disposal procedures for this experiment.

Sign here: ______________________________

Experiment 23

Theory of Color in Organic Compounds: Preparation of Organic Dyes

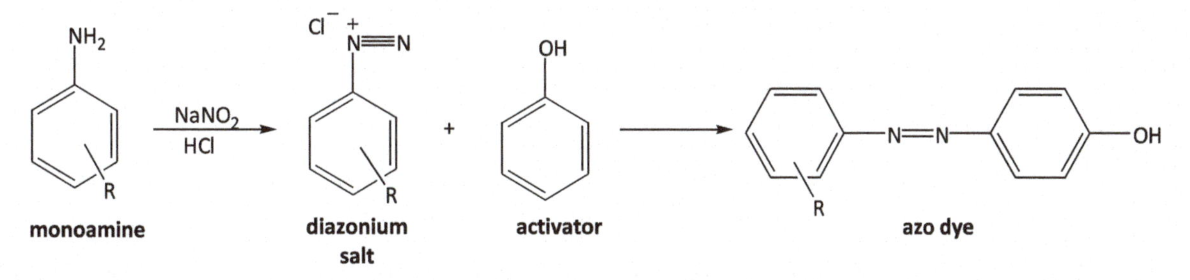

Figure 23.1 Synthesis of an azo dye.

Introduction

The preparation of azo dyes involves two reactions: diazotization and coupling. Both reactions are very simple operations that are carried out in aqueous solution. In this experiment you will synthesize azo dyes with various amines (Figure 23.1). You will study the relationship between the extent of conjugation and the color that something appears. You will also gain knowledge of how conjugation and color are affected by pH.

Formation of Diazonium Salts

Reactions of amines with nitrous acid (HNO_2) are particularly useful in organic synthesis. Nitrous acid is relatively unstable, therefore is typically generated in situ by mixing sodium nitrite ($NaNO_2$) with cold, dilute hydrochloric acid (Figure 23.2). This reaction produces the nitrosonium ion ($^+NO_2$) which can then be reacted with an aromatic amine to generate a diazonium salt. Diazonium salts are relatively stable, and are used as intermediates in several important reaction schemes. This is a process called diazotization.

$$\underset{\text{amine}}{R{-}NH_2} + \underset{\text{sodium nitrite}}{NaNO_2} + \underset{\text{hydrochloric acid}}{2\,HCl} \longrightarrow \underset{\text{diazonium salt}}{R{-}\overset{+}{N}{\equiv}N\ Cl^-} + 2\,H_2O + NaCl$$

Figure 23.2 Synthesis of diazonium salt.

The aromatic amines used in today's synthesis (Figure 23.3) are prepared in dilute hydrochloric acid. When reacted with sodium nitrite, the arene diazonium salt will be prepared.

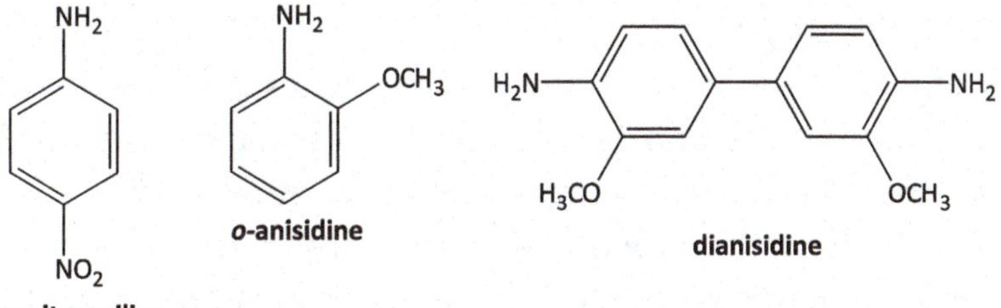

Figure 23.3 Aromatic amines used in dye synthesis.

Diazo Coupling: Azo Compounds

Arene diazonium ions undergo a reaction called diazo coupling with strongly activated aromatic rings (called activators) to produce azo compounds with the structure Ar-N=N-Ar' (Figure 23.4). These azo compounds are highly conjugated and strong chromophores, resulting in highly colored compounds. For this reason, many azo compounds are used as dyes, called azo dyes.

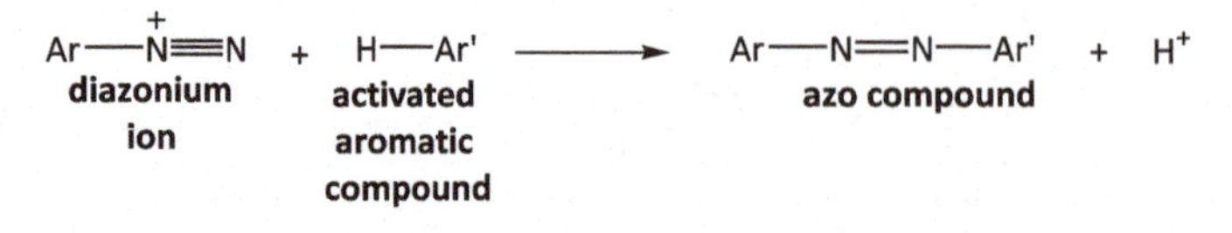

Figure 23.4 Synthesis of azo compound.

Many common azo dyes contain sulfonate groups ($-SO_3^- Na^+$) on the molecule to make the dye more water soluble. To obtain water-soluble azo dyes, a naphthol containing such groups is used for coupling. A typical reaction involves coupling with the salt of 2-naphthol-3,6-disulfonic acid, commonly referred to as "R-salt" (Figure 23.5).

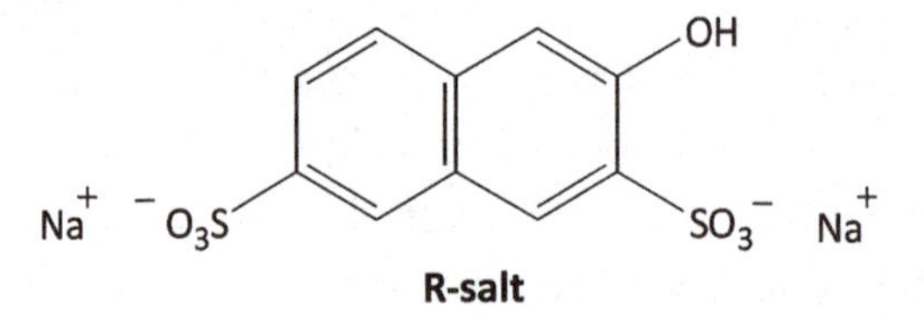

Figure 23.5 Structure of "R-salt" activator used in coupling step.

Dyes formed with "R-salt" contain these sulfonate groups on the molecule, which not only increases water solubility, but also improves the affinity of the dye for the surfaces of certain fiber types containing polar surfaces such as cotton and wool. The structural units for these types of fibers are shown in Figure 23.6.

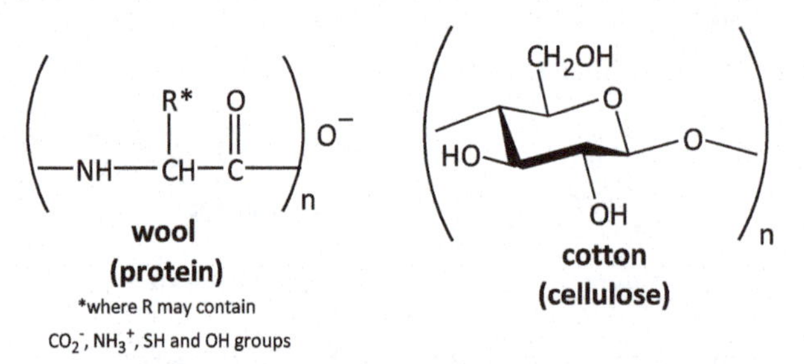

Figure 23.6 Structural units of wool and cotton fibers.

Theory of Color in Organic Compounds

Molecular orbital (MO) theory provides the simplest explanation for the color of organic compounds. Only π electrons need to be considered, since they are higher in energy than σ electrons. Consider the MO description of the bonds in ethene (Figure 23.7). The p_z and p_z' atomic orbitals combine to form the π (bonding) and $\pi*$ (antibonding) orbitals, and the electrons occupy the lower energy π (bonding) orbital. The energy gap between the two orbitals is large (about 630 kJ/mol) in ethene. When high energy, short wavelength light (corresponding exactly in energy to the difference between the bonding and antibonding orbital energy levels) shines on ethene, one of the bonding electrons absorbs a quantum of energy and is promoted to the higher energy level.

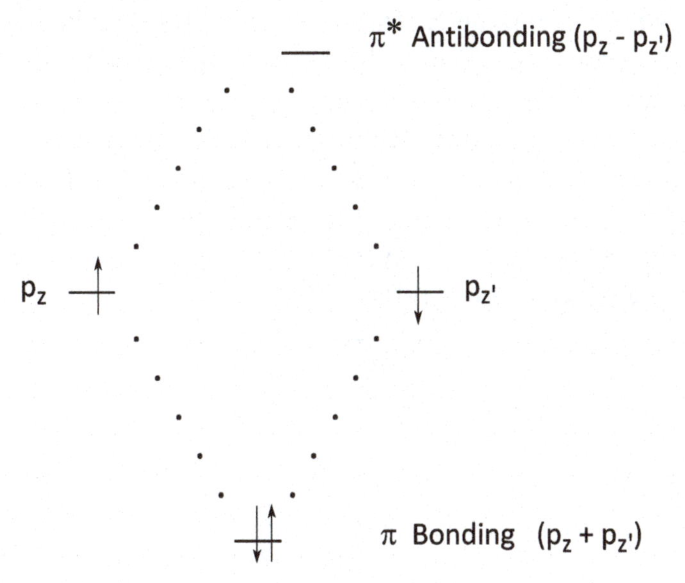

Figure 23.7 Molecular orbital description of π bonding in ethene.

The situation is similar in 1,3-butadiene (Figure 23.8), but the energy gap is smaller (about 555 kJ/mol), so the energy of the light absorbed is less (and therefore the wavelength is longer). In 1,3,5-hexatriene the energy gap is still smaller. As the length of the conjugated system of electrons increases, the wavelength of light absorbed also increases, as illustrated in Figure 23.8 below. When the wavelength of light absorbed is greater than 400 nm, it enters the visible region of the electromagnetic spectrum and the absorbing substance appears colored.

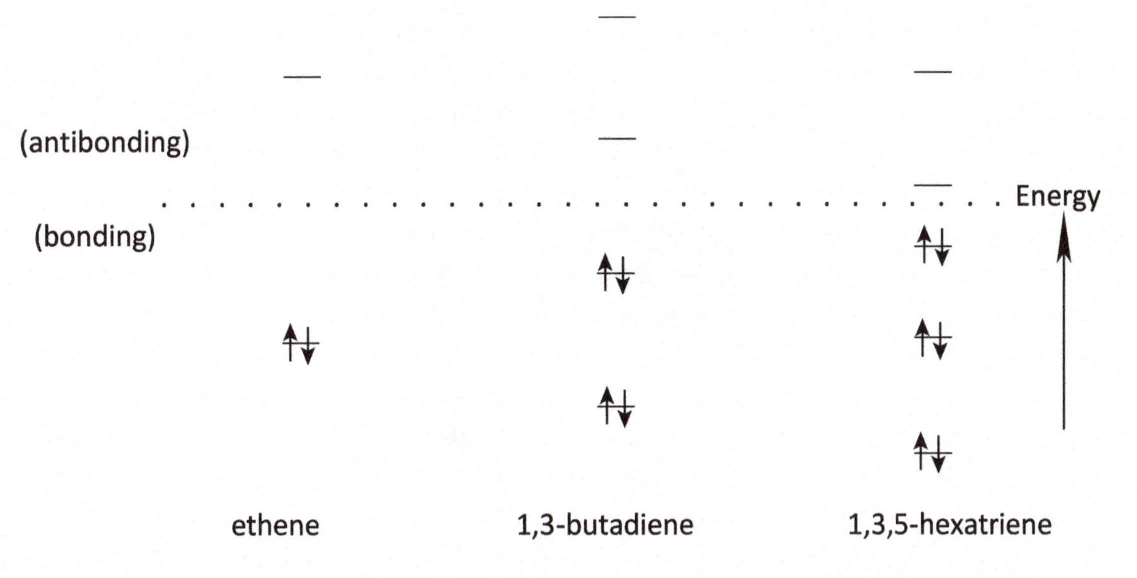

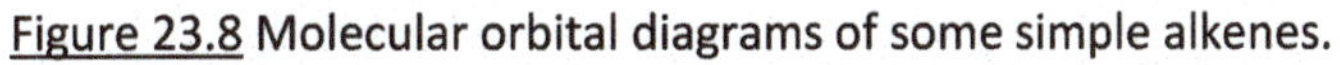

Figure 23.8 Molecular orbital diagrams of some simple alkenes.

Effects of Conjugation on Color in Organic Compounds

Substituents that carry nonbonded electrons (*n* electrons) can cause shifts in the wavelength absorbed by organic compounds. The *n* electrons can increase the length of the conjugated system through resonance. Examples of groups with *n* electrons are the amino, hydroxyl, and methoxy groups, as well as the halogens.

In compounds that are acids or bases, pH changes have very significant effects on the positions of the absorption bands. At high pH, the phenolic group of an "R-salt" dye (see Figure 23.5) is converted to a phenoxide anion, which through resonance increases the extent of conjugation of the dye. This increases the wavelength of light absorbed and observed, so significant color changes are observed. This is referred to as a bathochromic, or red shift.

	n	λ (nm)
	1	180
$CH_3(CH{=}CH)_nCH_3$	3	270
	5	345
	7	400

Figure 23.9 Effect of the length of the conjugated system on the wavelength of light absorbed.

The Complementary Nature of Color

The color that something *appears* is the ***complement*** (opposite on the color wheel, Figure 23.10) of the color light it *absorbs*. For example, if light of wavelength 430 nm (indigo) is absorbed, the object appears orange (650 nm).

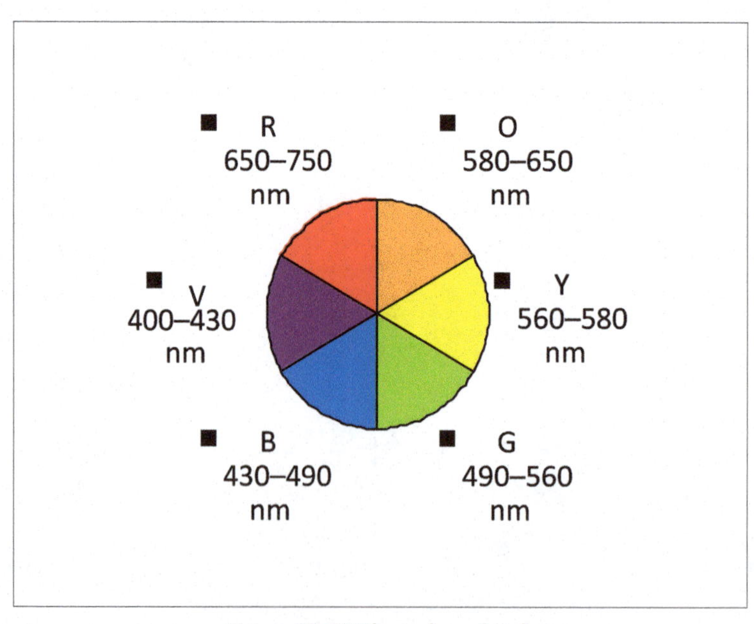

Figure 23.10 The color wheel.

Objectives

In this experiment you will synthesize azo dyes with various amines. In part A, you will use the monoamine *p*-nitroaniline to compare the effects of pH on the color that the dye solution appears. In part B, you will use a monoamine (*o*-anisidine) and a diamine (dianisidine) to compare the color that the dye solution appears versus the conjugation of the dye structure. You will study the relationship between the extent of conjugation and the color that something appears. You will also gain knowledge of how conjugation and color are affected by pH.

Experimental Procedure

Part A: Comparison of pH effects on color:

- Place 5 mL of *p*-nitroaniline in a large test tube. Chill several minutes in an ice bath.
- Add 1 mL of 0.5 M $NaNO_2$ to the test tube. Keep this solution in the ice bath.
- Add 0.5 mL of "R-salt" to the tube. Mix well. Note any change of appearance.
- Divide the *p*-nitroaniline dye solution evenly into two small test tubes.
- Add ten drops of 5 M NaOH solution to one test tube. Stir gently with a glass rod to mix. Note any change in appearance in the observations section of your laboratory notebook.
- Using plastic pipettes, add five drops of each dye solution to clean test tubes.
- Using pH-Hydrion paper, measure the pH of the dye solutions.
- Predict the wavelength of the color that ***appears*** using a color wheel.
- Predict the wavelength of the color that each dye solution ***absorbs*** using a color wheel.
- Record this data in the laboratory notebook and complete Table 23.1 on the POST LAB Assignment (provided online).

Part B: Comparison of effect of conjugation on color:

- Place 5 mL *o*-anisidine and 5 mL dianisidine in separate large test tubes.
- Chill several minutes in an ice bath.
- Add 1 mL of 0.5 M $NaNO_2$ solution to the test tubes. Keep these solutions in the ice bath.
- Add 0.5 mL of "R-salt" to each test tube. Note any change in appearance.
- Predict the wavelength of the color that each dye solution appears using a color wheel.
- Predict the wavelength of the color that each dye solution absorbs using a color wheel.
- Compare this data to the length of conjugation of the dye structure. Record this data in the laboratory notebook and complete Table 23.2 on the POST LAB Assignment (provided online).

SAFETY

All experiments should be performed in a fume hood with appropriate safety glasses. Gloves will be provided by request.

All organic compounds used in this experiment are irritants, can be toxic if ingested or absorbed through skin, and are also suspected carcinogens. Hydrochloric acid and sodium hydroxide are extremely corrosive.

WASTE MANAGEMENT

All of the dye solutions and any aqueous rinses should be placed in the container labeled "LIQUID WASTE—DYES". This mixture will be concentrated to reduce the volume, and then disposed of commercially.

Reference

Klein, David. (2015). *Organic Chemistry*, 2nd ed. Hoboken: John Wiley and Sons.

Name:			
	Max	Score	Total Grade
Tech	10		
Pre	20		
In	20		
Post	50		

Exp. 23 Theory of Color in Organic Compounds: Preparation of Organic Dyes

PRE-LAB ASSIGNMENT: *(**EACH STUDENT** will complete and submit an original copy at the beginning of the lab period. **Without a complete pre-lab assignment, you will not be allowed to perform the experiment, and will receive a zero for the lab.**) .....max score = 20 pts.*

1. **Objective** *(Write a brief purpose of the experiment in **complete** sentences, addressing all of the following points.)*
    - Name the *specific* reactants combined during the synthesis and the type of product formed.
    - What *two* relationships will be studied during this experiment?

2. **Chemical Equation** *(Draw the chemical equation for the synthesis of an azo dye using actual chemical structures.)*

3. **Compound Structures** *(Complete the following table before coming to lab.)*

p-nitroaniline	dianisidine	*o*-anisidine	R-salt

4. **Experimental Outline** *(Give a brief description of the procedure that will be followed in this experiment in 5 lines or less.)*

1	
2	
3	
4	
5	

5. **Pre-Lab Questions** *(Answer the following questions prior to lab.)*

A. Based on the molecular orbital theory (MO), as the conjugation of a molecule _______ the energy gap between the highest occupied molecular orbital (HOMO) and the lowest unoccupied molecular orbital (LUMO) _______, resulting in a _______in the wavelength light absorbed.

a. decreases, decreases, decrease.
b. increases, decreases, increase.
c. increases, increases, increase.
d. none of the above.

B. _____ is required for a substance to absorb in the visible region and exhibit a color:

a. extended conjugation
b. high molecular weight
c. an amine functional group
d. all of the above

I have read and understood the experimental procedure for this experiment. I am familiar with the hazards and the required disposal procedures for this experiment.

Sign here: ____

Experiment 24

Reductive of Amination

Introduction:

Amines are abundant in nature and play an important role in the survival of life—they are involved in the creation of amino acids, the building blocks of proteins in living beings. Many vitamins are also built from amino acids. Besides the amines that the human body is composed of, humans have found a range of other uses for amines (Figure 24.1). Medicines based on amines such as Morphine and Demerol are commonly used analgesics. The anesthetic Novocaine is an amine. Tetramethyl ammonium iodide is used in the disinfection of drinking water. Amines are used in industry for pest control and dye manufacturing. Some amines are popular recreational drugs, such as amphetamine and methamphetamine.

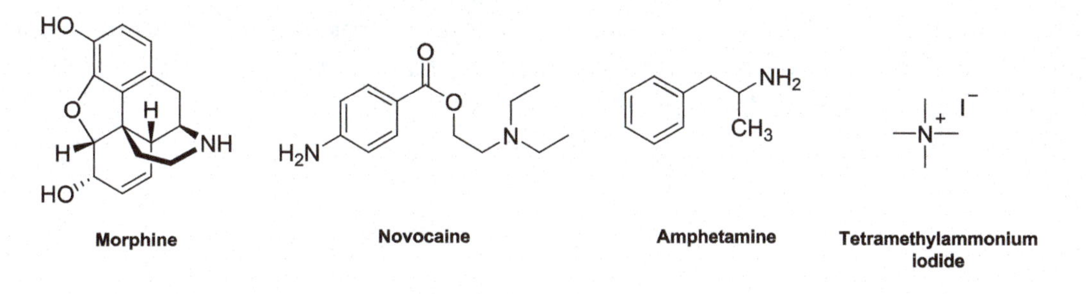

Figure 24.1 Commonly known amines.

In this experiment, you will synthesize an amine using an important synthetic method known as reductive amination (Figure 24.2). Reductive amination involves the reaction of an aldehyde or ketone with an amine to form an imine, followed by in situ reduction to a more substituted amine product. Initially, an aldehyde known as *o*-vanillin will be reacted with *p*-toluidine to generate an imine. This reaction will take place without the need for a reaction solvent, similar to the aldol condensation experiment performed previously in lab. Once formed, the imine is reduced to form a secondary amine using sodium borohydride. Finally, the secondary amine will be acetylated with acetic anhydride to produce a solid amide derivative. The entire reaction sequence can be performed in one hour using an open beaker.

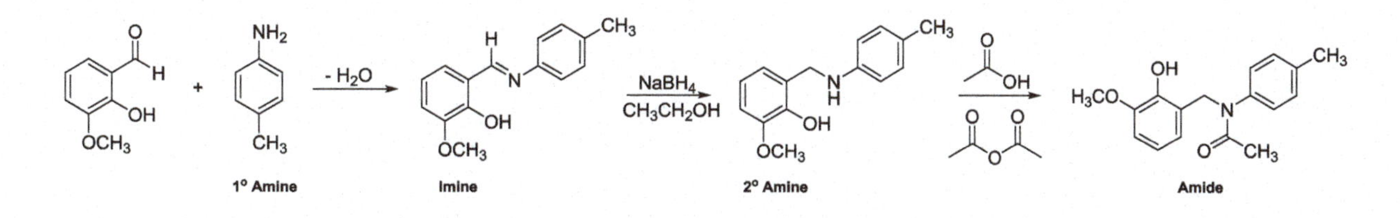

Figure 24.2 Reductive amination reaction scheme.

Reductive Amination

It can be difficult to add amines to organic compounds due to the nucleophilicity of the amine. In order to achieve this goal, an amine is added to a carbonyl to form an imine (nitrogen double bonded to carbon), then the imine can be reduced to an amine using a hydride reagent. Usually imine formation requires the use of an acid catalyst, but in the case of *o*-vanillin, the phenolic group is acidic enough to act as an internal catalyst.

In the first step of the reaction, the amine nitrogen attacks the carbonyl carbon of the aldehyde to form an ionic intermediate. A series of proton transfers occurs, then the electrons on the nitrogen will form the π bond of the imine, as water is eliminated.

The second step of the synthesis includes the reduction of the imine to a secondary amine using $NaBH_4$. Initially, the borohydride forms a complex with the imine. The carbon gains a hydrogen in the form of a hydride ion, while the boron remains with the nitrogen. A proton transfer followed by the elimination of $-BH_3^-$ results in the secondary amine.

In the final step of the synthesis, the secondary amine is acetylated to an amide using acetic anhydride. The nitrogen performs a nucleophilic attack on the anhydride carbonyl, producing a protonated amine. After a proton transfer, the nitrogen electrons form an iminium ion, as an acetyl group is eliminated. A final proton transfer yields the desired amide product. The entire mechanism is shown in Figure 24.3.

Step 1: Imine formation

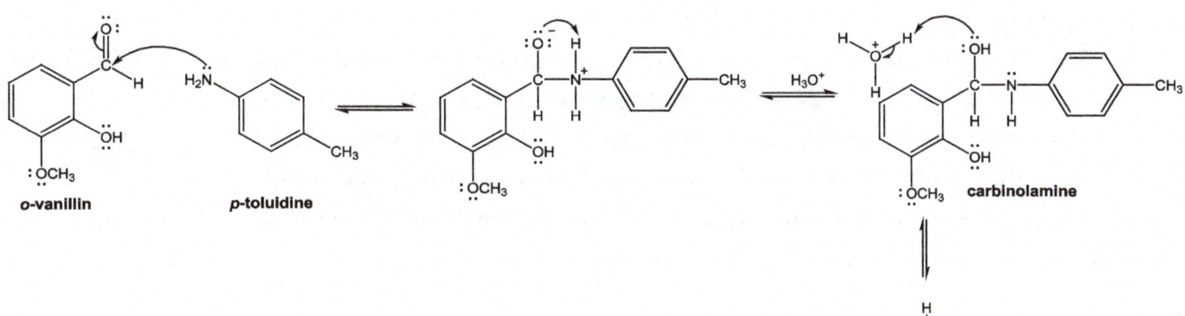

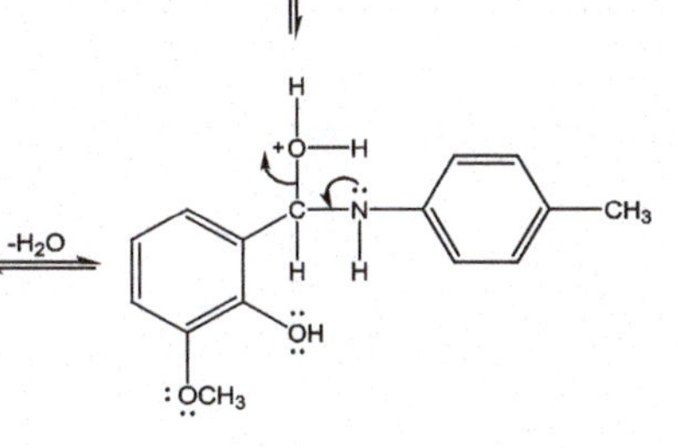

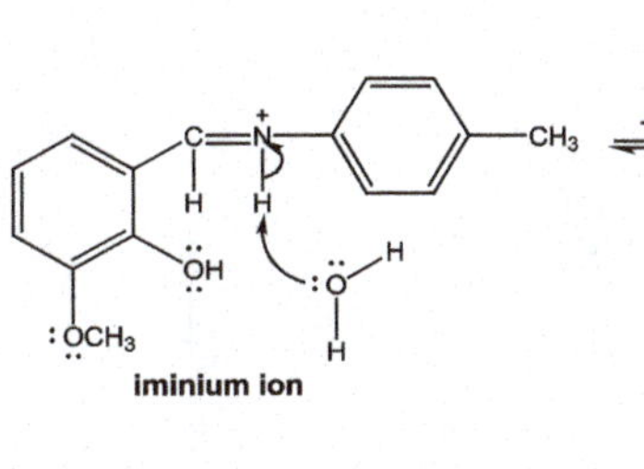

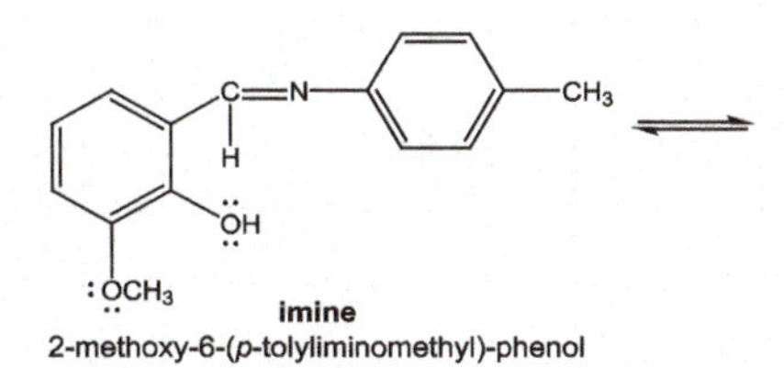

Step 2: Reduction of imine

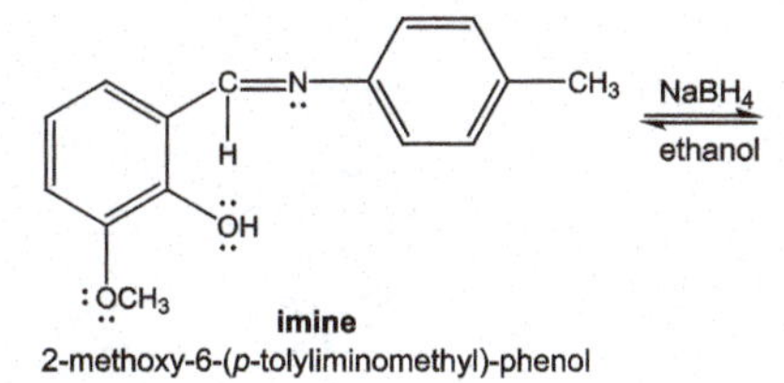

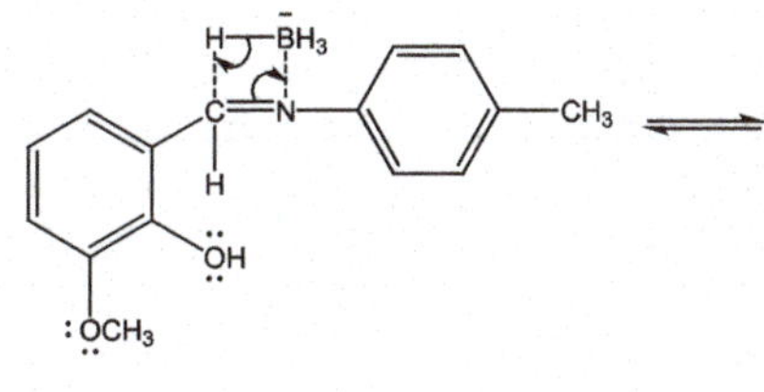

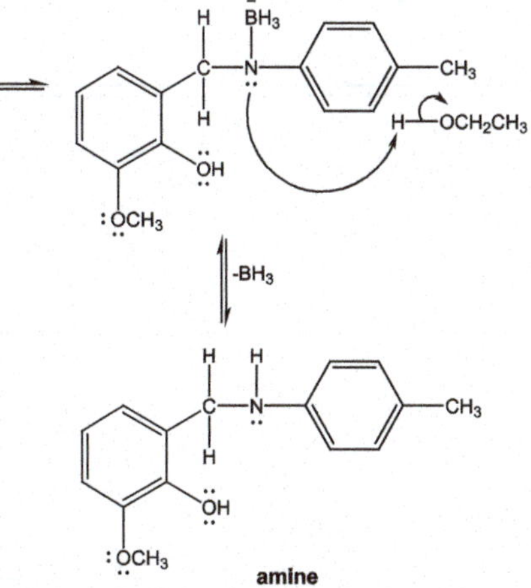

Step 3: Acetylation of amine

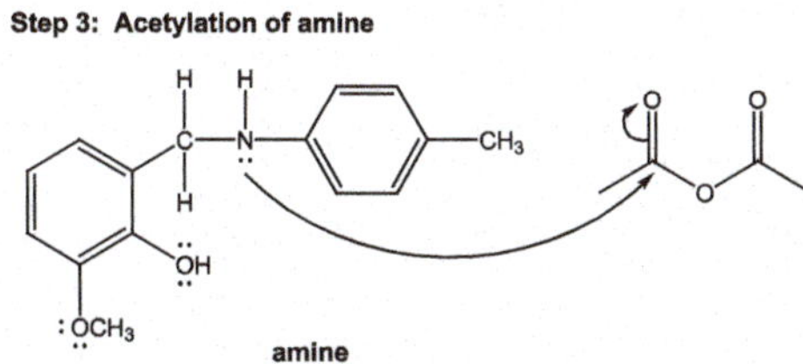

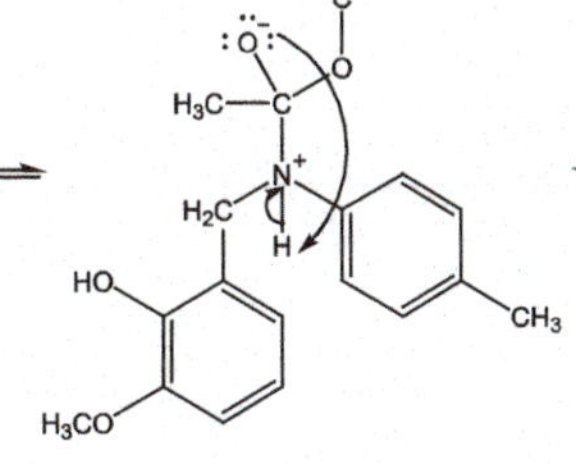

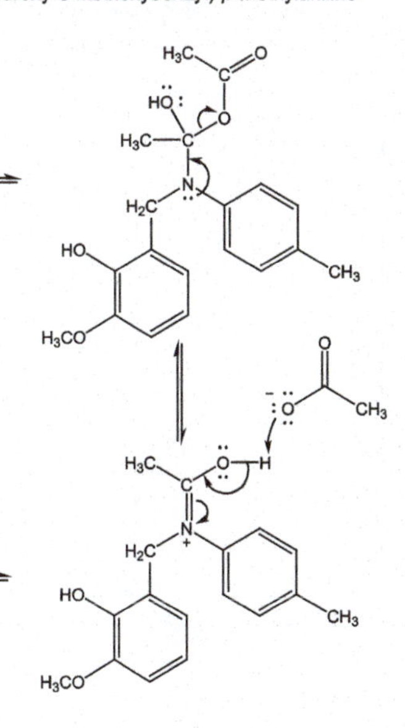

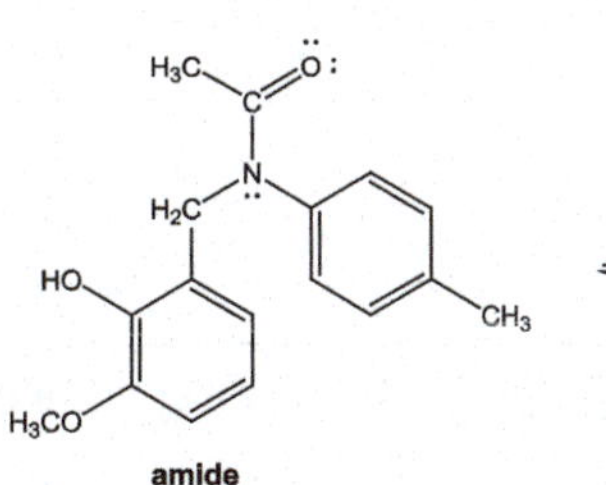

Figure 24.3 Mechanism for three step synthesis of amide.

NMR Spectroscopy

The ^{1}H-NMR spectra of the reactants *o*-vanillin and *p*-toluidine, and the final amide product are shown in Figure 24.4. One notable difference which indicates the conversion of the *o*-vanillin to the product amide is the absence of the aldehyde proton in the product spectrum. The shift in the NH protons between the reactant amine and the product amide are another noticeable difference. Finally, the appearance of the methyl proton signal in the amide product. This signal was not present in the spectra of either reactant.

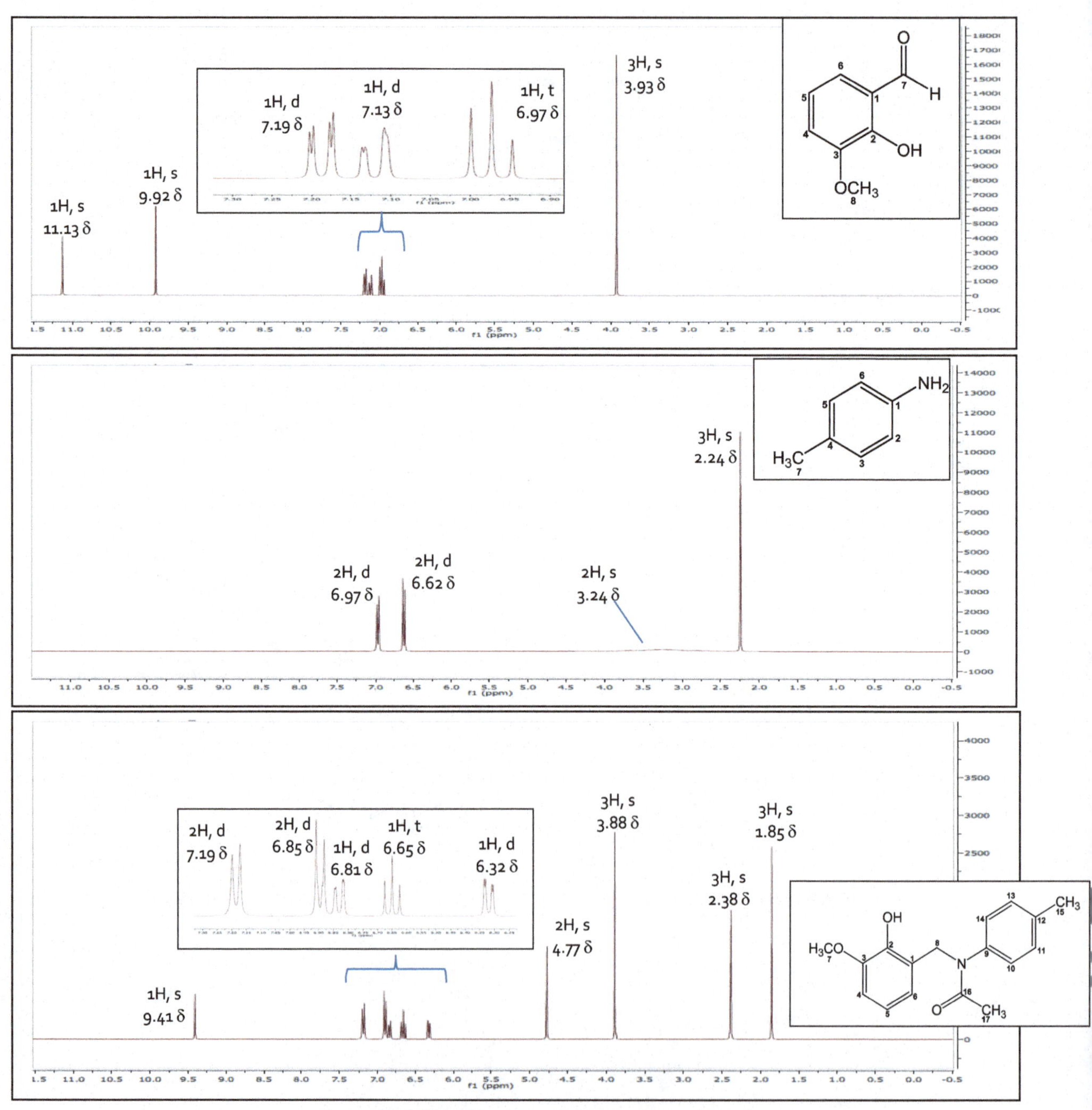

Figure 24.4 ^{1}H NMR spectra of reactants and product.

Objectives

This experiment is an example of a solvent free synthesis in which a primary amine will be converted to an imine. The imine will be reduced to a secondary amine, followed by conversion to an amide via acetylation of the amine. All three steps will be performed in an open beaker during a single lab period. The initial imine product and the final amide product will be analyzed by melting point analysis and TLC analysis. The reactants and final amide product will then be characterized using ^{13}C-NMR spectroscopy.

Experimental Procedure

Synthesis:

Imine formation:

- Weigh an empty 250 mL beaker. Add ~0.76 g of *o*-vanillin directly to the beaker.
- Weigh ~ 0.54 g *p*-toluidine onto a piece of weigh paper. Add this solid to the beaker containing the *o*-vanillin. Observe and record any physical changes in your lab notebook.
- Mix solids together using a glass rod until a dry, homogenous solid forms. Reweigh beaker to determine product yield. Calculate percent yield and record data in Table 24.1 on the POST LAB Assignment (provided online).
- Calculate atom economy, experimental atom economy, cost per synthesis and cost per gram for the reaction. Record data in Table 24.3 on the POST LAB Assignment (provided online).
- Prepare a melting point sample of the solid imine and set aside for *Product Analysis*.

Imine reduction:

- Add ~15 mL 95% ethanol to the reaction beaker.
- Using a magnetic stir bar, stir the reaction on a warm hotplate (setting of 2) until the majority of the solid is dissolved.
- While the solid is stirring, weigh ~ 0.10 g of $NaBH_4$ onto a piece of weigh paper.
- ***SLOWLY*** add the $NaBH_4$ in small increments, while stirring. Stir for 5 minutes.
- Observe and record any changes in your laboratory notebook.

Acetylation of amine:

- Add 2 mL of acetic acid to the beaker while stirring. Stir for 5 minutes. This will neutralize the phenoxide ion.
- Add 2 mL of acetic anhydride. Stir for 5 minutes.
- Add 2-3 boiling chips, then place the reaction beaker in a shallow hot water bath prepared in a 400 mL beaker. Allow the reaction to heat in the water bath for 10 minutes. This will acetylate the amine.
- Move the beaker to a cool stir plate and begin to stir the mixture rapidly.
- ***SLOWLY*** add 75 mL of water, while stirring. This will result in precipitation of the desired amide product from the diluted acetic acid and alcohol.
- Cool the mixture in an ice bath for 10 minutes. Suction filter to isolate the solid product. Rinse the solid twice with 5 mL of ice cold deionized water. Leave the product under vacuum for 10 minutes. ***PROCEED TO PRODUCT ANALYSIS.***

Product Analysis:

TLC Analysis

- Prepare a TLC sample of the solid product by placing a few crystals into a small test tube and adding 1 mL of reagent acetone.
- Perform a TLC analysis on the previously prepared sample against the provided standards. Prepare a single TLC plate with 4 lanes.
- Apply the provided standards of *o*-vanillin, *p*-toluidine, and the desired amide product. Apply the previously prepared sample in the far right lane.
- Check the TLC plate under a UV lamp prior to development to ensure that samples will be visible.
- Develop the TLC plate in 4:1 hexane /ethyl acetate. Visualize spots using UV lamp.
- Circle all spots, and determine the TLC R_f values. Sketch a diagram of the TLC plate in the lab notebook, with all spots identified and cm measurements of all spots and solvent front given. Complete Table 24.4 on the POST LAB Assignment (provided online).

Melting Point Analysis

- Prepare a melting point sample of the solid amide. The solid imine sample was prepared in a previous step.
- Determine the experimental melting range of the imine produced during the first step of the synthesis and the final amide product for comparison. Compare these values with the literature melting points of the imine and amide product in order to determine the degree of purity. Record the melting ranges in Table 24.1 on the POST LAB Assignment.

NMR Characterization

- Using the provided ^{13}C-NMR spectra of the reactants and final amide product (Figure 24.4), identify signals of the assigned carbons. Complete Table 24.2 on the POST LAB Assignment (provided online).

SAFETY

****All experiments should be performed in a fume hood with appropriate safety glasses. Gloves will be provided by request.****

All organic solvents used in this experiment are flammable, irritants, and can be toxic if ingested or absorbed through skin. Acetic acid and acetic anhydride are flammable, corrosive, and acutely toxic if inhaled. Vanillin is toxic if swallowed. Toluidine is toxic if swallowed or inhaled, and is a suspected carcinogen.

WASTE MANAGEMENT

Place all liquid waste in the bottle labeled "LIQUID WASTE—REDUCTIVE AMINATION". Place solid product in bottle labeled "SOLID ORGANIC WASTE". Place all used TLC spotters and melting point capillary tubes in the broken glass container.

Reference

Klein, David. (2015). *Organic Chemistry*, 2nd ed. Hoboken: John Wiley and Sons.

Name:			
	Max	Score	Total Grade
Tech	10		
Pre	20		
In	20		
Post	50		

Exp. 24 Reductive Amination

PRE-LAB ASSIGNMENT: *(**EACH STUDENT** will complete and submit an original copy at the beginning of the lab period. **Without a complete pre-lab assignment, you will not be allowed to perform the experiment, and will receive a zero for the lab.**) .....max score = 20 pts.*

1. **Objective** (*Write a brief purpose of the experiment in **complete** sentences, addressing all of the following points.)*
   - Name the *specific* compounds combined during all steps of the synthesis and name the final amide product formed.
   - What analytical techniques will be used during the experiment to identify and determine purity of product?
   - What type of spectral analysis will be used to characterize the reactants and product?

2. **Chemical Equation** *(Draw the chemical equation showing the reaction scheme for the reductive amination and acetylation using actual chemical structures.)*

3. Physical Data *(Complete the following table before coming to lab.)*

Compound	MW (g/mol)	mp (°C)	bp (°C)	d (g/mL)
o-vanillin			X	X
p-toluidine			X	X
2-methoxy-6-(*p*-tolyliminomethyl)phenol	241.29	103	X	X
sodium borohydride		X	X	X
ethanol		X		
N-(2-hydroxy-3-methoxybenzyl)-*p*-toluidine	243.30	X	X	X
acetic anhydride		X		
acetic acid		X		
N-(2-hydroxy-3-methoxybenzyl)-*N*-*p*-tolylacetamide	285.14	125-127	X	X

4. Experimental Outline *(Give a brief description of the procedure that will be followed in this experiment in 5 lines or less.)*

1	
2	
3	
4	
5	

5. Pre-Lab Questions *(Answer the following questions prior to lab.)*

A. Usually an acid catalyst is used in imine synthesis. Why is it unnecessary to use one in this synthesis?

B. Which of the following compounds are used as the reaction solvent in the synthesis of the imine?

a. water

b. ethanol

c. acetic acid

d. none of the above

I have read and understood the experimental procedure for this experiment. I am familiar with the hazards and the required disposal procedures for this experiment.

Sign here: ____

Experiment 25

Synthesis and Analysis of Commercial Polymers

Introduction

A large number of materials of biological and economic importance are polymers. Polymers are very large molecules and are made by adding many small molecules, called monomers, to form large chains. These high molecular weight macromolecules consist of repeating units of much smaller molecular weight.

Polymers may be natural or synthetic. An example of a naturally occurring polymer is a protein. Proteins are extremely important biologically occurring polymers of amino acids that make up a major portion of our hair, skin, tissue, and cellular contents. A polysaccharide is another example of natural polymer, which is made from smaller carbohydrate units. Starch and cellulose are also examples of naturally occurring polysaccharides. Starch and cellulose are natural polymers, made from repeating units of glucose. In starch the glucose molecules are linked through α–glycosidic linkages. In cellulose, the monomers are linked by β–glycosidic linkages. Both starch and cellulose can be extensively branched (Figure 25.1).

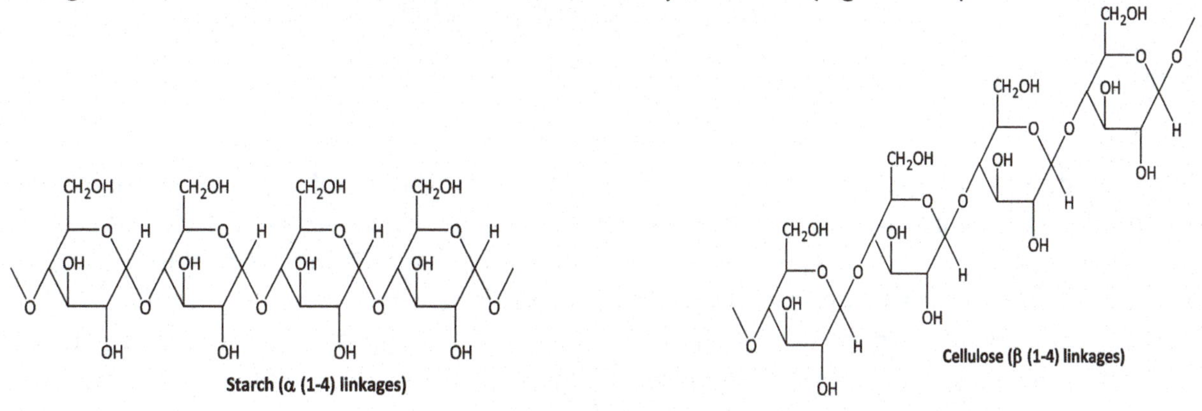

Figure 25.1 Glycosidic linkages of starch and cellulose.

A third class of biologically important polymers consists of the polynucleotides such as deoxyribonucleic acid (DNA) and ribonucleic acid (RNA). The materials that contain all the genetic information of a living cell, DNA and RNA, consist of linear chains of smaller molecules called nucleotides.

Synthetic polymers are of vast economic importance in manufacturing processes. Some of the best-known types of synthetic polymers are nylons, polyesters, acrylics, polyvinyls, and polystyrenes. Nylons, polyesters, and acrylics are used mainly in the clothing industry. Polyvinyls are used to make plastic sheeting and plumbing materials. Polystyrenes are used extensively for insulation materials. Synthetic polymers can be classified in a variety of ways, such as the type of reaction used to make the polymer, the mode of assembly, the structure, or the properties of the polymer.

Classification by Reaction type

Two common types of reactions used to build polymers are addition reactions and condensation reactions. An addition reaction can occur in which two molecules can combine to form a long chain (linear or branched) polymer. These reactions typically occur in monomers which contain π bonds and are also classified as chain-growth polymers due to the method in which the polymerization occurs. For example, in the synthesis of polystyrene, also known as Styrofoam, the linear polymer is produced by a chain-reaction polymerization in which an initiator adds to a carbon-carbon double bond of styrene to yield a reactive intermediate (Figure 25.2). This intermediate reacts with a second molecule of styrene to yield a new intermediate, which reacts with a third unit, and so on. Polystyrenes are used extensively for insulation materials. Tiny bubbles of HCFC gases are trapped between the flat polymer chains during the production of polystyrene. Heat must travel along the polymer chains, making a number of detours around these trapped gases, which are poor thermal conductors thus increasing the usefulness of polystyrenes for insulation materials.

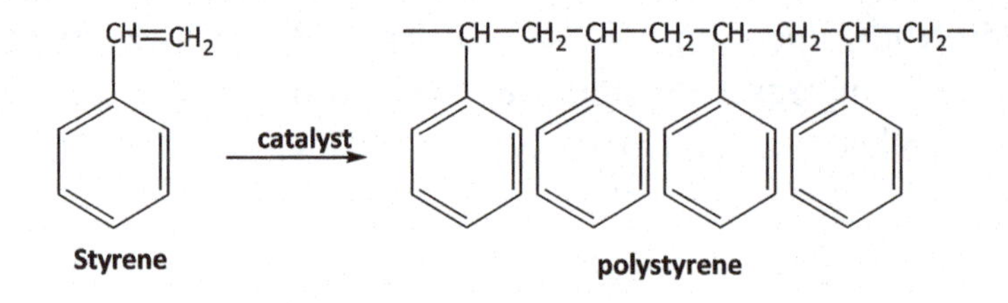

Figure 25.2 Synthesis of polystyrene.

Condensation reactions can also be used to form synthetic polymers. A condensation reaction is one in which two molecules undergo addition accompanied by the loss of a small molecule as a byproduct. One example of a condensation polymer is the polyamide Nylon 6,10 (Figure 25.3). In this reaction, a diacid chloride is reacted with a diamine, resulting in the formation of a polymer and the release of hydrochloric acid as a byproduct. Each diacid chloride and each diamine is capable of reacting twice, enabling the formation of a polymer. Notice that HCl is generated as a byproduct, characteristic of a condensation reaction. Since each bond in the polymer is formed independently of the others, these reactions can also be classified as step-growth polymerizations.

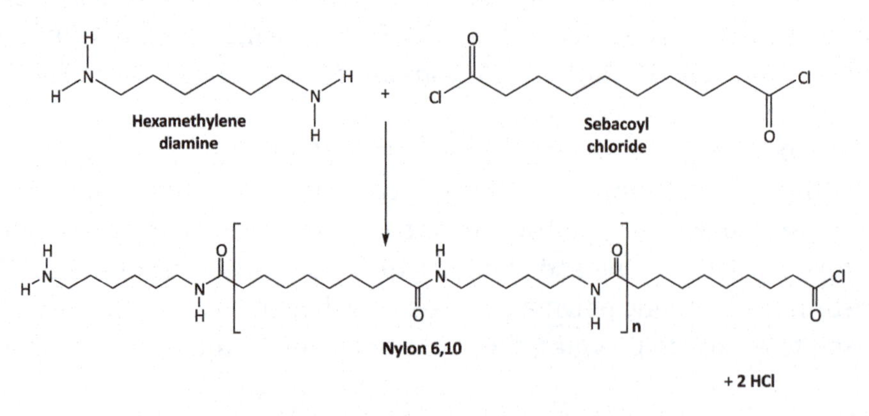

Figure 25.3 Synthesis of nylon 6, 10.

Another example of a condensation polymer is a type of linear polyester, formed by the reaction of phthalic anhydride and ethylene glycol (Figure 25.4). In this type of reaction, an anhydride is reacted with a diol to generate linear polyester, with water formed as a byproduct. This linear polyester is an example of a thermoplastic, which is a type of polymer which is hard at room temperature, but softens upon heating.

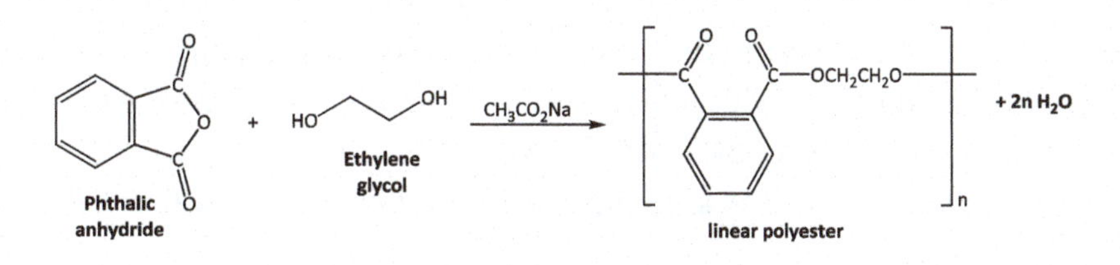

Figure 25.4 Synthesis of a linear polyester.

If more than two hydroxyl groups are present on one of the monomers, the chains can be linked together to form a large, strong three-dimensional structure. These structures are more rigid than their linear counterparts, are often used for paints and coatings, and are referred to as thermosets. Thermosets are very hard and insoluble, even at high temperatures. An example of a cross-linked polymer is that of Glyptal resin, shown in Figure 25.5. This polymer is made using the same phthalic anhydride monomer used to prepare the linear Dacron isomer, however the glycerol monomer contains an additional hydroxyl group (a triol), whereas the ethylene glycol only contains two hydroxyl groups (a diol). The addition of this simple hydroxyl group allows for extensive branching during polymerization.

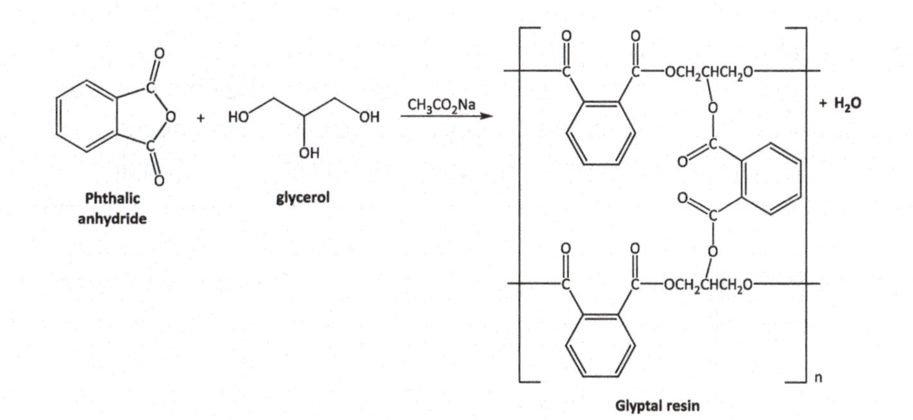

Figure 25.5 Synthesis of Glyptal resin.

Polymers formed via condensation reactions are also classified as step-growth polymers. Step-growth polymers are produced by reactions in which each bond in the polymer is formed independently of the others. The best known step-growth polymers are the polyamides and the polyesters which will be synthesized in this experiment.

Another interesting polymer is the cross-linked polymer commonly known as "slime". Slime is the result of the polymerization of poly (vinyl alcohol), commonly referred to as PVA, with sodium borate decahydrate (Borax) in aqueous conditions. PVA itself is an example of an addition polymer built on a carbon-carbon backbone with a hydroxyl (-OH) group on every other carbon. The presence of these hydroxyl groups give it the ability to form hydrogen bonds with other molecules such as the borate ion ($B(OH)_4^-$).

The borate ion results from the dissolution of Borax (Na_3BO_3) in water. When hydrolyzed, the Borax dissolves to form boric acid (H_3BO_3). The boric acid will then undergo a condensation reaction with PVA. The hydrogen bonds between PVA and the borate ion lead to a three dimensional network with extensive cross-linking (Figure 25.6). These hydrogen bonds are weak, however, allowing them to form and break easily. For this reason, slime is often referred to as a *reversible cross-linking gel*. Evidence of this is shown when a sample of slime is placed on a flat surface. It will slowly flatten out as the bonds break, rearrange, and reform, but not fully break apart. If a great deal of force is applied to the gel, however it will break, as some of the bonds are broken permanently.

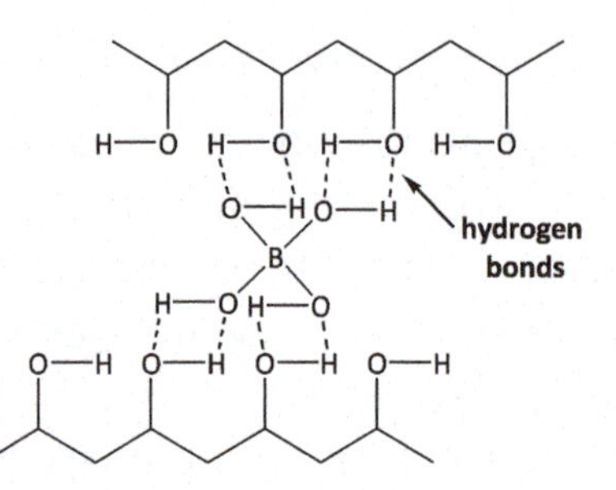

Figure 25.6 Borate ion cross-linking with poly (vinyl alcohol) to form slime.

The physical properties of a polymer, such as melting point and solubility, depend mostly on the intermolecular and intramolecular forces in the polymer. The monomers used to form polymers are generally gases, liquids, or solids, due to their weak intermolecular forces. During the polymerization reaction, the monomers are joined together to form larger molecules, the attractive forces between molecules increase, and the mixture becomes more viscous. Eventually, the molecules become so large that their chains become entangled, and it is this entanglement which gives the polymers their characteristic properties.

A common structural feature in all polymer materials is the presence of chain entanglement, but there are other features which give polymers unique properties such as flexibility of individual polymer chains, strength of the forces between the chains, and the presence of cross-linking between chains. Cross-linking restricts the ability of individual polymer chains to slide past one another and depending on the degree and type of cross-linking, various properties result. Materials with a low degree of cross-linking are elastic and deformable, such as rubber bands. Highly cross-linked materials are more rigid and brittle, as in the case of Bakelite, used in cooking pot handles and billiard balls. Temporarily cross-linked materials behave as viscous, liquid-like gels, as seen in the case of slime (Figure 25.7).

Figure 25.7 Linear, branched, and cross-linked polymer chain structures.

Objectives

In this experiment, three different synthetic polymers will be prepared. Once synthesized, the length of the polyamide strand prepared during synthesis will be determined. The elasticity of the slime prepared will be investigated. Finally, the physical properties of the polyester formed will be studied through solubility testing.

Experimental Procedure

Synthesis:

Nylon 6, 10

- Pour 3 mL of the hexamethylene diamine solution into a 50 mL beaker.
- Pour 6 mL of the sebacoyl chloride solution into a 100 mL beaker.
- *Slowly* pour the sebacoyl chloride solution into the beaker containing the hexamethylene diamine solution. A white film should form at the interface of the two layers.
- Reach into the beaker with forceps and grasp the film in the center of the beaker. Slowly pull straight up. Try not to let the thread of nylon touch the sides of the beaker.
- Pull the nylon from the beaker and wrap the thread around the tip of a large test tube. Rotate the tube slowly, counting revolutions, until no more nylon can be obtained.
- Record the number of revolutions of the nylon, along with the circumference and diameter of the test tube used in the data section of your laboratory notebook.
- Cover the nylon on the test tube *entirely* with a paper towel, and carefully remove the nylon from the test tube. Place the nylon in a small beaker of tap water for 2-3 minutes, then remove and blot the nylon several times to remove excess water. Continue to blot the nylon until the sample is fairly dry. ***Set sample aside for product analysis.***

Slime

- Place 50 mL of 4% polyvinyl alcohol solution into a 150 mL beaker. Add one drop of food coloring (optional), and stir with a wooden stick.
- Pour 5 mL of 4% sodium borate solution into the beaker and stir until smooth.
- Remove the polymer from the beaker and transfer to a small piece of aluminum foil.
- ***Set sample aside for product analysis.***

Glyptal Resin

- Weigh between 1.2-1.3 g phthalic anhydride into an aluminum weigh boat.
- Weigh between 0.5-0.6 g anhydrous sodium acetate onto a piece of weigh paper.
- Return to fume hood and transfer the sodium acetate to the weigh boat containing the solid anhydride. Carefully mix the solids together using a wooden stick to form a homogenous sample.
- Using a glass dropper, transfer 15 drops (0.5 mL) of glycerol to the solid sample. Stir to mix.
- Place the aluminum weigh boat directly on a hotplate (setting of 6), and heat while stirring with a wooden stick until a white paste forms.

- Continue to heat until the solution appears to boil. Allow the solution to boil for 5 minutes.
- Remove the weigh boat from the hotplate using tweezers. Allow the sample to cool to room temperature.
- To remove the polyester polymer from the weigh boat, simply fold the weigh boat backwards until the polymer separates from the foil. ***Set sample aside for product analysis.***

Product Analysis:

Nylon String Length

- During the synthesis of nylon, the number of revolutions made as the nylon string was wrapped around the test tube was recorded. Using this information, along with the diameter of the test tube, calculate the length of the nylon string using the following formula:

Nylon produced (mm) = (Diameter of test tube) * (π) * (# test tube revolutions)

- Record this data in the laboratory notebook and complete Table 25.1 on the POST LAB Assignment (provided online).

Physical Properties of Slime

- Separate the slime into two separate portions.
- Knead the two slime portions into separate balls. Hold a small part of one slime ball and slowly pull the polymer to stretch it without breaking it. Measure the length of the slime.
- Using the other slime ball, pull apart quickly to see if it breaks when pulled apart.
- Record this data in the laboratory notebook and complete Table 25.2 on the POST LAB Assignment (provided online).

Solubility Testing of Glyptal Resin

- Identify all of the IMF present in the Glyptal resin and solvent molecules. Based on the similarities or differences in the IMF present between the solute and solvent, predict the solubility of the polymer in each solvent.
- Verify the solubility of the polymer in each solvent. Add 3 mL of acetone to the first test tube, 3 mL of toluene to the second test tube, and 3 mL of deionized water to the third test tube.
- Add a very small amount of the Glyptal resin polymer to each test tube. No weight is necessary, however only an amount equivalent the inside of the circle shown should be used (—**O**).
- Mix the contents in the tube with a glass rod. Clean the glass rod between solvents by simply wiping it thoroughly on a paper towel.
- If the polymer does not dissolve completely in the solvent, record the solubility as ***INSOLUBLE*** in the laboratory notebook. If the polymer does dissolve completely, record the solubility as ***SOLUBLE***. Complete Table 25.3 on the POST LAB Assignment (provided online).

SAFETY

****All experiments should be performed in a fume hood with appropriate safety glasses. Gloves will be provided by request.****

All organic compounds used in this experiment are irritants, and can be toxic if ingested or absorbed through skin. Sebacoyl chloride, hexamethylene diamine, and sodium hydroxide are corrosive. Hexane and toluene are suspected teratogens.

WASTE MANAGEMENT

Pour the remaining contents of the beaker in which you synthesized the nylon and any liquids used during the solubility testing into the container labeled "LIQUID WASTE—POLYMERS" located in the waste hood. Place solid nylon product and aluminum foil in yellow solid waste can.

References

Klein, David. (2015). *Organic Chemistry*, 2nd ed. Hoboken: John Wiley and Sons.

Hart, Harold, Craine, Leslie E., and Hart, David J. (1999). *Organic Chemistry Laboratory Manual: A Short Course*, 10th ed. Boston, MA: Houghton Mifflin Company.

Name:			
	Max	Score	Total Grade
Tech	10		
Pre	20		
In	20		
Post	50		

Exp. 25 Synthesis and Analysis of Commercial Polymers

PRE-LAB ASSIGNMENT: *(**EACH STUDENT** will complete and submit an original copy at the beginning of the lab period. **Without a complete pre-lab assignment, you will not be allowed to perform the experiment, and will receive a zero for the lab.**) .....max score = 20 pts.*

1. **Objective** *(Write a brief purpose of the experiment in **complete** sentences, addressing all of the following points.)*
    - Name the *three* polymers formed in this experiment and the *specific* chemicals used to synthesize them.
    - What physical property will be determined for the polyamide polymer?
    - What physical property will be investigated on the polyester polymer?

2. **Compound Structures** *(Draw the compound structure of the repeating unit for each polymer.)*

Nylon 6,10

Slime

Cross-Linked Polyester

3. Experimental Outline *(Give a brief description of the procedure that will be followed in this experiment in 5 lines or less.)*

1	
2	
3	
4	
5	

4. Pre Lab Questions *(Answer the following questions prior to lab.)*

A. A common structural feature in all polymer materials is the presence of chain entanglement, but there are other features which give polymers unique properties. Name **3** other features.

a.

b.

c.

B. Describe the difference between a ***step growth polymer*** and a ***chain growth polymer***. Give an example of each.

I have read and understood the experimental procedure for this experiment. I am familiar with the hazards and the required disposal procedures for this experiment.

Sign here: ____

Experiment 26

Synthesis and GC Analysis of Fatty Acid Methyl Esters

Introduction

A fatty acid is a carboxylic acid with a long aliphatic carbon chain, which can be either saturated or unsaturated. Chemically, fats and oils are triacylglycerols, or TAGs, which are triesters of glycerol with three long-chain carboxylic acids. Hydrolysis of a fat or oil with a base (saponification) results in a glycerol and three fatty acid salts. Acidification then gives the fatty acids (Figure 26.1).

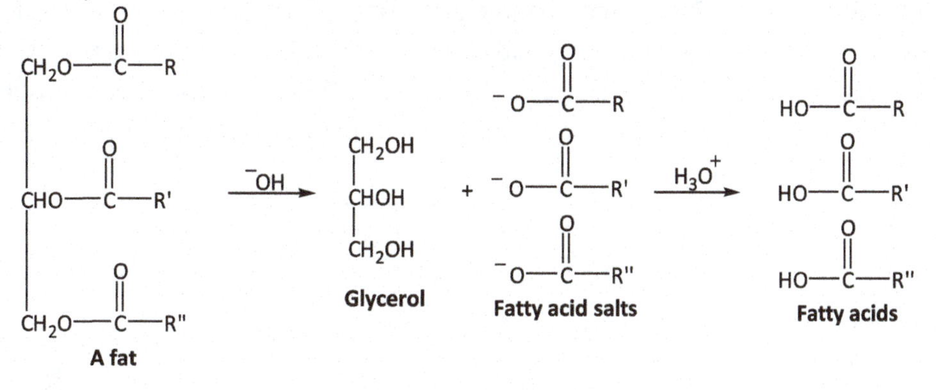

Figure 26.1 Hydrolysis of fat.

Fatty acids derived from natural fats and oils have at least eight carbon atoms. The most common fatty acids in nature have an even number of carbon atoms, typically between 14 and 22 carbons. There may or may not be double bonds within these long carbon chains. Saturated fatty acids (SFA) do not contain any double bonds. Most saturated fats in nature tend to be solids. Saturated fats are straight chain molecules, which can pack together very closely. This close contact between fat molecules results in greater London dispersion forces between them; therefore, they have a higher melting point.

An unsaturated fatty acid has at least one double bond, if not many. These double bonds occur almost exclusively in the *cis* configuration. Those containing only one double bond are called monounsaturated fatty acids (MUFAs). Those containing more than one double bond are called polyunsaturated fatty acids (PUFAs). A high percentage of unsaturated fatty acids tend to be liquids or oils. The greater the number of *cis* double bonds present, the more bent the molecules will be (Figure 26.2). This "kink" in the molecule prevents close packing of the molecules. The London dispersion forces between the molecules are reduced; thus, the melting point is lower than a saturated fatty acid. Most fatty acids in the *trans* configuration (trans fats) are not found in nature, but are actually the result of human processing such as hydrogenation. *Trans* fatty acids have been implicated in a variety of conditions, such as heart disease, cancer, and diabetes. This shows the importance of the differences in the geometry between various types of unsaturated fatty acids, as well as between saturated and unsaturated fatty acids. They play a very important role in biological processes.

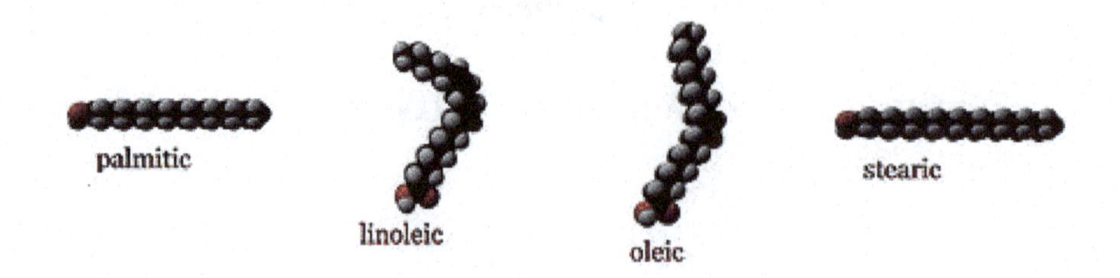

Figure 26.2 Shapes of common fatty acids.

Many fatty acids are known best by their common names, such as palmitic, oleic, and linoleic acids (Figure 26.3). The IUPAC names for those three fatty acids are hexadecanoic acid, 9-octadecenoic acid, and 9, 12-octadecadienoic acid. These fatty acids are often abbreviated by noting the number of carbons in the fatty acid chain followed by the number of double bonds in the chain. The location of the double bonds is identified with a Δ, followed by the number of the first carbon of the double bond(s). The abbreviations of the fatty acids listed above are 16:0, 18:1 Δ_9, and 18:2 $\Delta_{9,12}$, respectively.

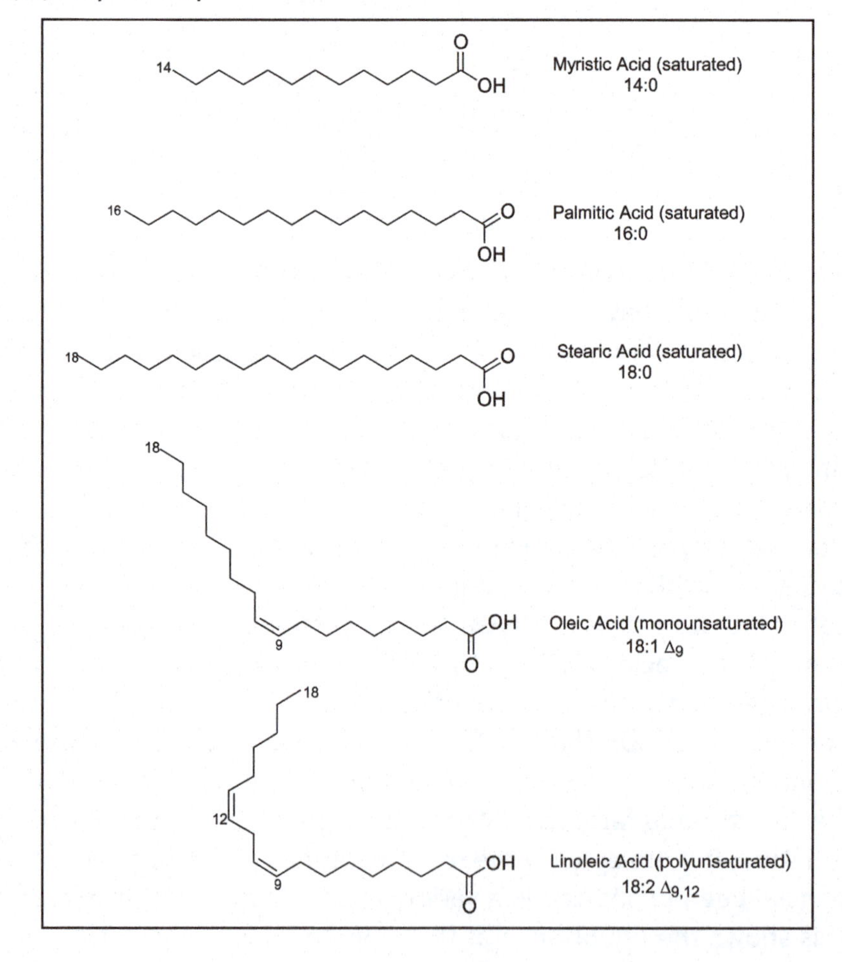

Figure 26.3 Structures of fatty acids.

Certain oils are considered to be nutritionally better than others.[1] Essential fatty acids are important in several human body systems, including the immune system and blood pressure regulation. An essential fatty acid in our diet is linoleic acid (18:2 $\Delta_{9,12}$) because we cannot synthesize it ourselves. Fish oils are now thought to be very healthy for us in that they have significant amounts of omega-3 fatty acids, such as docosahexaenoic acid (DHA = 22:6 $\Delta_{4,7,10,13,16,19}$) and eicosapentaenoic acid (EPA = 20:5 $\Delta_{5,8,11,14,17}$).

Vegetable oils are also one source of biodiesel. Soybean oil is a major component in biodiesel, a fuel for trucks, cars, and buses. Biodiesel fuel has fewer harmful emissions compared to petroleum diesel. It emits up to 100% less sulfur dioxide, a major component of acid rain, and 80–100% less carbon dioxide than traditional diesel fuel. Pure biodiesel is biodegradable and breaks down four times faster than regular diesel, and has a much higher flash point than traditional petroleum, making it much safer to transport and handle. Since biodiesel can be used in conventional diesel engines, this renewable fuel can directly replace petroleum products, thus reducing the country's dependence on imported oil.

Biodiesel can be made by mixing methanol and sodium hydroxide (lye) to make sodium methoxide ($CH_3O^- Na^+$). The sodium methoxide is then mixed with vegetable oil to form the fatty acid methyl esters and glycerol through transesterification. Due to the difference in density, the glycerol sinks to the bottom, while the fatty acid methyl esters (biodiesel) float to the top. The biodiesel can then be siphoned off easily and washed with water to remove any residual catalyst or soap, a common by-product formed during the transesterification process.

The following chart lists the relative fatty acid composition of several oils. Based on the identification and quantification of the component fatty acid methyl esters through gas chromatography, the relative percentages of saturated (SFA), monounsaturated (MUFA), and polyunsaturated (PUFA) fatty acids can be determined. The unknown vegetable oil can be identified using Table 26.1.

	Saturated Fatty Acids				**Monounsaturated Fatty Acids**	**Polyunsaturated Fatty Acids**
Name of Oil	Myristic **C_{14}** **(14:0)**	Palmitic **C_{16}** **(16:0)**	Stearic **C_{18}** **(18:0)**	**Total % SFAs**	Oleic **C_{18}** **(18:1Δ_9)**	Linoleic **C_{18}** **(18:1$\Delta_{9,12}$)**
Corn	**0**	**11**	**2**	**13**	**28**	**58**
Olive	**0**	**13**	**3**	**16**	**71**	**10**
Peanut	**0**	**11**	**2**	**13**	**48**	**32**
Soybean	**0**	**11**	**4**	**15**	**24**	**54**
Safflower	**0**	**7**	**2**	**9**	**13**	**78**
Cottonseed	**1**	**22**	**3**	**26**	**19**	**54**
Palm	**1**	**45**	**4**	**50**	**40**	**10**

Table 26.1 Relative fatty acid composition of common oils.

Preparation of FAMEs from Oils

Fatty acids in fats and oils exist primarily as triglycerides. To produce the methyl esters of the fatty acids, you will perform a base-catalyzed transesterification of the glycerol–fatty acid esters (Figure 26.4). The nonpolar FAMEs can then be extracted into hexane, leaving polar impurities in the polar methanol layer. After extraction to isolate the nonpolar FAMEs, you will purify the sample by column chromatography using a miniature SiO_2 column. TLC analysis will be used to ensure that all of the triglycerides have been converted to FAMEs, and that your sample is free of polar (and therefore nonvolatile) impurities. The sample must be free of nonvolatile impurities before injecting it into the GC system, since the mobile phase in the system is a gas.

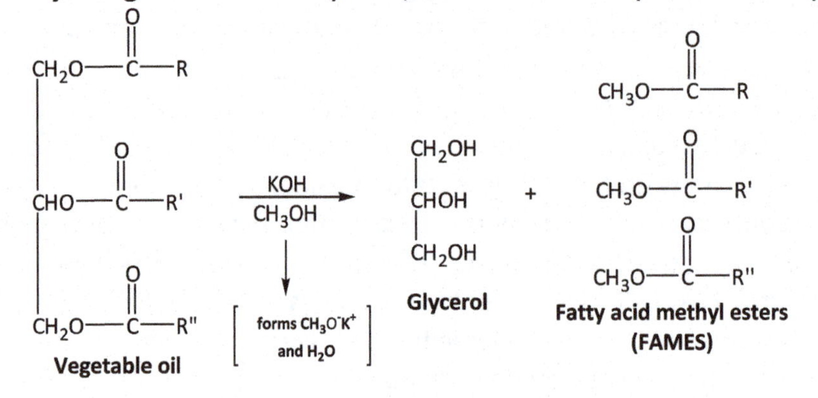

Figure 26.4 Transesterification of vegetable oil.

GC Analysis of Fatty Acid Methyl Esters

A powerful technique for identification and quantification of fatty acids in natural oils is through gas chromatographic analysis of the methyl esters of fatty acids. Fatty acid methyl esters (FAMEs) are more volatile than the original fatty acids, since the polar carboxylic acid has been converted to a relatively nonpolar methyl ester. The methyl esters also have better chromatographic characteristics. When underivatized fatty acids are analyzed by GC, the dual nature of the fatty acid (nonpolar tail vs. polar carboxylic acid group) often results in peak tailing and broadening, making efficient separation of fatty acid mixtures difficult. These problems are reduced to such an extent for FAMEs, that separation of FAMEs, which differ only in position or geometry (*cis* vs. *trans*) of double bonds, can be separated by capillary GC.

The mechanism of separation of components in a mixture depends on the chemical nature of both stationary phase and the components present. Typically, when a nonpolar packing material is used, components with *similar polarity* elute in order of volatility. The volatility of FAMEs depends on both the size and the unsaturation of the FAME. Shorter FAMEs are more volatile than longer FAMEs and polyunsaturated FAMEs are generally more volatile than monounsaturated and saturated FAMEs.

In gas chromatography, separated components are detected as they elute from the GC column by a flame ionization detector (FID) and a GC chromatogram of the mixture is obtained. The flame ionization detector is a very sensitive detector for organic compounds, especially those with a lot of saturated carbon atoms, such as FAMEs. As analytes elute from the column they pass through a flame that pyrolyzes the organic compound, producing ions and electrons that conduct electricity through the flame. The number of ions produced is roughly proportional to the number of reduced carbon atoms in the analytes.

When compared to the GC chromatogram of a standard solution, the identification of the FAMEs present can be determined by comparison of sample retention times to standard retention times. As shown in Figure 26.5, the smaller FAMEs, such as methyl palmitate, precede longer FAMEs, such as methyl linoleate, methyl oleate, and methyl stearate. Within the C18 FAMEs, more volatile polyunsaturated fatty acid methyl esters precede monounsaturated fatty acid methyl esters, which precede the less volatile saturated fatty acid methyl esters. In general, this will be the elution order of any fatty acid methyl esters in your unknown oil, but as always, you must use the chromatogram of the standard solution run with YOUR sample set, as the retention times may vary slightly. You will identify FAMEs in your unknown fats and oils by comparing retention times of standard FAMEs with those from your preparations.

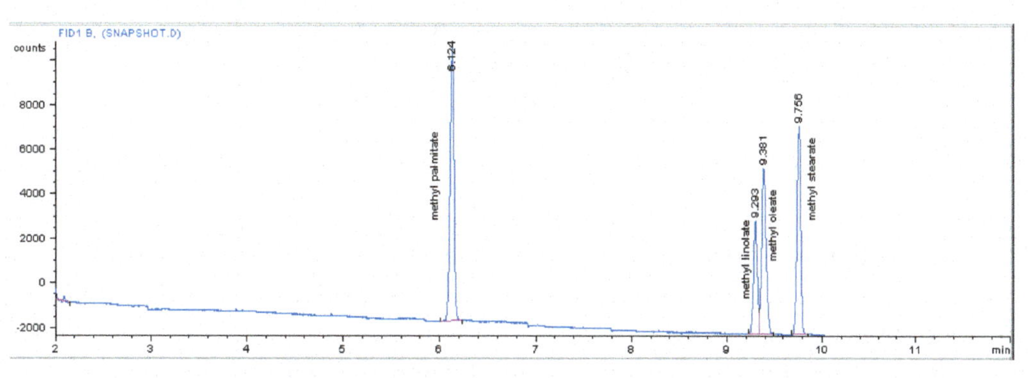

Figure 26.5 GC chromatogram of a FAMEs standard solution.

Objectives

In this experiment you will perform a transesterification on an unknown vegetable oil and purify it through extraction and column chromatography. You will perform TLC analysis on the sample to ensure purity. Using GC analysis, you will identify and quantify the FAMEs contained in your unknown oil. Based on the identity and quantity of the FAMEs in your unknown, you will identify the unknown oil using Table 26.1.

Experimental Procedure

Synthesis:

- Add 10 drops of an unknown oil and 1.0 mL methanol to a 5.0 mL conical reaction vial.
- Add a small stir bar and 1.0 mL of 1M KOH solution to the vial.
- Place the vial in the aluminum heating block on a hot-plate stirrer, and begin stirring the solution.
- Attach water hoses to a small reflux condenser, and then place the condenser atop the conical reaction vial, leaving the top of the condenser open. Begin water flow.
- Place a thermometer in the appropriate slot in the heating block, and set the hot-plate stirrer on a setting of 2.5–3. Once the solution begins to reflux, adjust the temperature as necessary to maintain a reaction temperature between 70°C and 80°C. ***Do not allow the temperature to exceed 100°C!***
- Reflux the solution for 30 minutes.
- Remove the condenser, and place the vial in a 50 mL beaker.
- Allow the reaction flask to cool to room temperature.

Purification:

- Add 2.0 mL of hexane to the reaction vial. Cap the vial and shake to mix. Allow the layers to separate. It may take several minutes for the separation to occur.
- Remove the top organic layer with a Pasteur pipette and run this solution through a mini SiO_2 gel column (prepared for you). Collect the eluent directly into a small test tube.
- Add an additional 1.0 mL of hexane to the reaction vial (which still contains the aqueous layer). Cap vial and shake to mix. Allow the layers to separate.
- Remove the top layer with a Pasteur pipette and run this solution through the mini SiO_2 gel column. Collect the eluent into the same small test tube as before.
- Proceed to *PRODUCT ANALYSIS* with this eluent.

Product Analysis:

TLC Analysis

- Prepare a TLC plate with three lanes. Apply the original unknown oil provided, the eluent from above, and the provided standard solution of the FAMEs. Develop the TLC plate in 85:15 hexane/ diethyl ether.
- Since the FAMEs cannot be seen under the UV lamp, you will need to stain the TLC plate by dipping it in a solution of 10% H_2SO_4 (in ethanol). ***USE EXTREME CAUTION WHEN HANDLING THIS SOLUTION! TWEEZERS AND SAFETY GOGGLES ARE REQUIRED!***
- Holding your TLC plate with a pair of tweezers, quickly immerse the entire plate in the stain solution.
- Remove the plate and remove excess stain solution by tipping the plate on the corner, blot it on a paper towel.
- Place the plate directly on a warm hot plate and heat until spots appear.
- Circle all spots and sketch a diagram of your TLC plate in the laboratory notebook and in Table 26.1 on the POST LAB Assignment (provided online).
- Have the instructor inspect the TLC plate for impurities prior to preparing the GC sample for analysis. Your FAME solution is pure enough for GC analysis if your TLC shows no traces of unreacted triglycerides (biggest component of starting material).

GC Analysis

- If TLC analysis indicates that your product is free of triglycerides, prepare a GC sample by transferring five drops of your FAME's solution to an autoanalyzer vial.
- Add 1.0 mL of GC solvent to your sample and place in the sample tray for analysis.
- Once the sample GC chromatogram has been returned to you, compare the retention times of the FAMEs in your sample with those in the provided standard chromatogram in order to identify the components in your sample.
- Tabulate the retention times of the FAMEs in the standard, those in your sample chromatogram, and the adjusted area percent of the FAMEs peaks (the solvent is not visible on the chromatogram but it is still detected, so calculation of adjusted area percent is still necessary).
- Complete Table 26.2 on the POST LAB Assignment (provided online).

- Once the FAMEs in your unknown oil are identified and quantified, use the information in Table 26.1 in the lab manual to identify your unknown oil. Structures of the individual FAMEs are located in Figure 26.3.
- Complete Table 26.3 on the POST LAB Assignment (provided online).

SAFETY

All experiments should be performed in a fume hood with appropriate safety glasses. Gloves will be provided by request.

All chemicals used in this experiment can be toxic if ingested or absorbed through skin. Potassium hydroxide and sulfuric acid are corrosive. Use extreme caution when handling these chemicals. Hexane, methanol, and diethyl ether are flammable solvents.

WASTE MANAGEMENT

Place all liquid waste in the container labeled "LIQUID WASTE—FAMES" Located in the waste hood. Place TLC plates and filter papers from the TLC chamber in the yellow solid waste can. Place used TLC capillaries, used SiO_2 columns, and used glass Pasteur pipettes in the broken glass container.

References

Christie, W. W. (1983). *Lipid Analysis: Isolation, Separation, Identification and Structural Analysis of Lipids*. Oxford: Pergammon.
Klein, David. (2015). *Organic Chemistry*, 2nd ed. Hoboken: John Wiley and Sons.
Quigley, Michael N. (1992). "The Chemistry of Olive Oil". *J. Chem D, 69,* 332–335.

Name:			
	Max	Score	Total Grade
Tech	10		
Pre	20		
In	20		
Post	50		

Exp. 26 Synthesis and GC Analysis of Fatty Acid Methyl Esters

PRE-LAB ASSIGNMENT: *(**EACH STUDENT** will complete and submit an original copy at the beginning of the lab period. **Without a complete pre-lab assignment, you will not be allowed to perform the experiment, and will receive a zero for the lab.**) .....max score = 20 pts.*

1. Objective *(Write a brief purpose of the experiment in **complete** sentences, addressing all of the following points)*

- Name the *specific* compounds combined during the synthesis and the name the *type* of product formed.
- What is the purification technique(s) used in this experiment?
- What analytical technique will be used during the experiment?

2. Chemical Equation *(Draw the chemical equation for the transesterification of vegetable oil using actual chemical structures.)*

3. Physical Data *(Complete the following table before coming to lab.)*

Compound	MW (g/mol)	mp (°C)	bp (°C)	d (g/mL)
potassium hydroxide			X	X
diethyl ether		X		
sulfuric acid		X		
hexane		X		
methanol		X		

4. Experimental Outline *(Give a brief description of the procedure that will be followed in this experiment in 5 lines or less.)*

1	
2	
3	
4	
5	

5. Pre Lab Questions *(Answer the following questions prior to lab.)*

A. The double bonds in unsaturated fatty acids occur almost exclusively in the ***trans*** configuration.

a. True
b. False

B. What does the symbol Δ refer to in the nomenclature of fatty acids?

I have read and understood the experimental procedure for this experiment. I am familiar with the hazards and the required disposal procedures for this experiment.

Sign here: ____

Appendix A

Typical Glassware Setups

Simple Distillation Apparatus

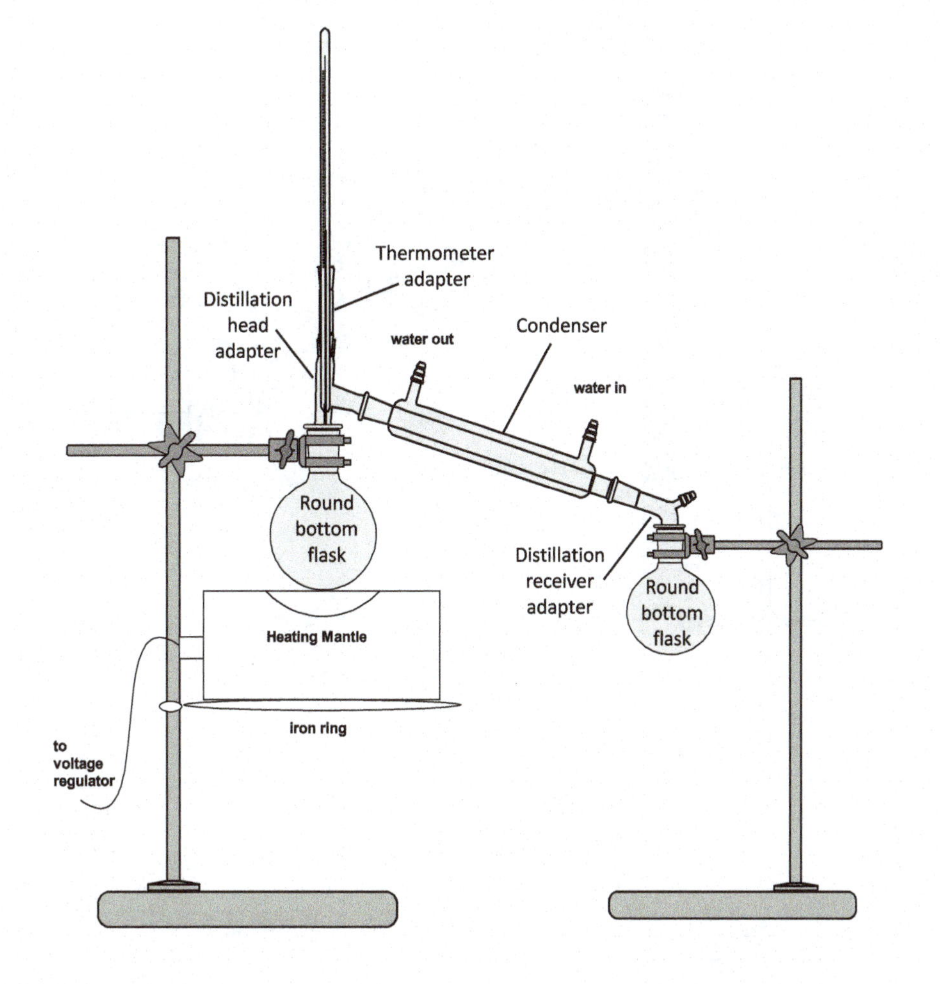

COMPONENTS:

Round bottom flask (reaction flask)
Distillation head adapter
Thermometer with adapter
Condenser (with clear hoses)
Distillation receiver adapter
Round bottom flask (receiving flask)
Heating mantle and voltage regulator

Fractional Distillation Apparatus

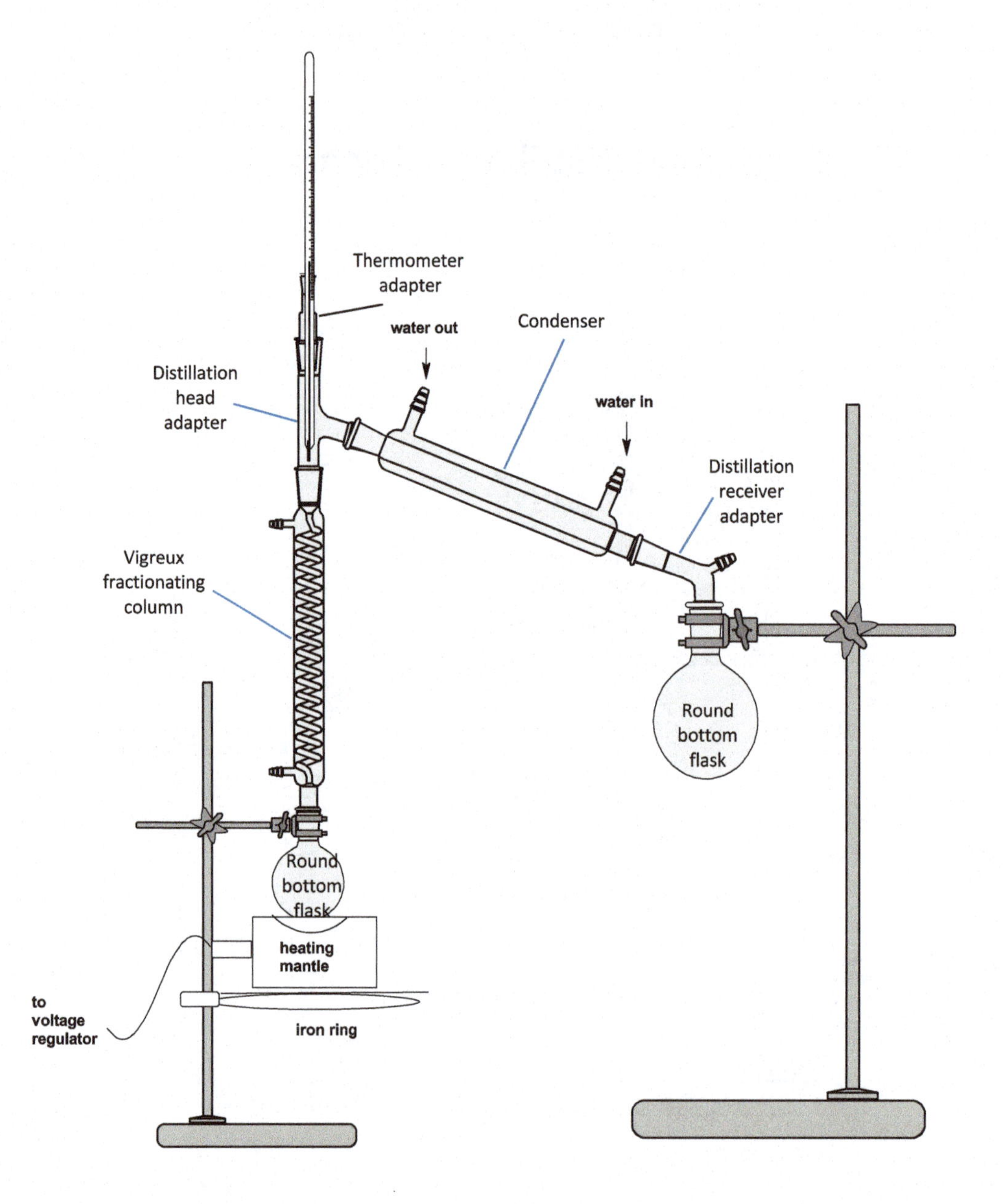

COMPONENTS:

Round bottom flask (reaction flask)
Vigreux fractionating column
Distillation head adapter
Thermometer with adapter
Condenser (with clear hoses)
Distillation receiver adapter
Round bottom flask (receiving flask)
Heating mantle and voltage regulator

Suction Filtration Apparatus

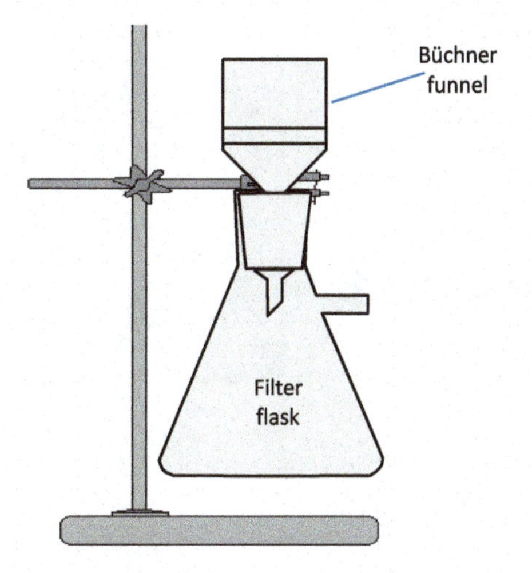

COMPONENTS:

Ring stand with clamp
Filter flask (with red vacuum hose, not shown)
Büchner funnel with filter paper

Extraction Apparatus

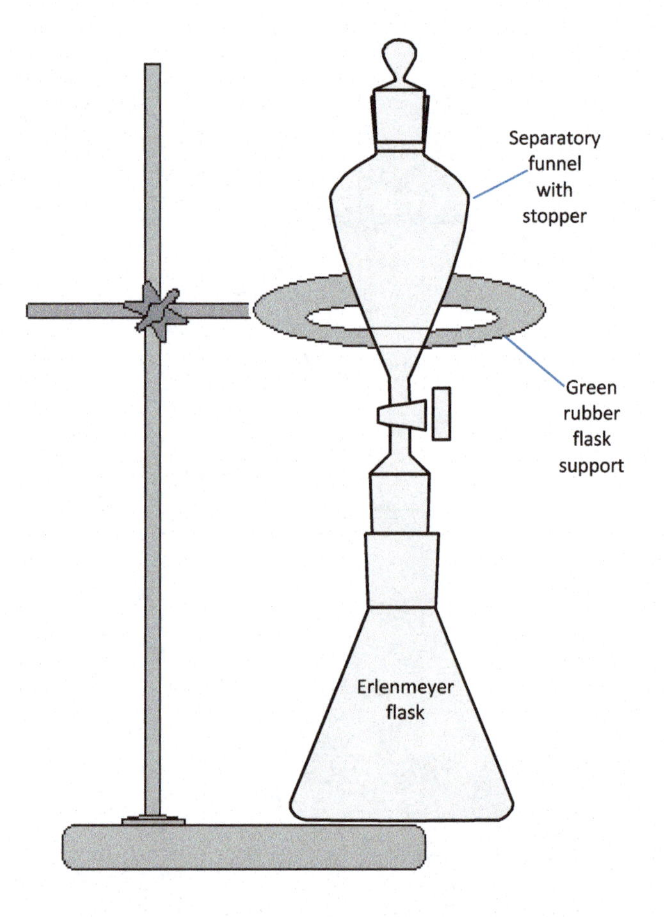

COMPONENTS:

Ring stand with iron ring and green rubber flask support
Separatory funnel with glass stopper
Erlenmeyer flask

Column Chromatography Apparatus

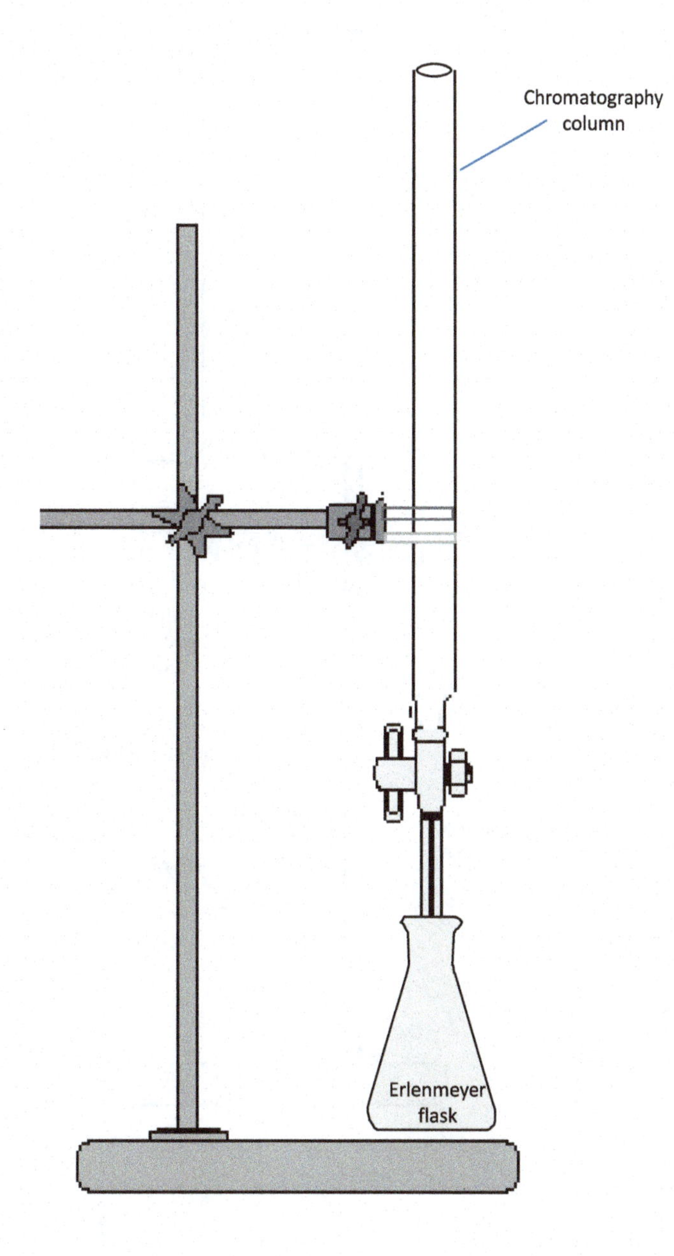

COMPONENTS:

Ring stand with clamps
Chromatography column (packed with adsorbent)
Erlenmeyer flask

Reflux Apparatus

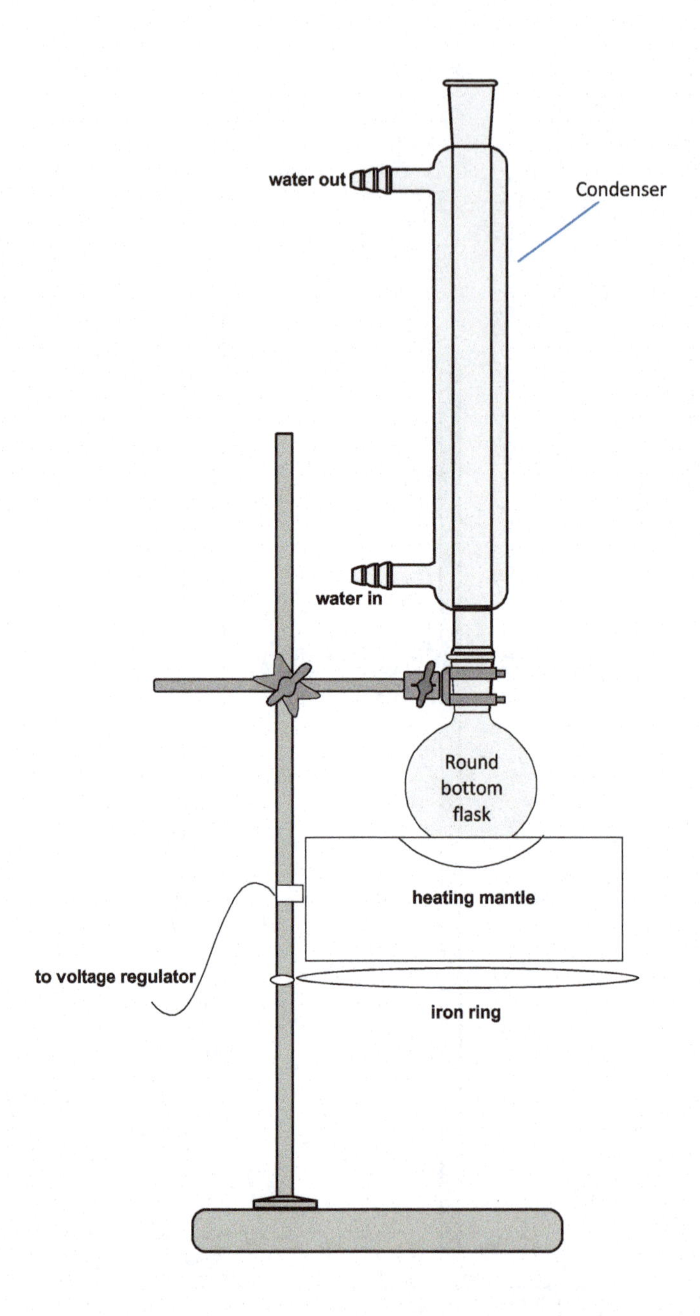

COMPONENTS:

Round bottom flask
Condenser (with clear hoses)
Heating mantle and voltage regulator

Reflux with Addition Apparatus

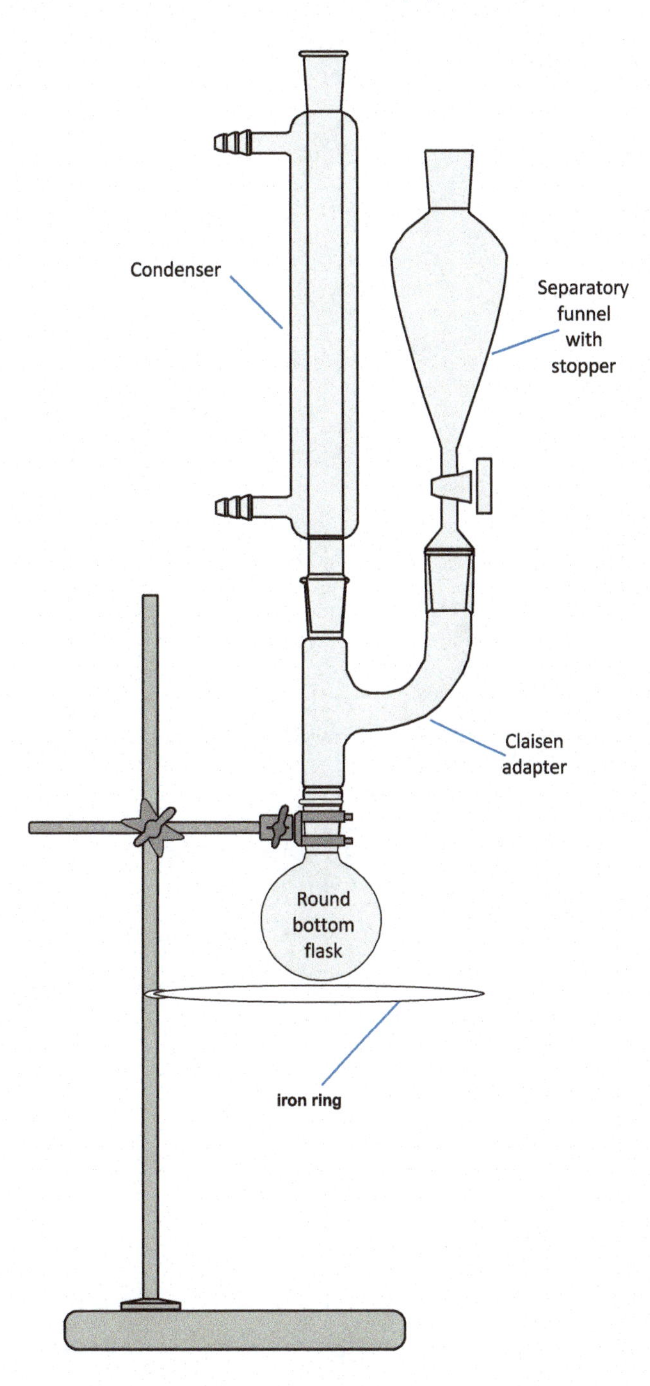

***COMPONENTS*:**

Round bottom flask
Claisen adapter
Condenser (with clear hoses)
Separatory funnel with glass stopper

Appendix B

Care of Equipment

Standard Taper Glassware

In addition to test tubes, beakers, Erlenmeyer flasks, and other familiar equipment, you will be using glassware fitted with standard taper ground glass joints. This equipment is VERY expensive, and it should be handled with care to avoid breakage.

The precision ground glass joints allow the pieces to be assembled interchangeably with a good tight fit, and therefore many types of apparatus can be constructed from only a few components. The joints will not "freeze" together if they are kept clean and free of chemicals, or lightly lubricated with stopcock grease.

The separatory funnel is equipped with a Teflon stopcock. Teflon has a low coefficient of friction and is chemically inert, but it deforms under stress, and if the stopcock is not loosened during storage, it will "creep" into the bore of the funnel, eventually causing leaks.

When assembling standard taper glassware, the two pieces to be joined should be given a slight twist while applying gentle pressure together; this will prevent them from coming apart. In a complex setup, it is essential to clamp only one place firmly (usually around the neck of the reaction flask); any other clamps that are used should be loose, providing gentle support only.

Glassware should be cleaned thoroughly after each use with detergent and water and a test-tube brush. Stubborn organic deposits may be more soluble in an organic solvent, such as acetone. Use acetone as a final rinse for your glassware to remove traces of water, but use it sparingly as it is quite expensive.

Electrical Heating Mantle and Regulator

In order to minimize the chances of a fire, an electrical heating device will usually be employed rather than Bunsen burners. This device consists of a ceramic heating mantle, which is regulated by connecting it to a voltage regulator. <u>Never</u> plug the mantle directly into the wall socket.

The ceramic heating mantle absorbs moisture from the air during storage, so it is advisable to preheat the mantle (while empty) at a voltage regulator setting of 5 for a few minutes to displace the moisture before inserting a flask. For the most efficient heating, the bottom of the flask should contact the ceramic surface of the mantle. Some heat is transmitted through the air space between the flask and the mantle (radiant heat), so that a perfect fit between the flask and the heating mantle is not necessary. In fact, the heating mantle provided will accommodate all of the different size flasks you will use.

Be careful not to spill chemicals, including acetone, on the heating mantle, regulator, or wires. In use, an iron ring on a ring stand should support the heating mantle. This allows it to be lowered to stop heating immediately if needed. Remember that even the gray metal cover will be hot, so be careful in moving the mantle when hot. When finished, turn the voltage regulator off and unplug the wire to your mantle. After the mantle has cooled sufficiently to allow handling, return it to its storage place with the cord wound around neatly, and push all equipment to the back of the lab hood.

NOTES

Appendix C

Final Lab Report Formatting

It is important to prepare lab reports in an organic chemistry laboratory course for several reasons:

- To help the student better understand the laboratory experience.
- To help the student improve analytical writing skills.
- To help the student learn how observations and conclusions based on laboratory experiences are recorded in the professional workplace.
- To help the instructor evaluate the student's laboratory understanding and ability.

The final lab report grade for each experiment in this course will consist of the following parts:

1. Pre Lab
2. In Lab
3. Post Lab
4. Lab Technique

Before Lab... **(Pre Lab Assignment)**

- Each student is expected to complete a pre-lab assignment before coming to the lab. The pre-lab assignments are located at the end of each experiment in the lab manual.
- The pre-lab assignment consists of the following:
    - Objectives
    - Chemical Equation/Chemical Structures
    - Physical Data
    - Experimental Outline
    - Pre-Lab Questions
- Each student will be expected to have a properly completed pre-lab assignment, which will be collected at the beginning of the lab period. Without a ***COMPLETE*** pre-lab assignment, the student will not be permitted into the lab, and will receive a zero for the experiment.
- Specific instructions for the pre lab assignments will be discussed during the first lab meeting, but a few basic rules for the pre lab assignments are:
    - Entries must be made in ***black or blue ink ONLY***. No colored ink or pencil will be accepted.
    - Only the original copy of the pre lab assignment (from the lab manual) will be accepted.

During Lab... **(In Lab Assignment)**

- Each student will maintain a laboratory notebook while they perform the experiment. A carbonless laboratory notebook will be a required text for this class, and this notebook can be used for both semesters of the lab course.

- Detailed lists of the data required in the laboratory notebook for each experiment will be available on the course website. Specific instructions for in-lab assignments (maintained in the laboratory notebook) will be discussed during the first lab meeting, but a few basic rules for the in-lab assignments are:
    - All entries must be made in ***black or blue ink ONLY***. No colored ink or pencil will be accepted.
    - The pre-printed top portion of each lab notebook page should be completed before submission.
    - All data should be entered in the lab notebook as it is gathered. No data should be written on scraps of paper or in the lab manual, then transcribed at a later time to the lab notebook.
    - Any mistakes made in the gathering of data should not be scribbled out, nor should white-out be used. It is best to cross out the mistake with a single line only.
    - Only the yellow carbon copy pages will be submitted for grading. The original white copy of the data should remain in the lab notebook, as those pages should be a permanent record of the work.

After lab... **(Post Lab Assignment)**

- This portion of the lab assignment is to be completed after the experiment has been performed. The post-lab assignment consists of the following:
    - Reference to Experimental Procedure
    - Results
    - Discussion
- Each student will submit an ***original*** copy of the post-lab assignment.
- The post-lab assignment for each experiment will be available on the course website. Specific instructions for the post-lab assignments will be discussed during the first lab meeting, but a few basic rules for the post-lab assignments are:
    - Tables must be completed in ***black or blue ink ONLY***.
    - Discussion questions must be answered in complete sentences.
    - All post-lab assignments are provided on the course website in Word format so that they can be edited with the data collected during the lab.
    - A student may download a copy of the post-lab assignment to their personal computer and bring it to lab to work on tables during the lab period. However, the final copy submitted for grading should be an electronic copy, with all tables completed and all discussion questions answered.

Appendix D

Finding the Limiting Reagent and Calculating Theoretical Yield

Limiting Reagent

In a chemical reaction, the starting substances in a chemical reaction are called the ***reactants***, and appear to the left of the arrow in a chemical equation. The substances produced in the chemical reaction are called the ***products***, and appear to the right of the arrow in a chemical equation. Typically, two or more reactants are used to synthesize a product, but only one reactant determines or limits the amount of product that can be formed. This reactant is referred to as the ***limiting reagent***, or limiting reactant, and is the chemical that determines how far the reaction will go before it stops forming the desired product. The reaction stops when all of the limiting reactant has been used. The other reactant(s) are referred to as excess reactants. Many reactions are carried out using an excess of one or more reagents, but the quantities of reactants consumed and the quantity of products formed are restricted by the quantity of the limiting reagent.

In order to determine which reactant is the limiting reagent, one must first determine the molar amount of product that can be formed based on the quantities of all reactants present. This must be determined for each of the reactants separately, assuming that all other reactants are in excess. For example, using the chemical equation below the molar amount of product is calculated using the starting masses of the reactants, their respective molecular weights, and the stoichiometry of the reaction. Suppose we started with 24.99 g of sodium iodide (NaI) and 8.47 g of dichloropropane (DCP):

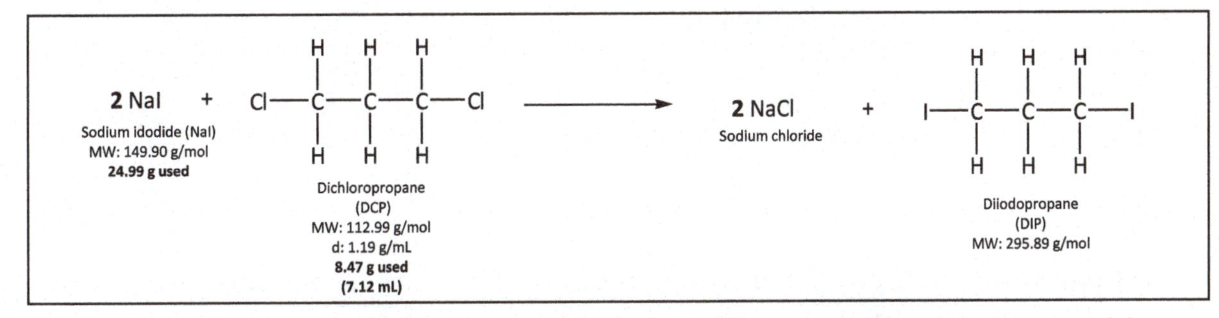

***__Be sure to always write down ALL units, as unit cancellation will help you do this calculation correctly!__

Moles of product = $\text{Mass}_{\text{reactant}}$ (g) * $\text{MW}_{\text{reactant}}$ * Stoichiometric ratio of products to reactants

$$\text{Moles of product}_{\text{NaI}} = 24.99 \text{ g NaI} * \left(\frac{1 \text{ mol NaI}}{149.90 \text{ g}}\right) * \left(\frac{1 \text{ mol diiodopropane}}{2 \text{ mol NaI}}\right) = \underline{0.083 \text{ moles}} \text{ of product formed (based on NaI)}$$

$$\text{Moles of product}_{\text{DCP}} = 8.47 \text{ g DCP} * \left(\frac{1 \text{ mol DCP}}{112.99 \text{ g}}\right) * \left(\frac{1 \text{ mol diiodopropane}}{1 \text{ mol DCP}}\right) = \underline{0.075 \text{ moles}} \text{ of product formed (based on DCP)}$$

Since the molar amount of product formed is less based on the DCP, it is identified as the limiting reagent. Had the ***volume*** (in mL) of DCP had been given instead of the ***mass*** (in g), we would first convert this volume to a mass using the density of the liquid.

$$\text{Mass of DCP} = 7.12\ \text{mL} * \left(\frac{1.19\ \text{g}}{1\ \text{mL}}\right) = 8.47\ \text{g DCP}$$

Theoretical Yield

One of the measures of the success of a chemical reaction is the ***percent yield***: the ratio between the amount of product actually isolated to the amount that (theoretically) could have been produced had the reactant present in the limiting amount been converted to product. Once the limiting reagent has been identified, the amount of product that can be produced, theoretically, can be calculated by simply converting the moles of product (based on the limiting reagent) to a theoretical mass (in g), using the molecular weight of the desired product.

The limiting reagent was identified as the DCP above. Based on the limiting reagent, the ***theoretical yield*** (or theoretical mass) for the product is:

$$0.075\ \text{mol}_{\text{DIP}} * \left(\frac{295.89\ \text{g}}{1\text{mol}_{\text{DIP}}}\right) = 22.19\ \text{g DIP is the theoretical mass}$$

In the event that some reactants never react, they do react, but form undesired side products, or the desired product is simply never fully recovered from the product mixture, the actual yield is almost always less than that theoretical yield, but never greater. The percent yield for the reaction is simply the actual mass of the product obtained divided by the theoretical mass of the product expected multiplied by 100:

$$\textbf{\textit{Percent yield}} = \left(\frac{\text{Actual mass (g)}}{\text{Theoretical mass (g)}}\right) * 100$$

Theoretical Yield Problems

1. You have six slices of bread and four slices of cheese. The recipe for a cheese sandwich is:

2 bread + 1 cheese ⟶ 1 sandwich

Question: How many sandwiches can you make, and which ingredient runs out first?

Answer: You can make three sandwiches and the bread runs out first:

$$6\ \text{slices of bread} * \left(\frac{1\ \text{sandwich}}{2\ \text{slices of bread}}\right) = 3\ \text{sandwiches}$$

$$4\ \text{slices of cheese} * \left(\frac{1\ \text{sandwich}}{1\ \text{slice of cheese}}\right) = 4\ \text{sandwiches}$$

In the example shown, the bread is the ***limiting reagent***. The number of slices of bread or cheese is analogous to the number of moles of each of the reactants in a chemical equation.

2. Shown below is the balanced chemical equation for the synthesis of 2,3-dibromo-2,3-dimethylbutane. Assume that you start with 47.27 grams of pinacol and 56.19 mL of aqueous HBr.

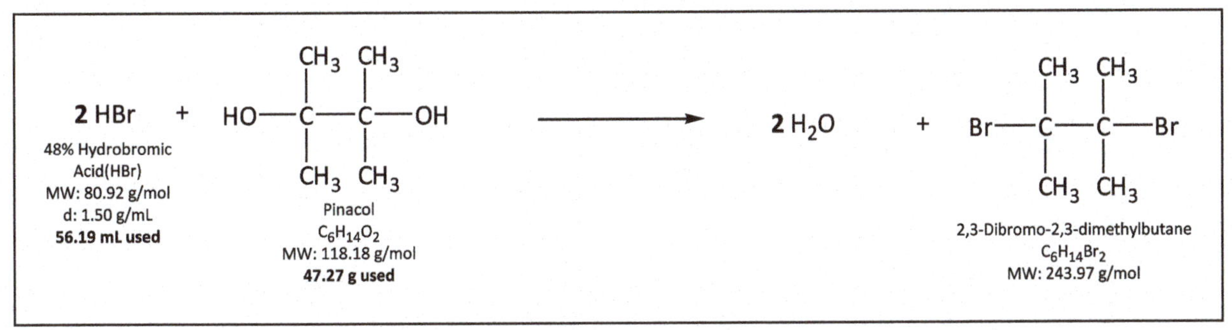

Question: Which compound is the limiting reagent? Calculate the theoretical yield based on the limiting reagent.

Answer:

Moles of product based on 48% HBr:

Concentration of aqueous HBr

$$\underline{56.19\text{ mL}}\text{ HBr} * \left(\frac{1.50\text{ g HBr}}{1\text{ mL}}\right) * \left(\frac{1\text{ mol HBr}}{80.92\text{ g}}\right) * \left(\frac{1\text{ mol } C_6H_{14}Br_2}{2\text{ mol HBr}}\right) = \left(\frac{0.521}{\text{mol product}}\right) * 0.48 = 0.250\text{ mol product}$$

Moles of product based on pinacol:

$$\underline{47.27}\text{ g pinacol} * \left(\frac{1\text{ mol pinacol}}{118.18\text{ g}}\right) * \left(\frac{1\text{ mol } C_6H_{14}Br}{1\text{ mol pinacol}}\right) = \frac{0.400}{\text{mol product}}$$

It appears that HBr is the limiting reagent, as it produces a smaller molar amount of the product. The theoretical yield is then:

$$\underline{0.250}\text{ mol product} * \left(\frac{243.97\text{ g}}{1\text{ mol product}}\right) = \underline{60.99}\text{ g product}$$

NOTES

Appendix E

Chromatography in the Organic Laboratory

Gas Chromatography

Gas chromatography (GC) is an important analytical technique that enables one to determine the composition of a mixture both qualitatively (**what** is present) and quantitatively (**how much** of each component is present). GC is used to analyze volatile liquids or solids that can be dissolved in volatile solvents.

In GC, the mobile phase, which carries the sample through the instrument, is helium gas. Using a small syringe, the sample is injected through a hot injection port, which vaporizes it. The vaporized sample is then picked up by the helium and carried into the column (stationary phase). If the sample is a mixture, the compounds separate on the column and reach the detector at different times because of variable strengths of interactions between the compounds and the column. Typically, a nonpolar column is used, and the compounds elute in order of increasing boiling point. As each component reaches the detector, the detector generates an electronic signal. The signal goes through an attenuator network, then to a computer where the signal is recorded, and a chromatogram is produced. ***Identity*** of the compounds is determined by comparison of the *retention times* to those of standards. ***Purity*** of the sample is determined by the *number of peaks* that appear in the chromatogram, other than the solvent peak. A single peak (other than the solvent peak) in the chromatogram indicates a pure compound.

High Pressure Liquid Chromatography

High Pressure Liquid Chromatography (HPLC) is one of the most widely used analytical methods available to the organic chemist. HPLC is routinely used to analyze pharmaceuticals and other organic compounds for purity and to identify substances in complex mixtures.

In contrast to GC, the samples need not be volatile. In principle, it is similar to GC, except that the mobile phase is a liquid forced through the column by high pressure. As in GC, the sample is injected through an injection port, where the sample is picked up by the liquid mobile phase and carried into the column. The components of a mixture spend more or less time in the moving liquid, therefore reaching the detector at different times. In CHML211/212, a polar silica gel column (stationary phase) is used, with a relatively nonpolar solvent system (mobile phase). This is referred to as *normal phase* chromatography, and compounds elute from the column in order of increasing polarity. The amount of time that the compounds spends retained in the column before it reaches the detector is defined as the *retention time*. The less polar compounds have shorter retention times, the more polar compounds have higher retention times.

The HPLC is equipped with a very sensitive UV-Vis absorption detector. Any sample that absorbs light in the 220-800 nm wavelength ranges can be detected, and this includes most compounds that have aromatic rings or other multiple bonds. This creates one of the main differences between the appearances of a GC chromatogram vs. an HPLC chromatogram. Most HPLC solvents have little ability to absorb UV/Vis light and are therefore not detected. The GC detector detects solvents, and thus an adjustment must be made to the area % values prior to quantification, since the solvent was simply added to ease injection of the sample. ***Identity*** of the compounds is determined by comparison of the *retention times* to those of standards. ***Purity*** of the sample is determined by the *number of peaks* that appear in the chromatogram. A single peak in the chromatogram indicates a pure compound.

Thin Layer Chromatography

Prior to running an experiment on an HPLC, test separations can be carried out on Thin Layer Chromatography (TLC) plates rather than on chromatographic columns. Compounds should separate the same way on a column as they do on a TLC plate if the same stationary phase and mobile phase are used. This will reduce solvent waste and the experimental time involved in method development.

TLC can be used for identifying compounds and determining their purity, as well as following the course of a reaction. TLC operates based on the same principles as HPLC, using a nonpolar liquid mobile phase and a polar silica gel stationary phase. The difference is that the stationary phase is applied as a thin layer on a plate, as opposed to a thin tube, which is packed with silica gel in HPLC. The sample is applied to the surface of the TLC plate and placed in a small jar containing a small amount of a solvent. The solvent begins to wick up the plate by capillary action, and as it comes in contact with the sample, it dissolves it and begins to carry it up the plate. If the sample is a mixture, the compounds appear as separate spots on the plate, which have traveled different distances, depending on their polarity. The distance that a compound travels from the origin divided by the distance that the solvent traveled from the origin is defined as the *R_f value*. The more polar compounds will stick to the silica gel, and travel less distance, resulting in a lower TLC R_f value. The nonpolar compounds will spend more time dissolved in the nonpolar solvent and travel further up the plate, resulting in a high TLC R_f value. In lab, the method of detection of the spots in TLC is a UV lamp, set to detect compounds that absorb UV light around the 254 nm range. ***Identity*** of the compounds is determined by comparison of sample *R_f values* to those of standards. The *number of spots* that appear in the sample lane determines ***purity*** of the sample. A single spot in the sample lane indicates a pure compound.

Column Chromatography

This method of chromatography is a *preparative* method, not an *analytical* method, since it is used to purify compounds from one another, and a separate analytical method must be used to identify the compounds once isolated from one another. Column chromatography works in a similar manner to TLC, although column chromatography is carried out on a column of silica gel, rather than on a thin layer, using a glass tube is as the support for the silica gel instead of a plate. TLC experiments are preliminary to separations using column chromatography. If the same mobile and stationary phases are used that sufficiently separated the components of a mixture in a TLC experiment, the components should separate the same way using column chromatography. As in TLC, the polarity of the mobile phase must me intermediate in strength. If the solvent is not polar enough, the compounds will remain in the column; if it is too polar, the compounds will emerge from the column too quickly to have a chance to separate. Those compounds which have a low TLC R_f value (polar) take longer to emerge from the bottom of the column than the compounds that have a high TLC R_f value (nonpolar).

Identity of Unknowns based on GC, HPLC and TLC

Identity of unknown compounds can be determined by comparing the retention time of the unknown to that of known compounds, called standards, under identical conditions. Three things to keep in mind:

1) If the retention times (or R_f values) are the *same*, the two compounds are probably *identical*.
2) The retention time of an unknown may be different from the retention time of a standard by roughly 0.1 min for a variety of reasons.
3) If the retention times *differ* by *>0.1 min*, the compounds are probably *not identical.*

Purity of Compounds based on GC, HPLC and TLC

We can determine the purity of the unknown compound by the number of peaks and their relative areas in the chromatogram. There are several guidelines than can be used:

1) If we see only *one peak (*or *one spot),* the unknown compound is probably *pure*. However, seeing one peak only tells us nothing about the percent yield.
2) If we see *two peaks*, and the larger peak is our product, then the size of the smaller peak correlates with the *degree of impurity*.
3) We always see a solvent peak when we run gas chromatography, but we ignore the solvent peak when we decide how pure our compound is. We added the solvent to the sample ourselves, and this is why we calculate the adjusted area percent on GC data WITHOUT including the solvent. This is the % composition of the sample, NOT the % yield of the reaction.
4) Purity or impurity has nothing to do with very slight differences in retention times between standards and unknowns run under identical conditions.
5) Not all impurities show up in GC or HPLC. Many impurities are not volatile enough to observe by GC, and therefore they would not emerge from the column. In HPLC the detector is sensitive only to compounds that absorb ultraviolet or visible light of a particular wavelength. Many impurities do not absorb light and would not be detected even if they did emerge from the column. The same is true for TLC. Compounds that do not absorb UV light will not be detected without the use of special stains. Therefore, it is best to have more than one method to judge the purity of your compound.

Appendix F

Chromatographic Solvent Effects on Analyte Mobility

Analytes are the compounds we are trying to separate from each other and identify. We can understand how the polarity of the solvent affects the mobility of the analyte by picturing a competition between solvent molecules and the analyte for binding sites on the silica stationary phase. Although the bulk solvent molecules (*mobile phase*) move, a few molecules of solvent bind to the silica. A given portion of the silica can either bind to a molecule of the analyte or a molecule of the solvent. When the analyte is bound to the stationary phase, it doesn't move. When a solvent molecule binds to the silica and displaces the analyte, the analyte is forced to move along with the bulk solvent.

Different analytes have different polarities. If the analyte is relatively polar (Figure A), it wins the competition for binding sites to the stationary phase most of the time, and the analyte moves slowly. If the analyte is relatively nonpolar (Figure B), the solvent wins the competition for binding to the stationary phase most of the time, and the analyte moves quickly.

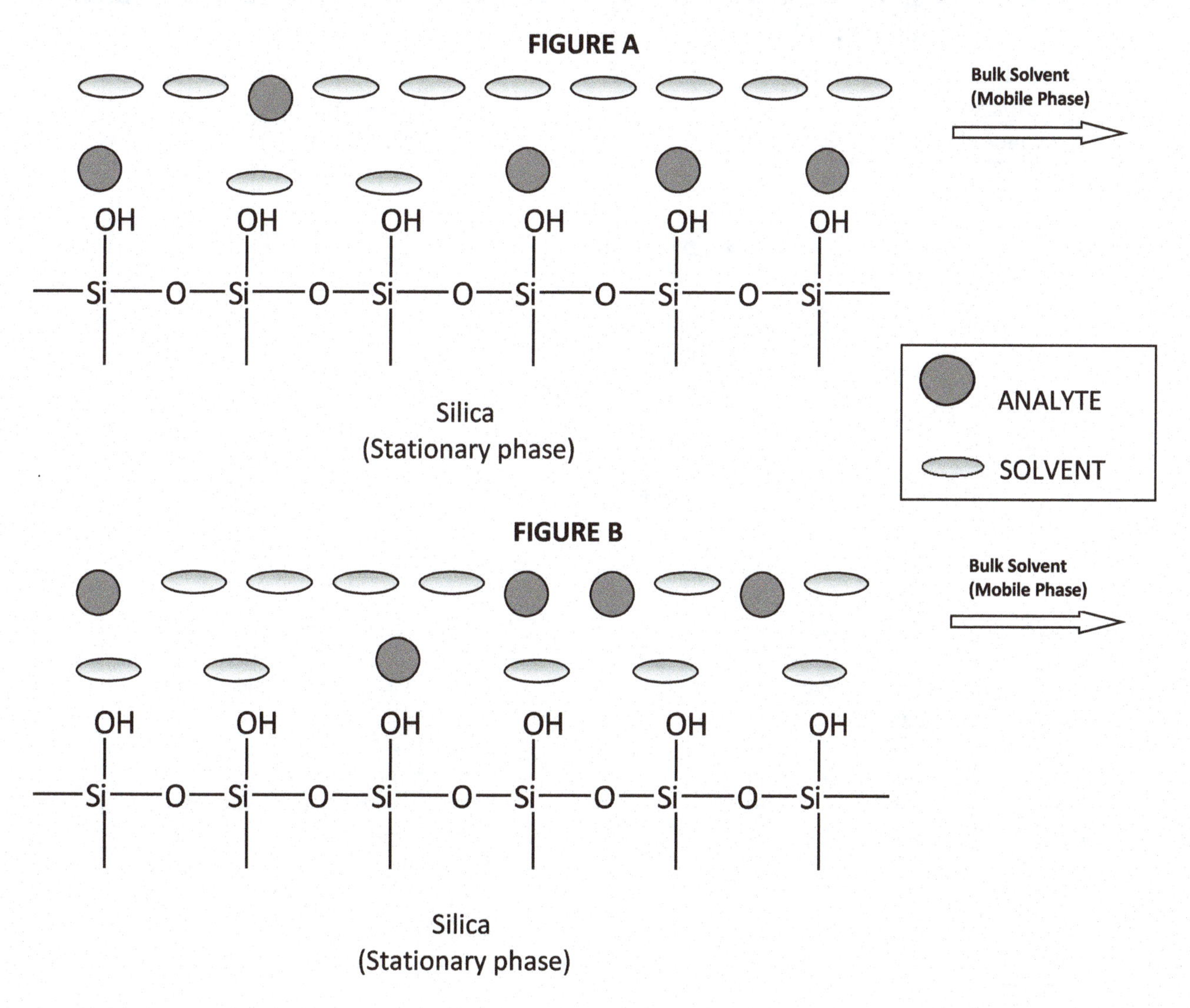

Thin Layer Chromatography

In thin layer chromatography, the analyte and the solvent compete for binding sites on the stationary phase. Silica is very polar and can form hydrogen bonds to the solvent or the analyte. A site on the silica can be occupied either by a molecule of solvent or a molecule of the analyte, but not both at the same time; therefore, we say that the solvent and analyte are competing against each other for binding to the stationary phase. If the analyte is polar (Compound A below), it wins the competition most of the time, and it stays bound to the stationary phase and is immobile most of the time. Compound A, therefore, has a low R_f value. If the analyte is nonpolar (Compound B below), solvent molecules win the competition most of the time, displacing B into the mobile phase most of the time. Since B spends most of its time dissolved in the solvent, it has a high R_f value.

If solvent polarity is increased, it can more effectively compete for sites on the silica against both Compound A and Compound B. Therefore, both of their R_f values will increase. If the solvent is too polar, both compounds will travel almost as fast as the solvent front, and there will be little to no separation between compounds. If the solvent is not polar enough, both compounds will have very low R_f values, and again, there will be little to no separation between compounds. The best solvent for TLC is one that gives the greatest separation between Compound A and Compound B.

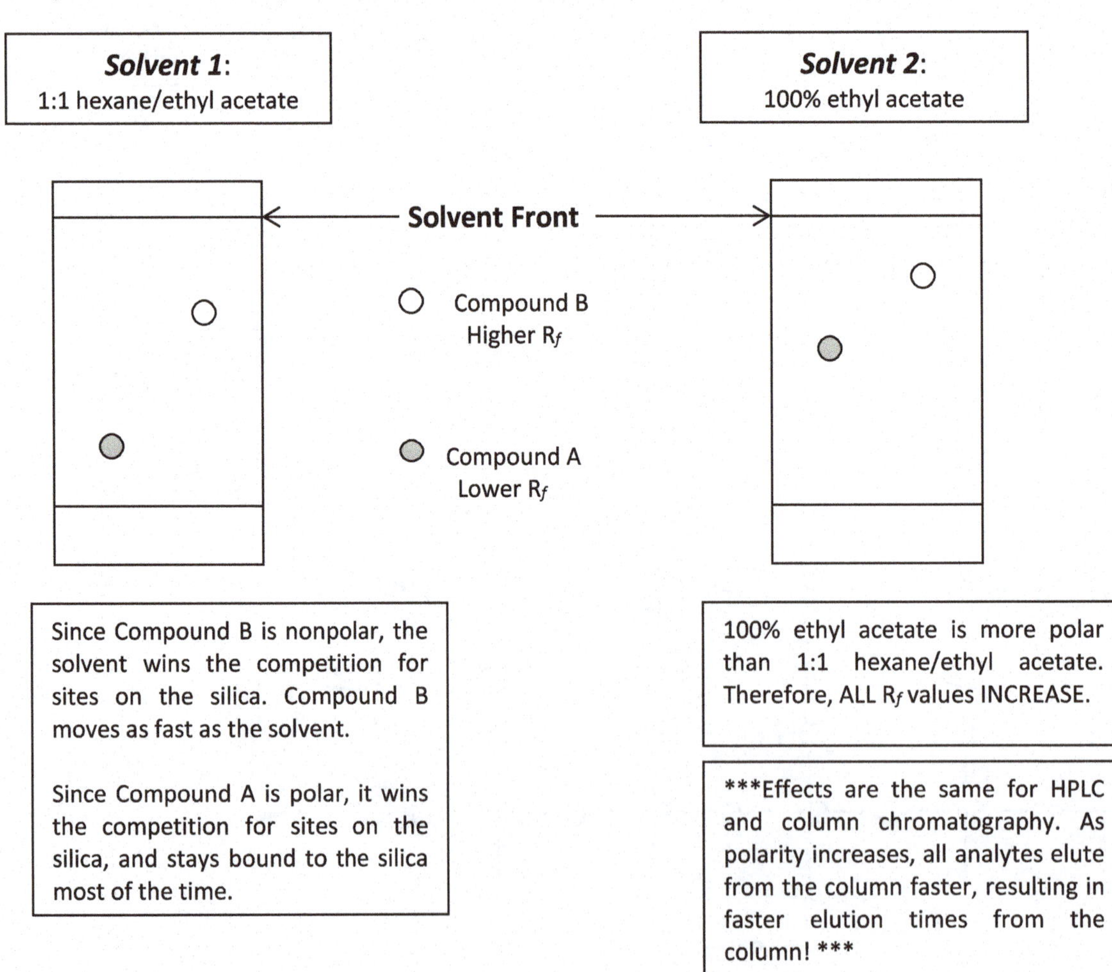

Solvent 1:
1:1 hexane/ethyl acetate
Solvent 2:
100% ethyl acetate
Solvent Front
Compound B
Higher R_f
Compound A
Lower R_f
Since Compound B is nonpolar, the solvent wins the competition for sites on the silica. Compound B moves as fast as the solvent.
Since Compound A is polar, it wins the competition for sites on the silica, and stays bound to the silica most of the time.
100% ethyl acetate is more polar than 1:1 hexane/ethyl acetate. Therefore, ALL R_f values INCREASE.
***Effects are the same for HPLC and column chromatography. As polarity increases, all analytes elute from the column faster, resulting in faster elution times from the column! ***

Appendix G

Definitions of Stereochemistry Terms

absolute configuration: A particular three-dimensional arrangement of a stereogenic center. Every stereogenic center can be classified by the Cahn-Ingold-Prelog sequence rules to assign priority, which are based on atomic number, as being R or S.

anomers: Two carbohydrate diastereomers that differ in configuration only at the aldehyde or ketone carbon. Unlike most diastereomers, anomers **can** easily interconvert in solution. When alpha-D-glucose and beta-D-glucose interconvert in solution, this process is called mutarotation.

chirality: Having handedness. A chiral molecule is not superimposable on its mirror image, has no plane of symmetry, and rotates plane-polarized light. All molecules are chiral when they have one stereogenic carbon, but not all molecules with two stereogenic carbons are chiral.

configuration: The three-dimensional arrangement of atoms bonded to a stereogenic carbon atom. The configuration around a stereogenic carbon atom cannot change unless covalent bonds are broken, a process requiring a great deal of energy.

conformation: The shape of a molecule in space. One conformation of a molecule can be turned into another one by rotations around single bonds. No covalent bonds are broken when a molecule changes conformation, and this process requires almost no energy. An example would be the *gauche* versus the *anti* conformations of butane.

diastereomers: Stereoisomers that are not mirror images of each other. Diastereomers differ from each other in the arrangement of atoms in space. Diastereomers may be either configurational isomers or *cis-trans* isomers. The comparison between *cis-* and *trans-*2-butene is an example of a pair of diastereomers that do not have any stereogenic carbon atoms. Configurational diastereomers must have at least two stereogenic carbons.

enantiomer: A type of stereoisomer. Enantiomers are non-superimposable mirror images of each other. Enantiomers have identical physical properties other than the direction that a solution of each enantiomer rotates plane-polarized light. The stereogenic center(s) in one enantiomer has (have) the opposite configuration(s), meaning R or S designation, as in the other enantiomer. *There is no simple relationship between the absolute configuration (R or S) and the direction of rotation of plane-polarized light (dextrorotatory or levorotatory).* A molecule with more than one stereogenic center can have many stereoisomers but only one enantiomer.

meso compound: A compound that has stereogenic centers but is not chiral. Meso compounds have an internal plane of symmetry (this plane may not be obvious unless one rotates a single bond to obtain the desired conformation). They do not rotate plane-polarized light.

optical activity: The ability to rotate (twist) plane-polarized light. There must be a chiral compound present to produce optical activity, but if two enantiomers are present in equal concentrations (a racematic mixture), no rotation is observed.

plane of symmetry: An imaginary plane that cuts an object into perfect (matching) halves.

racemic mixture: A 50/50 mixture of an enantiomeric pair. A racemic mixture has no optical activity.

specific rotation: A physical constant of a pure, chiral compound. Specific rotation, $[\alpha]_D$, is the observed rotation (α) divided by the concentration (C in g/mL) and the path length (l in dm). The direction of rotation is either to the right (+) or to the left (-). A pair of enantiomers have $[\alpha]_D$ values that are equal in magnitude but opposite in direction.

stereogenic carbon atom: A carbon atom connected to four non-identical groups. Also referred to as *chirality center, stereocenter, chiral carbon*.

stereoisomer: Isomers that have the same connectivity between atoms (same bonding pattern) but have a different three-dimensional arrangement of the atoms. Stereoisomers may either be enantiomers or diastereomers.

Appendix H

Common Organic Laboratory Calculations

TLC R_f Value

An R_f value can be used to aid in the identification of a substance by comparison to standards. The R_f value is not a physical constant, and comparison should be made only between spots on the same sheet, run at the same time. Two substances that have the same R_f value may be identical; those with different R_f values are not identical. Thus, TLC can serve as a rapid and simple tool for identification of various compounds if appropriate standards are available. Since the R_f value is a ratio, it is unitless and usually has only ***two*** significant figures.

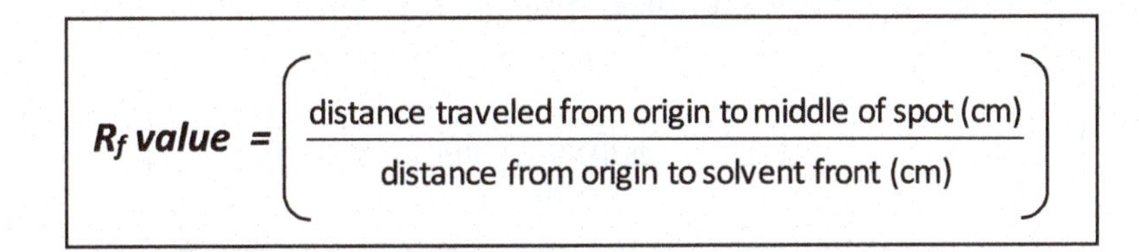

Rf value =
distance traveled from origin to middle of spot (cm)
distance from origin to solvent front (cm)

Adjusted Area Percent

Most often you will dissolve your compound or mixture in a low boiling solvent for GC analysis. The relative areas of the components of interest must therefore be adjusted to exclude the large percent area of the solvent peak. In general, to calculate the *adjusted area percent,* the area of each peak of interest is divided by the sum of the areas of all peaks of interest (excluding the solvent peak) and that value is multiplied by 100%.

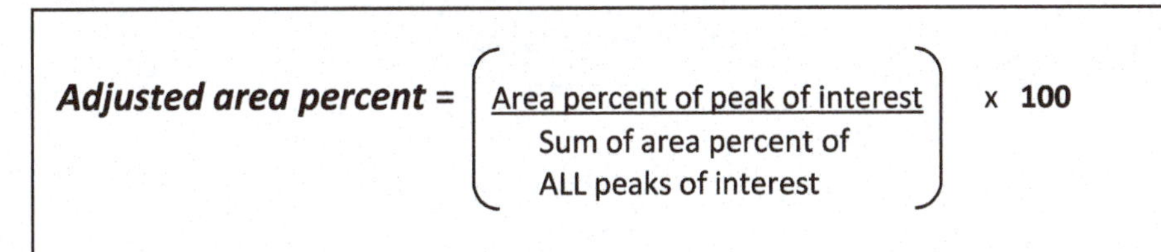

Atom Economy

Atom economy is a method of determining the efficiency of a reaction, and is typically calculated in green chemistry experiments. The atom economy reflects the success of the reaction based on how many reactant atoms are actually converted to product atoms, and how many result in side products or waste. This value is based on the percent of molecular weight of the product vs. the molecular weights of all of the starting materials appearing in the final product. The atom economy considers the success based on conditions that are *ideal*, with no side products or waste generated.

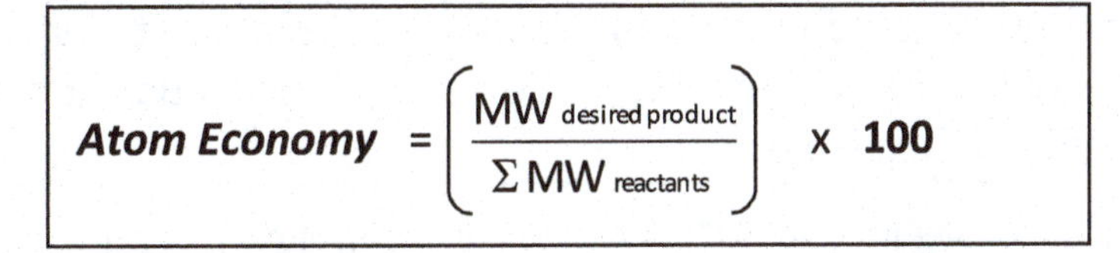

Experimental Atom Economy

Experimental atom economy is another method of determining reaction efficiency used in green chemistry. Since excesses of reagents are used at times to accelerate a reaction, and are therefore left over at the end of the synthesis, all of the reactant atoms are not typically used. Another possibility is that a side reaction may occur, generating waste that has to be disposed of at the end. Both scenarios reduce the actual utilization of reactant atoms. Experimental atom economy considers the conditions *actually used* in the synthesis.

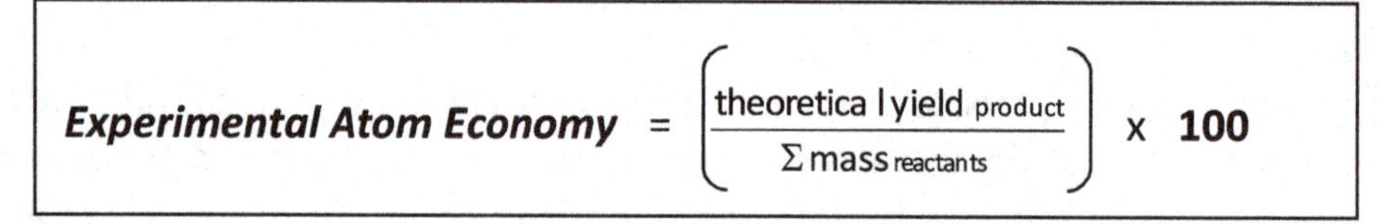

Experimental Atom Economy = (theoretica l yield product / Σ mass reactants) x 100

"$E_{product}$"

The "$E_{product}$" is the final method of determination of success of the reaction using the selected conditions. It is based on the final percent yield of the product and the experimental atom economy.

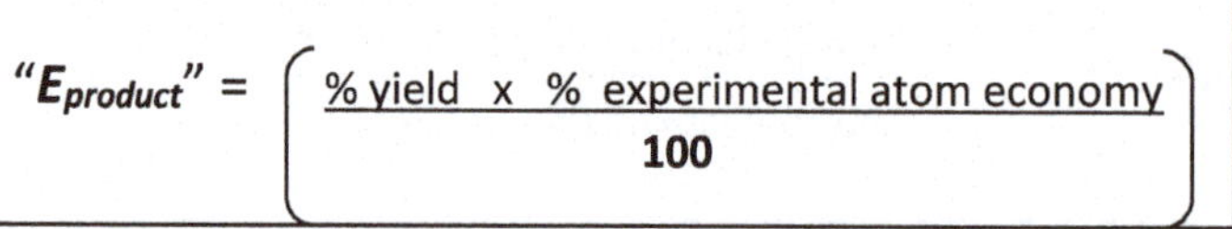

Cost Analysis

In order to determine the cost of a synthesis, one must first calculate the cost of each material required to perform the synthesis. Once the cost for each individual component is calculated, the costs are added together to reach the cost per synthesis (CPS). When calculating the cost of the synthesis, all reactants, solvents and catalysts must be taken into account.

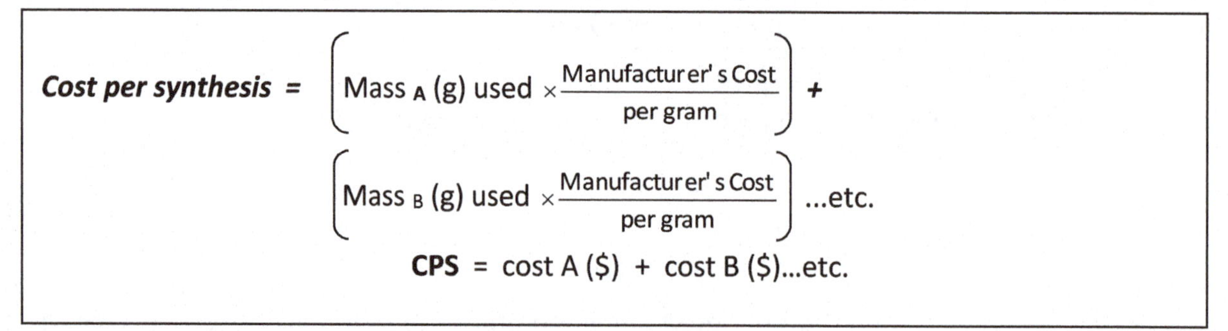

Cost per synthesis = (Mass A (g) used × Manufacturer's Cost per gram) +
(Mass B (g) used × Manufacturer's Cost per gram) ...etc.
CPS = cost A ($) + cost B ($)...etc.

Once the cost of the synthesis is determined, the cost per gram of your product can be determined by simply dividing the cost by the amount of the product that you actually obtained.

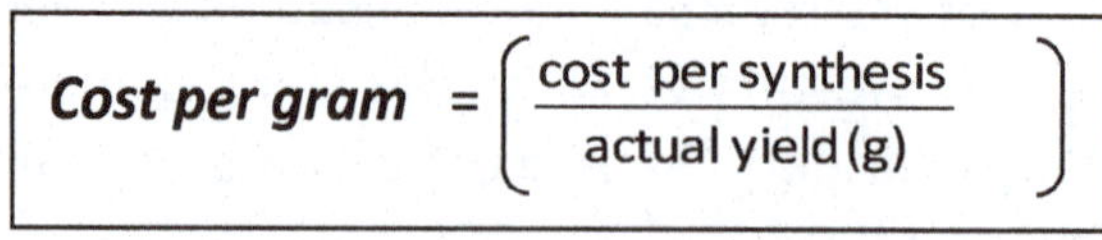

Cost per gram = (cost per synthesis / actual yield (g))

Degree of Unsaturation

The molecular formula can provide valuable information about the structural formula of a compound. The index of hydrogen deficiency, or degree of unsaturation, indicates the possible number or rings and pi bonds in a molecule.

Where $C_cH_hN_nO_oX_x$: $°unsaturation = \left(\frac{[(2c+2)-(h+x-n)]}{2} \right)$

Appendix I

Spectral Correlation Tables

IR Spectroscopy

IR Spectroscopy	
C-C and C-H Bonds	
sp^3 C—C	weak, not useful
sp^2 C=C	1600-1700 cm^{-1}
sp^2 C=C (aryl)	1450-1600 cm^{-1}
sp C≡C	2100-2250 cm^{-1}
sp^3 C—H	2800-3000 cm^{-1}
sp^2 C—H	3000-3300 cm^{-1}
sp C—H	3300 cm^{-1}
$C(CH_3)_2$	1360-1385 cm^{-1} (two peaks)
Alcohols and Amines	
O—H or N—H	3000-3700 cm^{-1}
C—O or C—N	900-1300 cm^{-1}
Ethers	
C—O	1050-1260 cm^{-1} (strong)
Ketones (saturated)	
C=O	1640-1820 cm^{-1}
Aldehydes	
C=O	1640-1820 cm^{-1}
C(O)—H	2820-2900 cm^{-1} (weak)
	2700-2780 cm^{-1} (weak)
Carboxylic Acids	
C=O	1700-1725 cm^{-1}
C(O)—OH	3330-2900 cm^{-1}
Acid Chlorides	
C=O	1800 cm^{-1}
Acid Anhydrides	
C=O	1740-1800 cm^{-1}
C—O	900-1300 cm^{-1}
Esters	
C=O	1735-1800 cm^{-1}
C(O)—OR	1100-1300 cm^{-1}
Amides	
C=O	1630-1680 cm^{-1}
C(O)N—H	3200-3400 cm^{-1}
Nitrile	
C≡N	2200-2250 cm^{-1}

^{1}H NMR Spectroscopy

Functional Group	Chemical Shift (δ) ppm
Primary alkyl, RCH_3	0.8 – 1.0
Secondary alkyl, RCH_2R	1.2 – 1.4
Tertiary alkyl, R_3CH	1.4 – 1.7
Allylic, R_2C=C—CH_2R	1.6 – 1.9
Benzylic, $ArCH_2R$	2.2 – 2.5
Iodoalkane, RCH_2I	3.1 – 3.3
Bromoalkane, RCH_2Br	3.4 – 3.6
Chloroalkane, RCH_2Cl	3.6 – 3.8
Fluoroalkane, RCH_2F	4.4 – 4.5
Ether, RCH_2OR	3.3 – 3.9
Alcohol, RCH_2OH	3.3 – 4.0
Ketone, RCH_2C(=O)R	2.1 – 2.6
Aldehyde, RCH(O)	9.0 – 10.0
Terminal alkene, R_2C=CH_2	4.6 – 5.0
Internal alkene, R_2C-CHR	5.2 – 5.7
Aromatic, Ar—H	6.0 – 9.5
Alkyne, RC≡C—H	1.7 – 3.1
Alcoholic hydroxyl, ROH	0.5 – 5.0
Amine, RNH_2	0.5 – 5.0
Acid hydroxyl, RC(O)OH	10.0 – 13.0

Note: Alcoholic hydroxyls, amines and acid hydroxyls show wide variability in chemical shift.

^{13}C NMR Spectroscopy

Functional Group	Chemical Shift (δ) ppm
Primary alkyl, RCH_3	10 – 40
Secondary alkyl, RCH_2R	15 – 55
Tertiary alkyl, R_3CH	20 – 60
Allylic, R_2C=C—CH_2R	20 – 60
Benzylic, $ArCH_2R$	20 – 60
Iodoalkane, RCH_2I	0 – 40
Bromoalkane, RCH_2Br	25 – 65
Chloroalkane, RCH_2Cl	35 – 80
Fluoroalkane, RCH_2F	60 – 90
Ether, RCH_2OR	40 – 80
Alcohol, RCH_2OH	40 – 80
Ketone, RCH_2C(=O)R	185 – 220
Aldehyde, RCH(O)	185 – 220
Acid, RC(O)OH	165 – 185
Ester, RC(O)OR	160 – 180
Amide, $RC(O)NR_2$	165 – 180
Alkene, R_2C=CR_2	100 – 150
Aromatic, sp^2 ring carbon	110 – 160
Alkyne, RC≡CR	65 – 85

Appendix J

Properties of Common Organic Solvents

	Physical Properties		Volatility		Solubility		Detection	Environmental Fate			Hazard Metrics				
Solvent	Mp (°C)	Bp (°C)	Fp (°C)	Evap rate	H_2O Sol (mg/kg)	Log K_{ow}	Odor Thres (ppm)	Atm half-life	Urban zone	Bio deg time	NFPA rating H,F,R	OSHA PEL (ppm)	Specific Hazards	LD_{50} (mg/kg)	EPA 33/50 listing
ALIPHATIC HYDROCARBONS															
Hexane	-95.3	68.7	-26	8.9	Negl	3.90	130	2.9d	0.12	d/w	1,3,0	50	M,T	28170	N
Ligroin		60-80	-26								1,4,0				
Pentane	-129.7	36.1	-40	>1	Negl	3.39	400	4.1d	0.11	d	1,4,0	600	-	-	N
Pet ether		35-60	-49												
CHLORINATED HYDROCARBONS															
Carbon tetrachloride	-23	76.7	-	6.0	793.4	2.83	96	47y	0.00	w	3,0,0	2	C,M,T	2350	Y
Chloroform	-63.5	61.2	-	10.45	7950	1.97	85	160d	0.00	w	2,0,0	2	C,M,T	908	Y
Dichloro-methane	-94.7	39.6	-	14.5	13030	1.25	250	110d	0.00	d/w	2,1,0	500	C,M,T	1600	Y
AROMATIC HYDROCARBONS															
Benzene	5.5	80.1	-11	5.1	1790	2.13	12	13d	0.03	w	2,3,0	1	C,M,T	930	Y
Toluene	-95.0	110.6	4	1.9	526	2.73	2.9	2.7d	0.24	d/w	2,3,0	100	T	636	Y
ETHERS															
1,2-dimethoxy-ethane	-57.8	82	-2		∞	-0.21					2,2,0				
Dioxane	11.8	101.3	12	2.42	∞	-0.27	24	1.5d	0.35	d/w	2,3,1	25	M,T	5700	N
Diethyl ether	-116.3	34.4	-45	33	69000	0.89	8.9	1.2d	0.45	d/w	2,4,1	400	M,T	1215	N
Methyl t-butyl ether	-109	55.2	-27	8.14	51000	1.24		5.7d	0.07	d/w	2,4,0	40		4000	N
Tetrahydro-furan	-108.4	66.0	-14	4.72	∞	0.46	2.0	1.0d	0.49	d/w	2,3,1	200	T	1650	N
KETONES															
Acetone	-94.7	56.1	-17	5.59	∞	-0.24	13	71d	0.01	d/w	1,3,0	1000		5800	N
Methyl ethyl ketone	-86	79.6	-9	3.8	223000	0.29	5.4	14d	0.04	d/w	1,3,0	200	T	2737	Y

Fp = flash point (°C) Evap rate = Evaporation rate (relative to butyl acetate) log K_{ow} = log (octanol-water partition coefficient) Odor thres = odor threshold (ppm) Specific Hazards: M = mutagenic, T = teratogenic, C = carcinogenic, Corr = corrosive
Atm half life = Atmospheric half-life (h = hours, d = days, y = years) Urban zone = Urban zone formation potential (relative to ethylene) OSHA PEL = Occupational Safety and Health Administration permissible exposure limit (ppm)
Bio deg time = Biological degradation time (d = days, w = weeks) NFPA = National Fire Protection Association ratings for health, flammability, and reactivity (0-4 scale, 0 = safest)
LD_{50} = lethal dose for oral ingestion (mg/kg)EPA 33/50 listing = Listing in Environmental Protection Agency 33/50 Program (33% reduction by 1992, 50% by 1995)

	Physical Properties		Volatility		Solubility		Detection	Environmental Fate			Hazard Metrics				
Solvent	Mp (°C)	Bp (°C)	Fp (°C)	Evap rate	H_2O Sol (mg/kg)	Log K_{ow}	Odor Thres (ppm)	Atm half-life	Urban zone	Bio deg time	NFPA rating H,F,R	OSHA PEL (ppm)	Specific Hazards	LD_{50} (mg/kg)	EPA 33/50 listing
ESTERS															
Ethyl acetate	-83.6	77.1	-4	3.90	80000	0.73	3.9	10d	0.04	d/w	1,3,0	400	M	5620	N
ALCOHOLS															
Diacetone alcohol	-42.8	168.1	58	0.12	∞	-0.098	0.28	12d	0.06	d/w	1,2,0	50	M?	4000	N
Ethanol	-114.2	78.4	13		∞	0.235	350	4.9d	0.12	d/w	0,3,0	1000	M,T	7060	N
Methanol	-97.7	64.6	11	2.10	∞	-0.77	100	17d	0.08	d/w	1,3,0	200	T	6200	N
isopropanol	-88	82.2	12	2.5	∞	0.05	28.2	3.1d	0.14	d/w	2,3,0	400	T	5045	N
MISCELLANEOUS															
Acetic acid	16.7	117.9	39	1.34	∞	-0.17	0.48	22d	0.02	d	2,2,1	10	Corr	3310	N
Acetonitrile	-45	81.6	12.8	5	∞		1143	>30d			2,3,0	40	?	2460	N
Carbon disulfide	-110.8	46.3	-30		2000	1.84	0.1				3,4,0	20	T?		N
Dimethyl-formamide	-61.1	152.8	58	0.17	∞	-1.01					2,2,0	10		3967	N
Dimethyl sulfoxide	18.5	189	95	0.026	∞	-1.35		6.2h	0.23	d/w	1,1,0		M?	14500	N
NMP	-24.4	202	92	0.06	∞	-0.11		0.78d	0.21	d/w	2,2,0		M,T	3914	N
Propylene carbonate	-54.5	241.7	132	0.005		0.017		3.7d	0.08	d/w					
pyridine	-42.2	208	20		∞	0.7	<1				3,3,0	5			

Fp = flash point (°C) Evap rate = Evaporation rate (relative to butyl acetate) log K_{ow} = log (octanol-water partition coefficient) Odor thres = odor threshold (ppm) Specific Hazards: M = mutagenic, T = teratogenic, C = carcinogenic, Corr = corrosive
Atm half life = Atmospheric half-life (h = hours, d = days, y = years) Urban zone = Urban zone formation potential (relative to ethylene) OSHA PEL = Occupational Safety and Health Administration permissible exposure limit (ppm)
Bio deg time = Biological degradation time (d = days, w = weeks) NFPA = National Fire Protection Association ratings for health, flammability, and reactivity (0-4 scale, 0 = safest)
LD_{50} = lethal dose for oral ingestion (mg/kg) EPA 33/50 listing = Listing in Environmental Protection Agency 33/50 Program (33% reduction by 1992, 50% by 1995)